From Additive Manufacturing to 3D/4D Printing 2

AF716310

To Laurent and Denis, who allowed me
to dedicate myself to additive manufacturing.

Series Editor
Jean-Charles Pomerol

From Additive Manufacturing to 3D/4D Printing 2

Current Techniques, Improvements and their Limitations

Jean-Claude André

iSTE

WILEY

First published 2017 in Great Britain and the United States by ISTE Ltd and John Wiley & Sons, Inc.

Apart from any fair dealing for the purposes of research or private study, or criticism or review, as permitted under the Copyright, Designs and Patents Act 1988, this publication may only be reproduced, stored or transmitted, in any form or by any means, with the prior permission in writing of the publishers, or in the case of reprographic reproduction in accordance with the terms and licenses issued by the CLA. Enquiries concerning reproduction outside these terms should be sent to the publishers at the undermentioned address:

ISTE Ltd
27-37 St George's Road
London SW19 4EU
UK

www.iste.co.uk

John Wiley & Sons, Inc.
111 River Street
Hoboken, NJ 07030
USA

www.wiley.com

© ISTE Ltd 2017
The rights of Jean-Claude André to be identified as the author of this work have been asserted by him in accordance with the Copyright, Designs and Patents Act 1988.

Library of Congress Control Number: 2017953144

British Library Cataloguing-in-Publication Data
A CIP record for this book is available from the British Library
ISBN 978-1-78630-120-8

Contents

Acknowledgments

Sincere thanks to 3A, BeAm, 3DCeram, 3D Systems, Fives, Nanoscribe, Poietis, Prodways, XtreeE, Alex (alias Alexandre Martel, co-founder of 3D Natives.com) and Laurent Dupont, head of the Lorraine Fab Living Lab® for their effective cooperation, and particularly for graciously providing certain images.

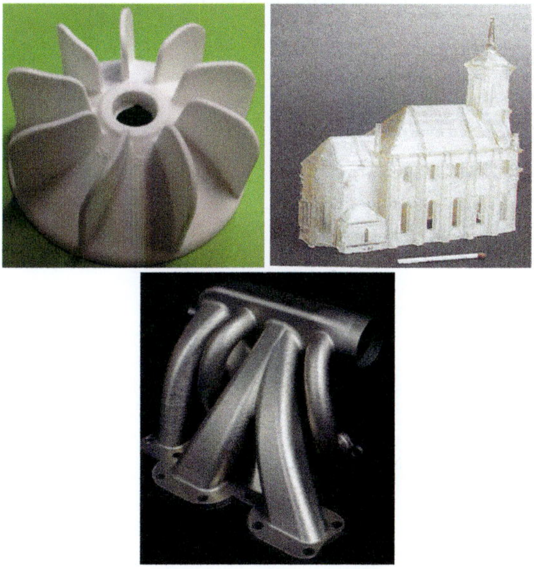

Prototype part (3DCeram, 2017), reproduction of the Église de Bonsecours in Nancy (LRGP, 1994) and metallic part (3A – Applications Additives Avancées, 2017)

Foreword

The evocative expression "3D printing" has been overtaken in everyday speech by the expression generally preferred by scientists and engineers, "additive manufacturing". In both cases, it is a matter of manufacturing objects in successive layers, and soon every workshop and every school will have a 3D printer and engage in additive manufacturing. Self-service workshops known as fab-labs already offer users the possibility to create their own objects. However, the adventure is not over, as "4D" is coming up over the horizon with materials that evolve over time, not to mention "bio-printing", which aims to create organs to be used to repair the living. Furthermore, the 3D printing of tomorrow, which will be performed without layers, threatens to make the term "additive manufacturing" obsolete, thereby making it possible to return to the initial concept of 3D printing. Whatever the case may be, we are faced with not only a very active and booming world, but also a complex world that calls on numerous skills in physics, engineering, chemistry of materials and mechanics with a resolutely multidisciplinary and convergent approach.

To understand the origin of the ideas in additive manufacturing/3D printing, learn about the current state of what is known and explore the developments to come, what could be better than to ask one of the inventors of the technology and one of the first French patent holders in the field, Jean-Claude André, to share his knowledge with us? This led to the idea of this 3-volume edition that I am pleased to present; a work that is both erudite and prospective, as its intention is to start at the genesis of the ideas that led to additive manufacturing to anticipate the impact and future of still "additive" technologies and, beyond this, to encourage reflection on the interactions between science and society of today and tomorrow.

If the first patents date back to 1984, an era where lasers, photo-materials and computer-aided design had already been mastered, was the idea of additive manufacturing completely disruptive as would be said today. What was creative was to put all of this knowledge together to come to something entirely new. Nevertheless, approval for the concept of additive manufacturing came rather quickly. It is on this basis that other additive methodologies, currently many of them with very specific niches, could be developed. These range from prototype and industrial parts to art, variable spatial scales – from the decameter to the nanometer –, from the inert to the living, from industrial organizations to very delocalized forms of manufacturing, etc.

On the basis of these works with varied applicative and societal spectrums, some of which are in the process of becoming stabilized, others to be invented, the principles of additive manufacturing can serve as an example, even as a "laboratory" to better understand how the interactions between research and society can (and must) develop, whether this is through new scientific concepts and the associated concepts of creativity, interdisciplinary scientific and technological operations, the popularization of public research, links with society in terms of the creation of new markets and jobs, and also forms of responsibility and ethics.

Throughout these three volumes, the author would like to invite you to reflect on the circuits between the applications that pose new scientific questions and prior research which opens the door to new applications or new products. The more we progress in the field of new niches, the more previously unasked scientific questions are considered, questions whose answers (if they exist) are supported and encouraged by public authorities and industry, which are gaining awareness of an immense industrial and/or medical market, as is the case for bio-printing. From dream to reality, scientists are often in the position to anticipate the length of the path; however, a dynamic is created. This leads to cultural changes and changes in practices, particularly concerning the importance of creativity, sharing enthusiasm for research, openness with others, the multiplying (and sometimes inhibiting) effect of public actors, on the one hand, and the economic world, on the other, as this work illustrates wonderfully.

This saga of additive manufacturing, told by one of its inventors, teaches us that creativity alone does not suffice; it is necessary to have a good dose of perseverance as well, and it is, of course, necessary to keep moving after the first failures. In addition, this shows us that sometimes the research structures and the environment are not entirely receptive to innovation, even when success comes relatively quickly.

Jean-Claude André also explains with great enthusiasm how we give shape to an idea to feed our intuition, which in turn increases creativity. On the whole, these three volumes provide a wealth of information on additive manufacturing, and additionally, they illustrate and encourage veritable reflection on the task of a researcher and research structures, as well as the role of creativity in research, and finally, they invite us to rethink and reinforce the relations between science and society.

Jean-Charles POMEROL
President of the Incubateur AGORANOV
and the ISTE Editions Scientific Committee

Preface

"We have too often forgotten that specialists are created from amateurs, just as soldiers are made from civilians". [LAT 07]

"In France, strangely enough, it is not those used to sailing the seas, the specialists of the real and tangible, who are asked for advice guiding the flagship, but the members of a caste who stay at port and who, for the most part, have only purely theoretic knowledge of the sea". [BEI 12]

"Technology has taken on a new breadth and organization. Here, I am searching for its specific structure, and I have noticed that it exists as a system, in other words, as an organized whole". [ELL 04]

"Those in the organization who have ideas to do things otherwise or better are divided into two categories: those who do not dare and those who dare. Those who do not dare understand very well the risks and the importance of new ideas, but they are paralyzed by risk taking and the fear of displeasing. Having never tried anything, they have not known failure and are thus unharmed by reproach [...], they are quitters. Those who dare, the innovators, move forward by challenging conventional ideas, organizations, and sometimes procedures. They stir up fears and a lack of understanding and are truly criticized...". [PHI 12]

"Science has largely renounced an interdisciplinary vision allowing the merits of different results to be faced". [THO 83]

"Theory is when everything is known and nothing works. Practice is when everything works and no one knows why. Here, we have united theory and practices: nothing works... and no one knows why!". [EIN 07]

"These creatures of man [machines] are exacting. They are now reacting on their creators, making them like themselves. They want well-trained humans; they are gradually wiping out the differences between men, fitting them into their own orderly functioning, into the uniformity of their own regimes". [VAL 57]

"Speaking of discipline is designating the scientific activity as a particular form of the division of labor in the social world". [FAB 06]

"The imagination is brilliant in that it produces images that enlarge reality and really invent it". [GUÉ 15]

"In cultural terms, no enterprise is built with dreams alone and none without. Action, if it is to be successful, is by necessity guided by practical circumstances. But the goal of any action is defined, implicitly or explicitly, by the deep nature of the human being, his dreams, his vision of life, his culture. The dynamics of life, the challenge of risk and uncertainty, today require from us a new creative effort leading to the reconstruction and to the re-conquest of the notion of progress, which the philosophies and the ideologies of certainty have shuttered almost to the point of destruction". [GIA 90]

"Researching is inventing the world, it is setting new rules of functioning for an ephemeral world. Not like the tyrants who also invent a new world for themselves, but impose it upon others. The researcher does not recreate the world, but rather unravels it to make it. He/She imagines one, then compares it with the real world to clarify it and not to exhaust it. Researching is an endless quest". [ROS 01]

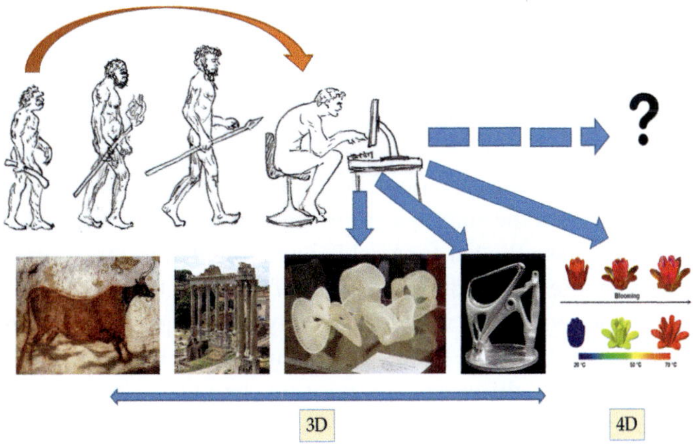

Figure 1. *From additive manufacturing to 3D/4D printing*

This book (in three volumes) is the result of a demand that has been repeated countless times for different reasons, notably among these, of the oversight and the reminder of the oversight to cite a French school that in 1984 succeeded to patent the first additive manufacturing process, stereolithography, several weeks before the Americans (who were working on the same subject, without either party knowing it). However, at the same time, thirty or so years later, it is a history lesson that can be told about a process concept, tossed out in France, without any malice of course, by "clairvoyant hierarchists", the explosion of the research team who felt their future was blocked and an American technical-economic development which has today led to several books and more than 50,000 scientific publications on additive manufacturing, because consequent applicative markets exist with profitable enterprises (and also because there is an immense attraction field around this subject that conditions the actions of a great number of researchers).

So why have we entitled these three volumes "From Additive Manufacturing to 3D/4D Printing"? First, it was about locally bringing material and/or energy to perform a transformation (e.g. from a powder to a solid or from a liquid to a solid). The expression "additive" then takes on its true meaning. But for a short time now, researchers have been developing (or working on) new processes that allow this change to be avoided through the additions mentioned at the start of this paragraph. It thus becomes possible to create an object in one go. Moreover, the use of so-called "smart" materials authorizes the introduction of a complementary parameter, i.e. time or functionality. The 4D aspect is thereby introduced.

The first volume on additive manufacturing is strongly linked to the existence of an effective economic market, one that is already significant, stemming from technological research in the engineering sciences connected to an essential component, that of materials (and of manipulating them to prepare them for manufacturing). It will take several decades for 3D technology to emerge and find its place as a robust technology for manufacturing objects in quite diverse domains. This situation, linked from the start to a strong attractiveness on the part of industrial R and R&D services, has allowed for "field" experimentation with competent users who are more and more demanding in terms of manufacturing qualities (without seeking in this preface to define what this quality, a true portmanteau, represents). Mastery by users, on the one hand, and competition between the bearers of knowledge pertaining to different 3D printing knowledge, on the other, are translated into new demands to be satisfied. In this framework, this demand has in fact made up one of the driving forces of incremental research, a "technology pull" described in Volume 2 (at least as much as is known (or published)).

A solution is good if and only if the concept, its demonstration with the right people, a culture of industrial innovation, and time and finances effectively come together. Maybe at that time, in 1984, there was a closed system of opinion and self-centered management that had not even thought of a possible debate on futuristic technological openings. This conformity to a manufacturing follower style of thinking was more and more often considered to be obsolete. But there was also, beyond socio-economic milieus, an incredible viscosity with many scientists: the most common attitude was not openness to other explicative schemata, but in the majority of cases, the ignorance and/or refusal to accept their existence. Tricks that only imperfectly fit into our ethics as researchers (at the time) must be made and likely developed.

According to estimation methods, the revenue from additive manufacturing lies somewhere between 5 and 40 billion euros (we could think that this is an estimation of the number of protesters in a claim by the police or trade unions!). Some speak of a revolution and others imagine senseless promises (which, according to Audétat [AUD 15], could put every emerging sector in danger); in short, things are booming at present with seven main stabilized technologies and a new kind of governance (Jeremy Rifkin's "makers"). This appreciative placement of the normalizers into categories is indeed rather artificial. Beyond a recent manufacturing technique that associates computer science and matter, 3D printing, with cheaper and cheaper home machines (down to a few hundred euros), constitutes a paradigm shift that impacts product design (which can even be defined, thanks to "open-source" systems), creation (from heavy industry to one's "garage"), consumption and the business models that result from them (from market activity, a new handicraft and DIY (Do-It-Yourself) to counterfeiting).

In fact, the progression rates are always in the double figures (between 20 and 40% per year), which leads some to believe that the additive manufacturing processes will continue to evolve for a long time to become a widespread technology, as they increasingly occupy ever-new applicative niches, quashing the other manufacturing methods that made up the skeleton of 20th Century industrial manufacturing. But what do tens of billions of euros per year represent for the world relative to France's "small" debt amounting to 2 trillion euros? It is therefore difficult to project a future which leads to a possible hegemony of additive manufacturing; besides, it would be more interesting to explore how intelligent synergies can be implemented with technology that emerged long before 1984. Yet, as is resurfaced in Volume 2, there are spaces, still relatively empty, where an attempt is made to challenge the very concept of adding material to processes.

The early 21st Century is marked by the "hegemonic" presence of the digital transition with the technological and practical complements of additive manufacturing processes likely to affect Western society in a quick and profound way. "In the face of radical innovation markets, where the first arrivers can acquire decisive, dominant positions and make the passage of other markets and the economic actors in place disappear, keeping a distance and watching things happen can lead to considerable social and economic costs" [FRA 17]. To go beyond this already uncertain space and become involved in disruptive innovations implies taking risks, thus accepting potential failure, facing their possible negative consequences, and being capable of learning all the lessons this teaches. "If we do not proactively incorporate innovation, this will end up being imposed all the same, in an even more disruptive manner" [FRA 17]. In short, it may be useful to anticipate.

In roughly a century, the number of researchers in Europe has gone from a few thousand to a few million, and despite some disturbances, this trend is continuing. Research activities have been the subject of reassuring discourse on the researcher's independence, on the one hand, and on the other, of a certain programming of research with the aim of achieving goals: security (before the fall of the "Iron Curtain", for example) and economic developments (from mass production with ECSC projects to information and communication sciences and technologies) participating in different forms of competition from France and the European Union.

On this basis, the stereotypical image of the scientist, responsible for the truth and good, is still part of the idealized image, which often positions him/her very highly in relation to a social reality of which he/she only has an imperfect mastery. The will to achieve the best "research efficiency" has led to the promotion of rather mono-disciplinary processes that are easier to manage from "peers", referents of a discipline. On the one hand, in-depth scientific study is maintained by actors from the same field provided that the guarantee of excellence is defined and respected; on the other, for the State, it is easier to realize international comparisons discipline after discipline. Indeed, and this is necessary to remember, without really noticing it, we have gone from a limited worldwide scientific elite to mass research (with tens of thousands of scientific journals) which represents a characteristic that is not discussed by developed nations: research must indeed allow society to respond to the great challenges that loom today: employment, progress, security, global warming, health and quality of life, sustainable development, etc.

Without seeking to speak of two worlds exploring different paradigms, one of in-depth study, the other of responding to social demand (even its anticipation), for this aim would be too limited, rather we look at evolutions translated by a research program that takes account of the different and sometimes antagonistic imperatives (see Volumes 2 and 3). This situation actually shows, at least in part, that the researcher is an element of society who is not independent, even if forms of "grand isolation" have long protected him. But, in the European Charter for Researchers signed by France at the CNRS (National Center for Scientific Research) in 2005, a reminder is given that "Researchers should focus their research for the good of mankind and to expand the frontiers of scientific knowledge, while enjoying the freedom of thought and expression, and the freedom to identify methods by which problems are solved, according to recognised ethical principles and practices."

Without this having been noticed by most of the research actors financed by the State, even if the notion of good is not easily defined (in any case, it does not simply mean the absence of evil), this sentence is a reminder of the role of research centers as a social (or societal) actor, implying new approaches like functioning through interdisciplinary projects and strategic reflections negotiated by stakeholders, stemming from a new prospective work. Considering their importance for the development of citizens' quality of life, research associated with technology is an element that is really starting to be discussed. Indeed, it has participated in the "natural" evolution of things and technological progress has long allowed man to be free from a number of material constraints. In this framework, the rhythm of implementing research results has been greatly modified and complicated, thanks to a more and more frequent coming-and-going between “manufacturing” and research, and thanks to a hybridization of technologies as well as, to a lesser extent, modes of research action (added value from the ability to interact). To work in the new economy of knowledge with greater partnership, there is a need for better reflection on creativity, innovation, and the societal impact of scientific and technological activities. So then, in today’s context of growing co-constructed and contractual research actions, must/can we break away from the “researcher’s temptation of innocence”, of the consoling illusion of “neutral” science, or of the simple transfer of responsibility to the deciders/financers? It will be understood that these are somewhat the stakes of the current evolutions/revolutions applicable to additive manufacturing, and particularly to its future.

With the concept of informed matter, there must be a possibility to modify the shape of objects in time (4D printing), to print living matter (bio-printing), etc. It is thus conceivable to come closer to life by flirting with its possible prolongation! This questioning, like 3D printing pushed to its limits (nanomanufacturing, micro-fluidics, electronics and robotics) associated with other domains, does not correspond to an economic market present today, but instead, if researchers, breaking with the traditions of incremental innovation, succeed (thanks to a bit of creativity and epistemic exploration), immense markets (relative to the “modest” market today amounting to 10 billion euros per year) should open up. The illustrative example of bio-printing which could correspond to a market worth several hundred billion euros per year is a great demonstration of the stakes linked to research concerning initial findings, presented in Volume 3.

If it is necessary to put some of this enthusiasm into perspective, the “classic” additive manufacturing technologies, which have already successfully demonstrated their numerous capacities of industrial development, offer application fields, some of which are very recent and possible, thanks, in particular, to disciplinary research, enabling existing manufacturing processes to be improved. This concentration on a clearly identified objective, process–material optimization, has limited more creative research leading to weaker programming and support for “divergent” researchers, whose numbers, for various reasons, are rather limited in the world of research. Nevertheless, these new applications called 4D printing, bio-printing, 5D printing, etc. result from more complex interdisciplinary activities that, if they succeed, could open markets, no longer in the 4-digit range (billions of euros around the world), but in all likelihood in the 5- or 6-digit range!

There are thus (at least) two types of challenge in additive manufacturing, one is the realization of 3D pieces which contribute a (the most) crucial input relative to the more traditional manufacturing techniques (prototyping, foundry, soldering, etc.) and the other is more prospective on openings in new fields with renewed approaches (and with the associated difficulties). For this reason, with the publisher (ISTE), there was a wish to present the 3D domain in three parts, one with validated scientific and technological bases (certainly with potential redundancies relative to other works on this subject) and the others based on a field of possibilities that offers new epistemological questions, terrible risk-taking, but considerable stakes.

In the first three volumes, it was actually about writing two open "scenarios" that were slowly constructed within a framework, but without a very strict preliminary plan, the scenarios in which the elements were to be introduced and discussed would be spread in an *a priori* graded manner. Each chapter has some degree of autonomy, which can be translated by possible repetitions (as few as possible, however), with a "history" that is progressively fed thanks to the in-depth reading of hundreds (thousands?) of publications, numerous times meandering through and delving into beautiful ideas and scientific meetings for debates, sometimes with success. The gray literature has been a vital source for what is happening in the field at times, which explains the numerous references to the websites in some chapters.

In Volumes 1 and 2, the reader is sensitively placed within the "summary table of disciplines" published in 1829 by Auguste Comte with an "institutional" organization for scientific disciplines, enabling incremental research and development in additive manufacturing. In Volume 3, the idea is to place the reader in a less programmable mode of functioning, with a recursive, systematic and self-organizing character of knowledge, a better willfulness in processes, which sets it apart from the first two, yet it is nevertheless complementary (because it is still constructed using what is known). However, a bit of naivety and/or ignorance may allow for progress to be made in the domain by tackling new paths of creation from a small amount of scientific and technical knowledge in a less "professional" manner, but full of enthusiasm towards a new world to be explored.

An intentional artifact (linked to the engineer and/or designer's work) may be considered a means of connecting an "internal" environment, the substance, the functioning, and the organization of the artifact itself and an external environment, the surroundings in which it is implemented. If the two environments are compatible, the artifact responds to the specifications. As underlined by H.A. Simon [SIM 04] in another framework, the knowledge of an artifact as an additive manufacturing machine "benefits from an advantage on the knowledge of nature, for it is based on valid, previous foundations whose ends will be perverted with a certain dose of new willingness to give projects intelligibility and openings on society." This notion can also be found within the facts in the three works, but with different interdisciplinary openings.

In Volume 3, for the researcher who studies the behaviors associated with the intrusion of temporal aspects and functionality in additive manufacturing, the systems operate for sufficiently long, entirely determined times. But, like "self-organization" phenomena, they can become very off-balance and sensitive to factors considered to be negligible near equilibrium. This is the intrinsic activity of the increasingly complex system, with an increasingly nonlinear behavior, which determines how it is possible to describe its relationship to the environment, which thus generates the type of intelligibility that will be pertinent to understand its possible stories. It is thus not only a matter of an applicative field with its constraints, but also of a theoretical domain to be approached and interrogated in order to resolve the end/means equation in a robust way so as to achieve it.

It will be understood that the epistemological foundations of the reflections in Volume 3 are based on the complexity paradigm, where interdisciplinarity is projected as one of the means of study. The disciplinary approach is too often divided, fragmentary and linear, hence a master idea aiming to know how to percolate through disciplinary borders so that the complexity paradigm can truly spread, notably because the recomposition of thought categories can no longer be based on borders and disciplinary subjects, but on boundary subjects based on the creative, the divergent, who, having no fear of recursiveness, hope to legitimately respond to the great risks society must face.

This change in delivering research for a more systematic approach does not hope to be the indicator of a field of scientific disciplines that, hoping to keep its power, loses its authority, even if current societal issues still cannot handle constructive forms of subordination well. It aims for a real, responsible integration of activities open towards society, bearers of meaning, allowing new research in additive manufacturing to be made to emerge as credible scientific evidence of movements that are materializing.

The evocation of different attractors of disruptive innovation in 3D manufacturing is the focus of Volume 3, in addition to its scientific and technical aspects. The author uses his experiences in this volume to recreate a bit of the history of new additive manufacturing processes, which could, in case of success, invade our daily lives in some years. It is in the spirit of creating a history, and interiorizing it by trying with the time and means available to re-establish them with a personal vision, with the risk of committing mistakes, of having failed with a promising idea. But this is the price to pay.

In the three volumes on the subject of additive manufacturing, it is shown that in relation to almost every problem, there is in fact a creative avant-garde with low inertia: this is carried out by groups of divergent researchers working in practice on the problem at hand. Then there are all the followers, who will structure the "paradigm" and engage it only in forms of conservatism authorizing research to improve processes or materials ("programmable" research). It will take years, even decades, for this paradigm to change positions – often with shoves (linked to the work of the creative by following information provided by the avant-garde). "Paths must be transformed into roads, the ground leveled, etc., so that the landscape will transform significantly until it becomes the main group's parking place" was written by L. Fleck in 1935. Could this context, in terms of research, be adapted to economic development? These characteristics of considering time, and its management, are the elements to be taken into consideration in a process of spatial and temporal transformation of matter that displays significant advantages.

Thus, beyond scientific aspects, indispensible techniques will be discussed to examine how the edifice of additive manufacturing was and is being built through its cultural filters and filters of understanding and interpretation. Anticipating the future of the field of additive manufacturing in the larger sense, to be in a position to prepare ourselves, is considered one of the keys for the long-term durability and competition of companies. This imperative to think of the future, to add to this divergent thought to create new devices for creating objects with the adapted material, devices that are functional, adaptive, "smart", etc., today seems even more significant considering the instability of the environment, the speed of evolution and the generalization of uncertainty. In such a context, research locations must be "offerers" of concepts, of their demonstration to anticipate the productive industrial future, not to mention the technological, economic and governance systems in which, on shorter and shorter reference times, companies evolve (undergoing nonlinear dynamics, splits and breaks). This mission is not only meant for individual researchers, but also for everything around them: research units, their administration and also (and above all else) the proactivity of economic milieus.

In terms of tomorrow and the future, can we not foresee new means of creating objects? At present, we have mastered synthesis, the way in which the objects are constructed. But we could also ask ourselves if it wouldn't be possible to develop systems in which we could give objects an intentionality, thus giving it the choice to look for itself for what changes it needs to make, thus moving onto self-organization with the selection of necessary elements that it would extract from a "bank" for the edification of the final object. This would go beyond the 4D printing that tackles the functional and evolutionary assembly of materials that should be able to come together to create an upgradeable object and that could be made easier through "programmable matter": "Programmable materials and objects that are themselves created would thus make assembly factors and heavy installation procedures superfluous… Robotization, the heart of progress in 20th Century productivity, could thus be integrated into the products themselves, with, as can be imagined, some ethical problems to be taken into consideration" [FRA 17]. Let us thus dream together of this future. The process attempted in these volumes therefore aims to try to question a present (it is impossible to know if this present will likely be able to achieve all its goals) and to determine the conceptual elements that could lead to an original future with access to new applicative niches by exploiting revisited paradigms.

Beyond the exhaustion of the reserves and consequences, it is also the way in which we understand scientific policy to be carried out by taking into consideration different world actors that should evolve to stimulate this nascent domain. In the reflection these books are aiming to create in its readers, it will likely be a matter of proposing changes to be undergone, which correspond to the conceptual displacement of the economy allowed by technology towards a new economy of creativity making a better effort to consider social, economic, organizational, geopolitical, even emerging environmental constraints. It is a form of "design thinking" that is thus to be considered. A reflection on the processes to help the integration of societal data, far from its disciplinary culture, would probably also be projectable (if only on the organizational aspects). In the end, it would be a matter of demonstration, through changes negotiated with the responsible authorities (some of whom are mute), leading to better exploration of the complexity, which can be done well, if not better, maybe with less equipment, but otherwise in a context of social and/or socio-economic demand that it would be advantageous to anticipate, if not follow. The paradigm shift would then take place thanks to scientific initiatives, which are marginal today, which remain aporetic in the paradigm in crisis, and which should be muted in a new scientific era, less framed, applied to 3D printing.

These three volumes can serve to think about the future in the domain that remains exciting for the author after more than 30 years since his 1984 patent, so that we can again find its place concerning its abilities of industrial creation and development in an ever more competitive environment. 3D, 4D, even 5D technologies constitute a path of promotion (among others that stem from the author's competence) of this desire for renewal.

NOTES.–

– For these three volumes, the search for the greatest possible number of specific or general visions concerning the subject of additive manufacturing, which can help the reader, has led to the presentation of the bibliography chapter by chapter and in alphabetical order. In fact, it was almost impossible to classify the bibliography through the numbering of entries.

– Some repetitions in the chapters of these three volumes may exist in an attempt to give them certain coherence and to provide them some degree of autonomy.

Jean-Claude ANDRÉ
Research Director at CNRS
August 2017

Bibliography

[AUD 15] AUDÉTAT M., *Sciences et technologies émergentes: pourquoi tant de promesses?*, Herrmann, Paris, 2015.

[BEI 12] BEIGBEDER C., *Puisque c'est impossible, faisons-le*, J.C. Lattès, Paris, 2012.

[EIN 07] EINSTEIN A., quoted by Debonneuil M., *L'espoir économique: vers la révolution du quaternaire*, Bourin, Paris, 2007.

[ELL 04] ELLUL J., *Le système technicien*, Le cherche midi, Paris, 2004.

[FAB 06] FABIANI J.L., "A quoi sert la notion de discipline?" in BOUTIER J., PASSERON J.C., REVEL J. (eds), *Qu'est-ce qu'une discipline?*, EHESS, Paris, 2006.

[FRA 17] FRANCE STRATÉGIE, "2017/2027 – Répondre à l'innovation disruptive – Actions critiques", available at: http://www.strategie.gouv.fr/publications/20172027-repondre-linnovation-disruptive-actions-critiques, 2017.

[GIA 90] GIARINI O., STAHEL W.R., *Les limites du certain: affronter les risques dans une nouvelle économie de service*, Presses Polytechniques et Universitaires Romandes, Lausanne, 1990.

[GUÉ 15] GUÉRIN M., *La croyance de A à Z; un des plus grands mystères de la philosophie*, Encre marine, Paris, 2015.

[LAT 07] LATOUR B., *L'espoir de Pandore; pour une version réaliste de l'activité scientifique*, La Découverte, Paris, 2007.

[PHI 12] PHILIPPE J., "L'innovation managériale, comment innover dans l'univers bancaire?", in EUROGROUP CONSULTING, *L'art du management de L'innovation dans le service public*, Eurogroup, Paris, 2012.

[ROS 01] ROSE J., *Profession quasi-chercheur*, L'Harmattan, Paris, 2001.

[SIM 04] SIMON H.A., *Les sciences de l'artificiel*, Folio-Essais, Paris, 2004.

[THO 83] THOM R., *Paraboles et catastrophes*, Champs Science, Paris, 1983.

[VAL 57] VALÉRY P., *Œuvres complètes*, La Pléiade, Paris, 1957.

Introduction

"In the first place, you learn that thinking too quickly might indeed be a very bad idea. So-called 'infobesity' surrounds us everywhere and it seriously undermines our creativity". [LEW 16]

"The adjustment of the distance to the subject through disciplinary remoteness and the implicit reference to scientific purity is slightly analogous to geographical and cultural remoteness, in the case of the ethnology of exotic worlds. The extreme cultural difference (indeed disciplinary here) allows us to empirically produce what resemble conditions of objectivation. In particular, this makes it appear as if the observer was, in fact, looking from another world from a point situated away from the subject, although the said subject moved towards this point". [LE 10]

"The practicality of creative genius is not realized until its existence is discovered". [FIM 17]

"Rapidity demands and creates insensitivity to everything which can slow us down, to friction and forms of hesitation, which make us feel that we are not alone in the world. The process of slowing down is to once again become capable of learning, to meet with and recognize both what we have and what defines us. When in such a physical state, we can think and imagine and, in the same process, create relationships with others, of which we would not otherwise be able to encapsulate the essence". [STE 13]

"It appears that comprehensive and systematic research is an inherent trend in the process of human thinking. One of the sources for this trend comes from the quest for certitude, a logical pursuit taking account of the precariousness of human existence. When we experience some fragments of reality, we wish to supplement them, so as to systematically give them meaning". [FRO 70]

"The answer is perceived as reassuring whilst the question gives cause for concern. In the truest sense of the term, it forces us towards activity and towards resourcefulness. Beginning to question necessarily compels us to shift the emphasis of our perspective, to take another look at the hypotheses, review the rules and unearth the hidden meaning". [LAG 12]

"The world is very complicated and it is clearly impossible for the human mind to totally understand it. This is why Man has found a device to criticize the complex nature of the world and its random character. Moreover, it thus permits him to abstract a sphere in which we find simple laws. We then describe the initial conditions for complications, and the spheres of uniformity within the laws of Nature". [WIG 49]

"Another major factor is that, for years, organizational management has been developing methods for increasing productivity and minimizing risks and errors that tend to stifle creative experimentation. The predominant approach to management that evolved during the industrial era, known as scientific management, broke jobs down into specific, sequential tasks, which could be allocated at appropriate times for completion in order to optimize efficiency". [SIM 11]

"In particular, the transformation of the research profession is seen through the inflation of administrative tasks, intended to 'make researchers accountable for everything'. These new administrative demands, often analyzed in accountability terms [NEY 07], and presumed to make organizations both transparent and responsible, weigh heavily upon researchers' daily tasks. This is so seeing that researchers devote a growing portion of their work time to it – an amount of time which one lady interviewed assesses as a third of her total work time". [HUB 11/12]

I.1. Introduction

Additive manufacturing is made up of seven standardized processes, according to the French standard NF: E 67-001 and international standard ISO 17296-2:2015E. It involves the following processes (MADEELI – the Agency for Economic Development, Export and Innovation based in Toulouse, France, 2017 [MAD 17]):

– VAT photopolymerization or stereolithography;

– Material jetting;

– Binder jetting;

– Powder bed fusion;

– Material extrusion;

– Direct Energy Deposition (DED); and

– Sheet lamination.

The themes in Volume 1 have already shown all the benefits of additive manufacturing within numerous spheres of application with the technological specifications that could be stipulated. The emergence of a small revolution translates through various inclusion difficulties within a manufacturing culture, which already has its own constraints. This leads to the promise of accomplishments whilst everything within the manufacturing environment is far from being stable. The performance itself is sometimes insufficient as is the time taken for manufacturing [DGE 16]. This is the case even if increasingly new materials, software and processes mature within their product lifecycle. Three-dimensional (3D) printing has been introduced in a slightly surreptitious manner by proposing local production of objects which would otherwise be practically impossible. Even though numerous outstanding issues still remain in this sphere, a transition to the growth phase still occurs with double-digit gains.

> "Additive manufacturing is often presented as being likely to radically alter how objects are designed as well as produced. However, the production system organization and the mass distribution of the given product still remains largely dependent upon a number of significant technical developments. The pace of such advances is difficult to anticipate. At the current time, machines and processes do not allow us to respond to all industrial production constraints. Progress may, in particular, have to be achieved to speed up the production rate" [COE 17].

At the same time, the arrival of a given technology removes numerous design phases by supplying a tool that (in principle) sits between the computer and the object concerned. This re-empowers designers and "makers" who can produce innovative objects "in their own garage workshops". Such objects would otherwise be out of reach to them. The enlargement of this new relationship between conception and the given tool may lead to cultural developments in the production process. This can be accompanied by the desired aspects of reintegration of some production nationwide and possible "Uberization" of the given activities.

These components will still increase or develop from the scientific, technological and social perspectives in equal measure. These comprise a unique sphere in which the author has "floundered" in a sense. In Volume 1, he has tried to share with you a little of his hopes, with suggestions and opinions that you may choose to ignore. From the author's perspective, the important aspect is to fuel reflections around the emergence of ideas for "informed" discussions.

In the second volume, the spirit of the text is not entirely the same. Indeed, it is envisaged that the sphere should be open to improvements, which are largely defined by their purposes. These improvements will have incremental innovations for methods. This is because the scales, constraints and other factors will change over time. In this setting, which remains vague and indeed yet to be constructed, there are still limited numbers of published knowledge sources. In these circumstances it is difficult to "explore" the sphere. This leads the author to simply mention the elements of information, which he has retained from his own reading. It will be interesting, in a few years, to examine how these spheres will develop a structure and nucleate so as to move towards preferred directions, which then appear more clearly. This analysis will be all the more interesting for France (but perhaps not France alone). This is because it will be necessary not only to establish a link between scientific and technological sets of themes, but also individuals, who are currently greatly "diluted" within the national research mechanism [FCM 17].

The so-called "criteriology" serves to measure scientific excellence within the current short-term accounting-productivist system. It falls more within the realms of a quantitative supermarket than a qualitative and detailed approach to research. In principle, the freedom to seek something is like the freedom to create. By definition, genuine research activity consists of uncertainties. Moreover, it is difficult to assess the "value" of a genuine new result due to various reasons. However, as we enter into a certain continuity of thought (the old paradigm of connection by point and tangent), there is no doubt that the increase in research in the sphere will be present. This is despite the constraints that will be discussed in terms of risk-taking (existing at various levels).

Indeed, to integrate the axes of support and seek grants for various activities, scientists should show the quality (which may also be questionable) of the project to the financing bodies and in some way illustrate with great conviction a potential return on investment. It is within this framework that it is necessary to have already published, if possible many times, in so-called "high-ranking" scientific journals. It is therefore necessary to have demonstrated in advance that the project will be placed in the right hands (having someone with a good brain, often a further questionable issue) if it is financed in such a way. Furthermore there is a need to show that it will become integrated, in keeping with being dynamic and having the promise of progress, which is associated with the given request for proposals.

The technologies which will be described are characterized by their major interdependence with other spheres: nanotechnologies, electronics, process engineering, chemistry and energy (and its storage). Of equal importance is the digital sphere, which must never be forgotten. The ruptures, which will be made within the spheres presented in the following chapters, are conditioned to a large extent by progress, which is both made (and long-awaited) in their cutting-edge research. "This interdependence of technologies makes the so-called foresight exercises regarding technological development fairly complicated endeavors. This is all the more true since innovations, in particular disruptive innovations, spring up at interfaces. Yet, these interfaces increase by reason of the possibilities of the information society, as well as the major interdependence of the most recent technological advances. These favor interactions between technical and professional spheres, which were previously less connected. Furthermore, the given technology's lifespan is uncertain and may be long. Sometimes many years pass between the first positive laboratory results and deployment of a given technology" [COE 17].

In 2016, the DGE (Direction Générale des Entreprises (the Directorate General for Enterprise)), division of the Ministry of Industry, published a report entitled "Technologies clés, 2020" (Key Technologies, 2020), with 47 targeted technologies. Within Volumes 2 and 3 of this series, the attentive reader is able to see that 10 of these 47 technologies will be directly or indirectly impacted by the disruptive technology activities laid out in the present work. These are the following technologies:

– Cell and tissue engineering;

– Health imaging;

– Collaborative robotics and human technologies;

– Microfluidics;

– Advanced and active (smart) materials;

– Autonomous robots;

– New hardware-software integrations;

– Modeling, simulation and digital engineering;

– The Internet of things;

– Not forgetting, of course, the subject of this work – additive manufacturing.

Lin *et al.*'s view [LIN 14] on the development of additive manufacturing is depicted in Figure I.1. This passes from the phase of producing monofunctional parts to the use of nanotechnologies in additive manufacturing. In this volume concerning the emerging activities with an essentially incremental origin, some of the recommendations of authors of other works will be included. These recommendations cover areas where additive manufacturing is pushed to its limits. In this figure, the fields on the left essentially correspond to the subject matter set out in Volume 2, and those on the right to that in Volume 3.

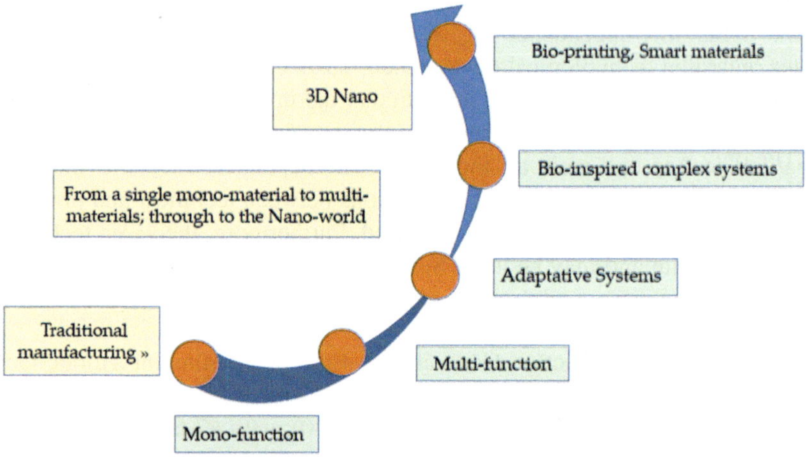

Figure I.1. *A medium- to long-term vision of additive manufacturing*

> "The engineer's role is to use the technical knowledge available to him/her to devise innovations which are an acceptable compromise for all stakeholders. It is said that technical perfection poisons the economy! It is necessary to be ready to 'tinker' to satisfy the greatest number of people. Technological resources are such that you cannot believe an engineer who assures you, without batting an eyelid, that there is only one efficient profitable solution and that you must choose it" [CAL 11].

A part of Volume 2 devotes a chapter to the developments and improvements of processes set out in Volume 1 (from the author's perspective). However, the "conventional technology" aspects of additive manufacturing will be pushed to their limits in fields where, until now, scientific activity has been modest. However, upon analyzing the development of scientific production over time, the reader is able to observe that we are currently within a period of high volumes of emerging technologies. Additive manufacturing, by its success and the pressure it exerts (its fascination to others) percolates within the available porosities and creates opportunities.

This slightly widespread, even loosely structured grouping corresponds to thematic nucleations, which are being designed. Moreover, these may be subject to the attention of researchers provided such researchers' culture and general environment allows it. Thus two chapters are devoted to discussing the subject of development of additive manufacturing and issues associated with risk-taking, innovation and creativity. With innovations, which rely upon scientific knowledge, we are dealing with boundary objects involving interdisciplinary approaches. These will be discussed in Volume 3, with an analysis of their integration under the term of "convergence". It goes without saying that these two chapters are adapted both to incremental innovations (Volume 2) and potential disruptive innovations (Volume 3).

The only way to recognize the scientific quality of a given interdisciplinary work may be to, in the end, translate it into as many versions as there are disciplines. With each translation, the work should be put into perspective in relation to the literature and research of each given discipline. Proving that it is not a question of an average, or indeed a mediocre, standard of research excellence, even if it is creative, is essential. This is because interdisciplinarity is generally viewed as a means to overcome "arbitrary compartmentalization of a given concept". It is also seen as a process, which goes beyond the simple juxtaposition of disciplinary standpoints. This involves implementation of cooperation and integration between specific disciplines around a common goal within a knowledge co-production perspective. It is clearly applicable to the various fields of 3D printing.

In other words, interdisciplinarity may be defined as the result sought by a conscious and controlled attempt of rational humans, despite having adverse effects. This involves time to understand, cross-fertilization, management of different timescales and methods, as well as other factors. Beyond its idealized character, doesn't this type of definition lead to an extremely restrictive view? In focusing on the result rather than the means to reach the result and in particular to implement disciplinary translations, doesn't it run the risk of compromising the given endeavor? Consequently, the eventual emergence of forms of innovation stabilization happens and even includes the bubbling phase of additive manufacturing. There is a chance that such additive manufacturing has not yet reached the maturity phase.

Researchers involved in different disciplines must often indeed communicate with each other, in the absence of a clear and common aim. They exchange and copy each other's ideas. This happens so much that it is no longer a question of them seeking to stand out from each other. They thus put in place the conditions for a future "active" (smart) form of interdisciplinarity. We could even argue that interdisciplinarity in the strictest sense of the term, that is to say conscious and active, assumes the prior existence of a form of "collateral" interdisciplinarity. This stems from the dissemination of a given concept, an instrument or a technical protocol within various disciplinary communities.

Hence, we consider the need to design structures whilst displaying a strong disciplinary identity. This is as long as additive manufacturing sciences have an identity, one of their missions being the purpose of allowing dialog and meetings between researchers from different disciplines and backgrounds. Such researchers are committed to shared research objectives. There are potential exchanges within these structures (possible examples being networks, laboratories and Joint Research Units between academic staff and industry). These should permit each stakeholder to discern conceptual or temporal discrepancies between representatives of various disciplines and to adjust these to make gains within scientific competition in terms of the application of their work.

With Kuhn's "cherished" scientific paradigms [KUH 72], a so-called "normal" science was invented, corresponding to an economy of thought, indeed to standardization of an epistemological simplistic procedure. So in order to create a discipline associated with additive manufacturing, we must go beyond the so-called "simple" intellectual requirements. Indeed, with possible external threats of vassalization and the need for greatness and legitimacy by some groups or individuals, it is necessary to demonstrate how the subject of 3D printing was at the crossroads of various social groups. All of these groups were indeed somewhat concerned by the transformation and shaping of the subject and the sciences, which work towards its development. This led to support from the so-called "aspiring/beginner scientific 3D printer specialists", an element of autonomy and seizing an element of (small) power (as it happens in politics) relative to other disciplines.

Within a creative pre-paradigm phase, social and economic realities are the determining factors leading to the cultural body that is currently stable. This together has recognition within the scientific system insofar as the sphere has defined its so-called "body of doctrines". It is simplistic through a form of, it could be argued, necessary disjunction, having a multitude of disciplines, which may participate in its development. However, as Edgar Morin pondered, is it not, in fact, necessary to acknowledge a principle of rational uncertainty? He thus wrote in [MOR 99], "Rationalization believes that it is rational because it makes up a perfect logical system, based around forms of deduction or induction. However, it is based around mutilated or mistaken bases. It closes its mind to the challenging of its arguments and to empirical testing".

As long as it does not voluntarily seek to deal with "unmanageable" problems the sphere remains alive, provided it maintains "acceptable" connection with anticipating society's needs (in various ways). This poses the delicate issue of voluntary withdrawal of habits perpetuated by the "system", its social organization and its ends. By analogy this is a little like the Tex Avery cartoons, in which the hero continues to run above the precipice before he falls. The issue of an unanticipated postmortem survival towards the opening up of new fields is however set out in a world which retrospectively supports excessive amounts of risk-taking, creativity and ruptures, but which claws back losses with ease.

As de Brabandere [DE 16] has stated, "It is very important to realize that we construct our thinking, our ideas, our judgments and our stereotypes more on the basis of ignorance than on knowledge. In order to design, we are, on the one hand, required to forget, and, on the other hand, to take a form of distance". Let us therefore endeavor to move forward in this direction.

> *"Projects and entrepreneurs appear to emerge more spontaneously within territories having far less resources and natural advantages. Human capital is all the more sought when there is nothing else to exploit". [GOD 11]*

> *"The mind formed by disciplines loses its natural aptitude to contextualize knowledge, thus integrating such knowledge as a natural whole within the mind's corpus. The weakening of the perception of the overall picture leads to a weakening of responsibility. In addition, it leads to the weakening of solidarity". [MOR 99]*

"Progress takes place when it appears in retrospect that it is difficult, even impossible, to go back. Progress is less about concepts moving forward than what prevents them from going back". [SON 05]

"The complexity of situations and the uncertain repercussions prohibits traditional sciences from definitively making decisions from above. The death of the right decision has already occurred. We are now living in the time of the satisfying decision". [GER 97]

I.2. Bibliography

[CAL 11] CALLON M., LACOSTE A., "Défendre l'innovation responsable", *Debating Innovation*, vol. 1, pp. 5–18, 2011.

[CAS 17] CASTILLO A., "La créativité ou l'art d'exhumer des analogies cachées", available at: https://www.letemps.ch/economie/2017/02/16/creativite-lart-dexhumer-analogies-cachees, 2017.

[COE 17] COE (CONSEIL D'ORIENTATION POUR L'EMPLOI), *Automatisation, numérisation et emploi – Tome 1 – Les impacts sur le volume, la structure et la localisation de l'emploi*, available at: http://www.coe.gouv.fr/IMG/pdf/COE_170110_Rapport_Automatisation_numerisation_et_emploi_Tome_1.pdf, 2017.

[DE 16] DE BRABANDERE L., "Les Big Data pourraient-ils nous rendre plus créatifs?", available at: http://www.ladn.eu/innovation/datas-datas/luc-de-brabandere-plus-dinformations-moins-de-certitudes/, 2016.

[DGE 16] DGE (DIRECTION GENERALE DES ENTREPRISES), "Technologies Clés 2020, Préparer l'industrie du futur", available at: http://www.entreprises.gouv.fr/politique-et-enjeux/technologies-cles-2020, 2016.

[FCM 17] FCM (FILIERE CARNOT MANUFACTURING), Cartographie des acteurs clés de la R&D en Fabrication Additive en France, Instituts Carnot, Paris, 2017.

[FIM 17] FIMECO, "Pourquoi la créativité est-elle le moteur de l'entrepreneuriat?", available at: http://www.blog-fimecobti.com/2017/02/pourquoi-la-creativite-est-elle-le-moteur-de-l-entrepreneuriat.html, 2017.

[FRO 70] FROMM E., *Espoir et révolution; vers l'huminisation de la technique*, Stock, Paris, 1970.

[GER 97] GERMINET R., *L'apprentissage de l'incertain*, Odile Jacob, Paris, 1997.

[GOD 11] GODET M., *Bonnes nouvelles des conspirateurs du futur*, Odile Jacob, Paris, 2011.

[HUB 11/12] HUBERT M., CHATEAURAYNAUD F., FOURNIAU J.M., "Les chercheurs et la programmation de la recherche: du discours stratégique à la construction de sens", Quaderni, pp. 85–96, available at: http://quaderni.revues.org/556, 2011/2012.

[KUH 72] KUHN T.S., *La structure des révolutions scientifiques*, Flammarion, Paris, 1972.

[LAG 12] LAGADEC P., *Du risque majeur au mégachocs*, Préventique, Bordeaux, 2012.

[LE 10] LE MAREC J., "Enquêtes et savoirs du contact dans les études de science: pour une réflexivité institutionnelle", in LE MAREC J. (ed.), *Les études de sciences: pour une réflexivité institutionnelle*, EAC, Paris, 2010.

[LEW 16] LEWIS C., *Too Fast to Think: How to Reclaim Your Creativity in a Hyper-connected Work Culture*, Kogan Page, Paris, 2016.

[LIN 14] LIN D., NIAN Q., DENG B. *et al.*, "3D printing of complex structures: man made or toward nature?", *ACS Nano*, vol. 8, pp. 9710–9715, 2014.

[MAD 17] MADEELI, "Fabrication additive: Enjeux et Impacts en Occitanie Pyrénées-Méditerranée", available at: http://www.madeeli.fr/wp-content/uploads/2017/01/Fabrication-Additive-en-Occitanie-Etude-2016.pdf, 2017.

[MOR 99] MORIN E., *Les sept savoirs nécessaires à l'éducation du futur*, Le Seuil, Paris, 1999.

[NEY 07] NEYLAND D., "Achieving transparency: the visible, invisible and divisible in academic accountability networks", *Organization*, vol. 14, pp. 499–516, 2007.

[SIM 11] SIMS P., *How Breakthrough Ideas Emerge from Small Discoveries*, Business Book, London, 2011.

[SON 05] SONDAG G., DUNS S., *La métaphysique de la singularité*, Vrin, Paris, 2005.

[STE 13] STENGERS I., *Une autre science est possible! Manifeste pour un ralentissement des sciences*, Les empêcheurs de penser en rond, Paris, 2013.

[WIG 49] WIGNER E.P., "Invariance in physical theory", *Proceedings of the American Philosophical Society*, vol. 93, pp. 521–526, 1949.

PART 1

Incremental Innovations and Technologies Pushed to their Limits

1

Incremental Developments of Processes, Machines and Materials

"When things are arranged so that description through using the senses allows us to easily imagine them, we say that they are well-ordered. Otherwise, we say that they are not well-ordered or confused. As the things which we can easily imagine are more pleasant than the others, Man therefore prefers order to confusion, as though, outside of our imagination, order were something found within Nature". [SPI 94]

"Knowledge… is not a series of coherent theories converging towards an ideal conception. It is not a gradual march towards the truth. Rather it is an ever-greater ocean of mutually incompatible (and perhaps immeasurable) options. Each singular theory, each fairy tale and each myth being part of this collection, forces the others to a state of greater flexibility. All of this contributes, via this rivalry to the development of our awareness". [FEY 79]

"Our post-modern societies seem to be characterized by a so-called 'dual loss': that of tradition and the certainties of pre-modernity. This is based, on the one hand around respect for the past, and that of the modernity myths anchored within a positive view of the future, and in faith in progress. On the other hand, it is based upon a Promethean concept of Man". [GUY 14]

"Bergson defined the laugh as the 'superimposition of the mechanical upon the living'. Likewise we could say of industry that it is the superimposition of machinery upon organic concepts, and it longs to become natural. The same can be said of digital being affixed to machinery". [MUS 14]

"The second stage is to divide each of the difficulties which I will look at into as many fragments as possible, and as may be required to best resolve these difficulties. The third stage is to put my thoughts into order, starting with the simplest and easiest subjects to understand to go up little by little, as if by degrees, to the most complex knowledge". [DES 92]

"It is important that we do not conceal our uncertainties. This awareness of the frailty of given structures is consubstantial to the scientific approach". [BAR 16]

"Make the present moment more intense so as to make the moment profitable in the extreme. We thus escape the routine of a life which is limited in space and time, such could be the sense of this present imprisoned within the moment". [KHE 10]

"The capitalist economy is not and does know not how to be stationary". [SCH 11/99]

"The entire issue lies within this opposition between gradual and disruptive innovations. Should innovation extend current cycles, support existing structures and make our way of life sustainable? Alternatively should it even open up radically new routes, substitute new technologies for those inherited from the past and revolutionize our societies? Owing to this ambivalence, the rhetoric of technological innovation takes the form of a paradoxical injunction: 'Let everything change, so that nothing changes!' As humans we recognize that this is hardly a stimulating approach'". [KLE 15]

"However, within many management teams, the appetite for, or the tolerance of risk, are perceived as interesting esoteric topics of conversation. This is especially the case for researchers and academics. However it has no actual bearing on their daily activities in terms of strategies, and no impact upon their decision-making processes". [LOU 17]

1.1. Introduction

The myth of the so-called "industrial revolution", promised for additive manufacturing, arises when the question concerns new technology creation. However, these myths have their significance. They permit the creation of the given movements and membership of those movements. Nevertheless, any production system only truly has direction if it is both explained and questioned at regular intervals. It can thus maintain a standard of technological excellence relative to a determined future view, preferably a technological view, for the purposes of this chapter. If we consider the innovation market, several principles are at work [GUZ 95, CAR 98]:

– an initial aspect corresponds to the principal of assimilation depending upon the success of a given new product, or identified as such (the launch of new products on traditional markets, or attacking new markets with advanced products, which have simply been re-tailored). Success can lead to the stabilization of ways of thinking and modes of action;

– a second aspect is linked to the principle known as plurality, which assumes that a new element (product, process or another aspect) can only be assimilated if it is integrated within a given existing set (linked with the notion of technological rupture);

– the principle of alignment stipulates that technical skills induced by a technical or organizational change should not exceed (in terms of training, hiring and other outlays) the maximum cost already agreed within the business in other circumstances;

– the principle of transfer to others is aimed at making the initial research and development effort profitable;

– the principle of expectation is linked to a response to demand without the business seeking to propose a given offer. Indeed, often the creative offer is only generally included because it flows from a technological innovation that disrupts the course of events.

These observations evidence accelerated production of new products using well-mastered methodologies. Under these conditions, there can be a connection between research temporalities and that of the company managing such research. This is so, as long as scientific "laboratory" expertise is effectively present and it is a question of "genuine" innovation.

Chapter 3 of Volume 1 of this series set out the majority of current technologies for additive manufacturing. We can imagine, within a world undergoing rapid development, that these few technologies grouped together by the authors within the 3D printing technology categories 3, 4 and 7 are only a small share of all publications and patents since 1984. However, doing everything possible may not necessarily lead to an industrial application, and several criteria come into play. These are criteria that range from types of obstructions on the part of interested businesses having a strong presence in the field to the purchase of start-ups and little promotion of entrepreneurship favoring the academic world. There are also difficulties in financing with the so-called "Death Valley crossing" and other factors. Moreover, not all ideas, even if their concept is proved, can be applied. This chapter therefore integrates an area in which there is a place for uncertainty with particular projects. This includes some ideas that are nearly at the end of the transition between research and application, with others still remaining at the stage of "academic" feasibility.

Intentional risk-taking by the inventor involves not siding too much with the fear of seeing the seeds of creative and original ideas killed too soon. Ideas should not be shelved before they have demonstrated their capacity to aid additive manufacturing in its development, "biodiversity" or abundance. If we take mechanical manufacturing into account, the number of modes of manufacture largely exceed the current 3, 4 and 7 technologies "acknowledged" in additive manufacturing. However, a personal choice criterion (which may be subject to criticism on the part of the reader) has been retained. This concerns the ability of ideas to emerge from forms of continuation of activities stemming from the sole "knowledge of prior art/knowledge". It is thus the case that this chapter does not describe how we may increase the size of a 3D machine to build houses (except perhaps for the "material" component), produce a 3D machine functioning on solar energy and with sand as the material, or spatialize a 3D machine.

Notwithstanding this, not taking this aspect into account does not mean that these activities are completely eliminated; after all, they are even considered by some to be industrious or exotic. It is simply that the development of such a machine takes time, necessitating a number of so-called "DIY (Do It Yourself) scientists". It also necessitates that technological, prescriptive and financial constraints, which are often difficult to surmount, are taken into account. What we note is that increasingly refined mechanisms and control procedures are being implemented all the time. These operations continue on a "business as usual" basis. Moreover, by being located upstream, no doubt fewer inertial constraints are experienced. However, clearly this is not the objective of this chapter.

Within a broad view, creativity, which conditions innovation, may be considered as the capacity to either provide or find original solutions to adjustment problems. In fact, every human being is confronted with such problems. The creative approach begins with the acknowledgment of a given problem [FOR 02]. From there, stakeholders become involved in a divergence process, which will finally perhaps end with convergence, in the form of a new solution to the problem. “Creativity and then innovation, is therefore ‘doing’ derived from ‘being’” [ANZ 88]. Innovation covers the aspects stated below. These are discussed by Schumpeter [SCH 04, SCH 11/99]; Gaglio [GAG 11a]; Larédo and Mustar [LAR 03]; Lesourne *et al.* [LES 04]; OCDE [OCD 05]; Ecrin [ECR 05]; Kuhlmann [KUH 01] and other works, they are:

– new goods or the same goods equipped with new and distinct original properties;

– new production methods with new potential markets;

– the opening up of new markets;

– conquering a source of raw materials, which has not yet been exploited;

– the creation of a new organization, such as the creation of a monopoly situation;

– an innovation assuming the emergence of new social practices within the wake of the new product [ALT 10];

– an innovation making the assumption that we are capable of imagining how it will be carried out for members of society [CHA 11];

– an innovation calls into question long-established positions and secure income. It upsets certainties and adaptations [GAG 11a].

Should not we consider creative innovation as governed, both by the search to discover fundamental principles on the one hand and by ideas stemming from divergent thought, and systemic research and development on the other, likely to favor the creation of economic and social value? If it is thus, these fields should have particular characteristics: one aspect is openings and the other is deepening understanding. Let us consider together what the author has selected as potential openings. We understand that these are essentially the foremost aspects of this list, mainly coming within the supply logic, and which are targeted below. Essentially, on the one hand, it is possible to evoke linear progress being predictable and not entirely used in every case. On the other, we may evoke a more spontaneous, reticular and chaotic innovation.

1.2. Undertaking non-layered stereolithography

The concept devised by the Institut Battelle in the 1970s [ADA 68, ADE 71, MCG 84] was the concept of the manufacture of three-dimensional (3D) objects within transparent fluids using multi-photon absorption. The principle used at that time involved the exploitation of two-photon and multi-photon absorption, such as that shown in Figure 1.1. In this figure, incident photons are shown by their hν energy, where h is Planck's constant and ν is the frequency of radiation. During that point in this particular field's development, sequential absorption was defined as follows: a chromophore absorbs the wavelength λ_1 (corresponding to the frequency ν_1 with the formula $\lambda = C/\nu_1$, where C is the velocity of light in a vacuum), which has a given lifespan (ranging from a few nanoseconds up to an hour). A second photon emitted at λ_2 allows conversion to a higher excited electronic state, leading to the fragmentation of the substance (or by energy transfer from a given acceptor). Thus, compared to the description in the previous volume concerning energizing a photon from a radical or ionic polymerization reaction, the formation of given species, which will initiate polymerization, occurs within a sequential or simultaneous two-photon process (see Figure 1.2).

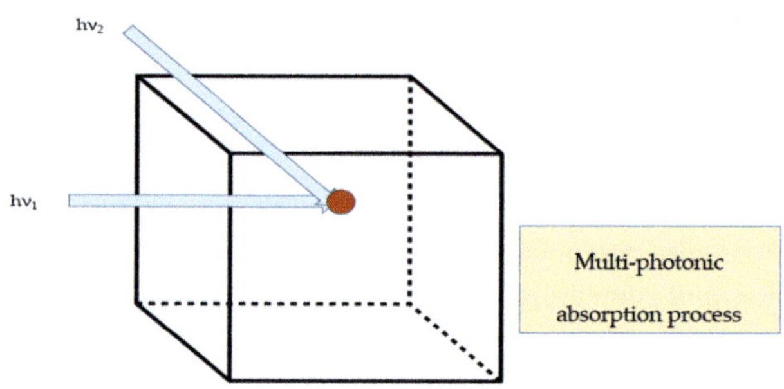

Figure 1.1. *Principle of two-photon simultaneous or sequential absorption*

In Figure 1.2, the principle of simultaneous absorption is shown on the right-hand side. Light beams are displaced within a transparent medium (which is generally not as simple to achieve when the thickness of the given medium to be penetrated is of the order of a few dozen centimeters). In principle, we might envisage carrying out a local-scale conversion of the medium. This process might include species involved in energizing, inhabiting an excited precursory electronic state. This may occur by either sequential absorption, necessitating the transition through an intermediate electronic excited state [JIA 14], or simultaneous absorption.

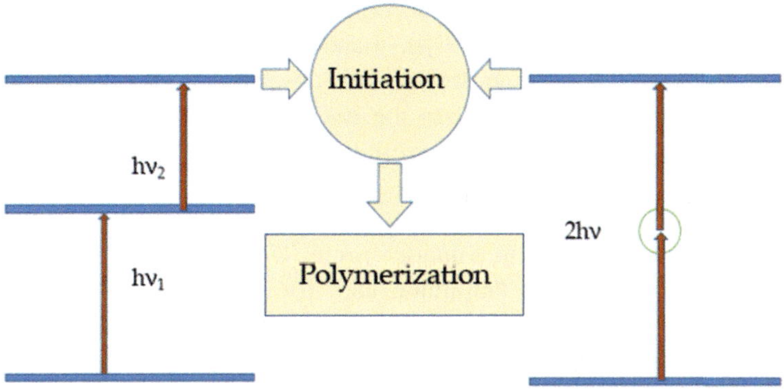

Figure 1.2. *Two-photon sequential absorption (left) and simultaneous absorption (right)*

To go further back in time, the access to picosecond lasers, or better yet femtosecond lasers, was not envisaged even though the photochemistry of two-photon processes had already been studied. This explains the origin of the patents proposed at the beginning of this section. Several possibilities might be mentioned at this point. These are photon absorption to inhabit an electronic singlet state, transition to a triplet state (with a lifespan which is potentially longer than that for a singlet state) and absorption of a second photon to change from inhabiting this state to that of a triplet reactive state. An example of the latter is acridine, referred to by Lewandowka-Andralojc *et al.* [LEW 12] and Kellmann and Tfibel [KEL 82]. Similarly, there is also the formation of a triplet state transferring its energy to another triplet, which itself absorbs a photon to create a reactive state (e.g. the acridine–rhodamine 123 couple). The significance of this system is that if linear transformation translates by a non-radiative return without a marked chemical transformation, there is a benefit. It is that the system can be reversible at low light intensities. On the contrary, triplet–triplet absorption allows for the chemical transformation of the medium, with apparent opportunities in terms of spatial resolution. The reader who is interested in this field may benefit from referring to several works. These include Bensasson and Land [BEN 71] who found coefficient values for increased molecular extinction coefficients for triplet–triplet transitions, with complements derived from Bagdasaryan *et al.* [BAG 72/73]. With regard to other substances, Nielsen *et al.* [NIE 98]; Brinen *et al.* [BRI 72]; Malval *et al.* [MAL 07, MAL 11]; Singh-Rachford and Castellano [SIN 09], and with regard to biacetyl (which has a lifespan of the high triplet at room temperature, of the order of 1 ms (as Almgren demonstrated in [ALM 67])), Land [LAN 68] is worth consulting. However, to deal with the question of sequential absorption, other than the choice of the right photochemical initiator, raises the issue of bringing together two beams within a parallelepipedic vessel (see Figure 1.1).

Simulations were undertaken, aiming to be completed within a plane, having a given number of voxels limited to a quarter circle. The two beams are considered to be perpendicular, of the same amplitude and, for simplicity, emitting the same wavelength ($\nu_1 = \nu_2$). They also have the same rate of absorption for all substances, taking part in single-photon absorption or sequential absorption. They therefore have a maximum flow for the parts subject to irradiation and nil for the non-polymerization zones. Through the selection of a six-step beam, Figure 1.3 shows the distribution of light for a given section by using false colors. In this case, the red area shows weakest intensities and the dark green area, the highest intensities. We can observe from this figure that where single-photon processes are involved, the quarter circle being produced is not visible and that the optimum position available is a square zone (green), which leads to polymerization.

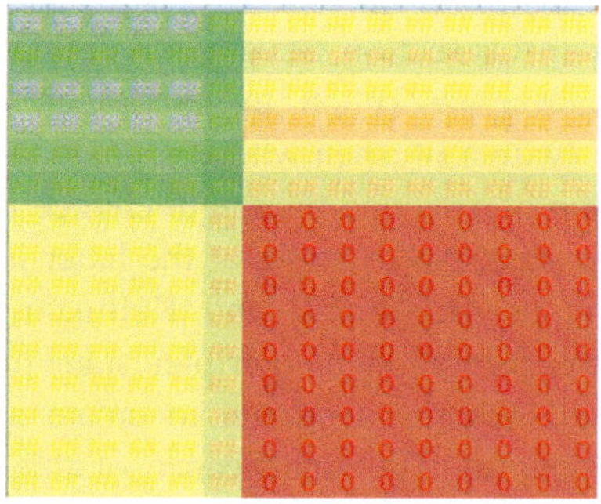

Figure 1.3. *R = 6 and the addition of photon flows. For a color version of the figure, see www.iste.co.uk/andre/printing2.zip*

The three diagrams in Figure 1.4 show the distributions using so-called “false colors” (on the same bases as those used above) for two-photon, three-photon and four-photon absorptions.

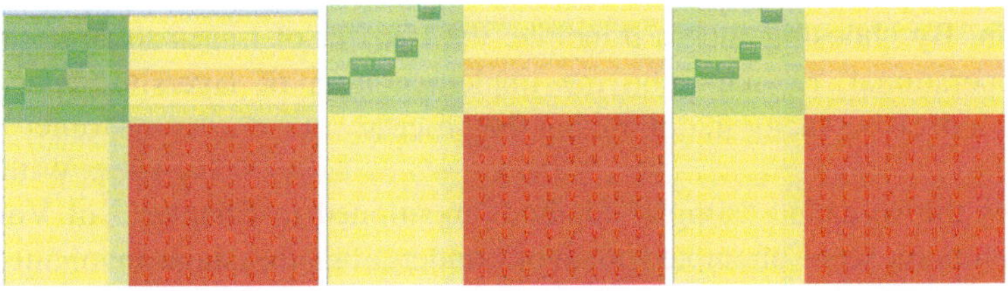

Figure 1.4. *Two-photon absorption (left); three-photon absorption (center) and four-photon absorption (right). For a color version of the figure, see www.iste.co.uk/andre/printing2.zip*

For these examples, the contrast (the dark green/yellow relationship) varies with the degree of multi-photon process. The results are 2.6, 5 and 6.8 for the two-photon, three-photon and four-photon processes, respectively. This is achieved with a very modest number of voxels. This technically non-exploitable situation has led scientists in this field to work in layers (see Chapter 3 of Volume 1).

Figure 1.5 shows the same phenomena for a four-photon process but for the case of R = 13. The size of the mesh is therefore appreciably reduced by a factor of 2, compared to that shown in Figure 1.4. It is worth noting that the contrast which was 6.8 decreases to 2 when the value of R is doubled (from 6 to 12). Under these conditions involving the intersection of two parallel and orthogonal beams in relation to each other, it is unrealistic to be able to produce an object with an accurate resolution using this process. This situation would be further accentuated if we were to take into account the third dimension of the space.

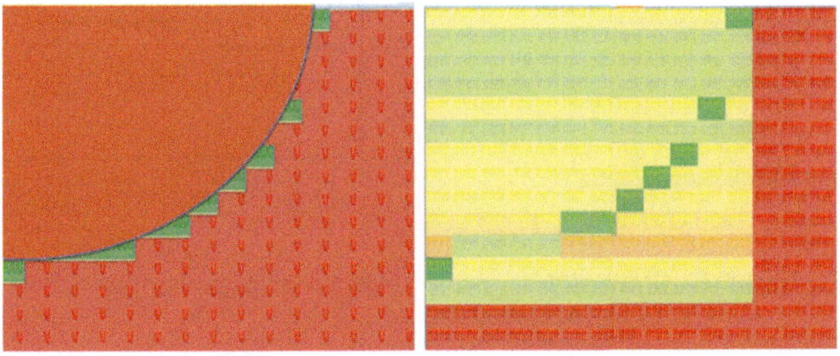

Figure 1.5. *Four-photon absorption; R = 12 (with the left-hand side diagram showing what is envisaged would happen with the quantification effect). For a color version of the figure, see www.iste.co.uk/andre/printing2.zip*

1.2.1. *Optimizing the light supply within a single-photon process*

Under these conditions, it was not necessary to find multi-photon absorption systems. Hence, at the time, the use of continuous lasers is necessary to "shape" the finish resin. By choosing such a principle, the transfer of issues so as to make the process operational, relying upon the implementation of layers, appeared fairly quickly as a decisive factor in optimizing the process.

Besides the so-called "simple" methods mentioned in Chapter 3 of Volume 1, several solutions were attempted to avoid undergoing the stage involving layers. These are described below.

1.2.2. *Transparent window*

Figure 1.6 shows the principle where a window, which is transparent to photons, is placed on an object in the course of its construction. This is done by mechanically imposing a layer between the object and the transparent window.

The process enables us to realize polymerization. However, the problem of polymer adhesion on the window, or the extremity of a given fiber optic, cannot be solved simply by using optical glass plates. This leads us to use flexible strippable materials or strive for another solution. The choice of such existing materials in the form of films allows for the effective production of a given object. However, the manufacturing time is longer than when using a skimmer (and in addition, it produces a lot of waste).

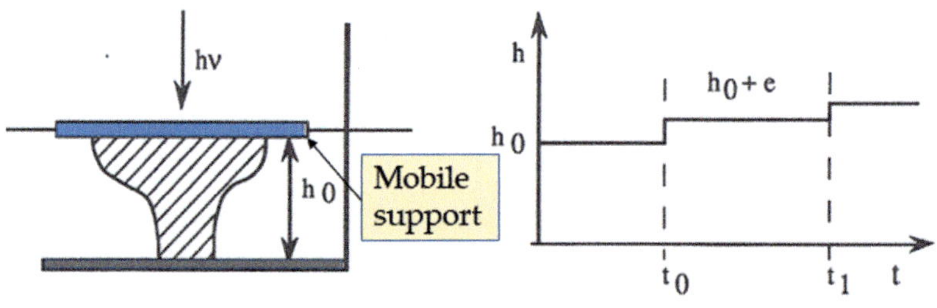

Figure 1.6. *Stereolithography using a transparent window*

1.2.3. *Gaseous interface*

The principle of such a procedure is shown in Figure 1.7. The basic concept is to protect the extremity of the optical fiber or the irradiation system by separating them using a gas bubble from the resin undergoing the polymerization process Although this system operates effectively, its size limits its displacement within the resin. This is particularly the case when the resin is extremely viscous. Once again, this potentially misguided "good idea" does not allow us to invalidate the basic process by adding successive layers of stereolithography.

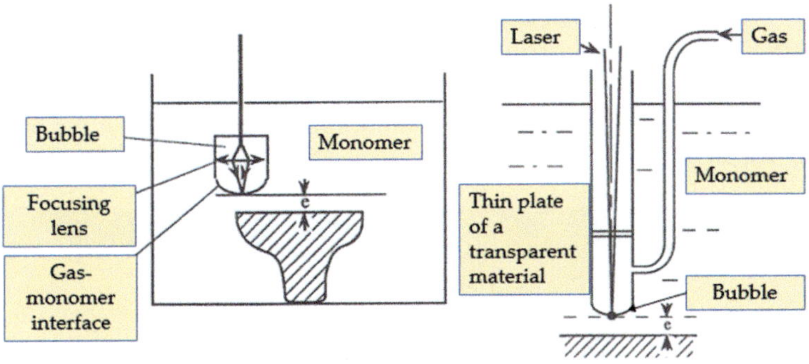

Figure 1.7. *Gas–resin interface (fiber optic (left) and direct irradiation (right))*

1.2.3.1. *Screw system*

A solution was proposed to avoid the use of layers by "screwing" the part being constructed within the appropriate resin. Instead of representing an object by the standard x, y and z coordinates, the part being printed is then defined by cylindrical coordinates (ρ, θ and z). In such a case, z is the height of the object undergoing light treatment, θ is the angle and ρ is the modulus (see Figure 1.8).

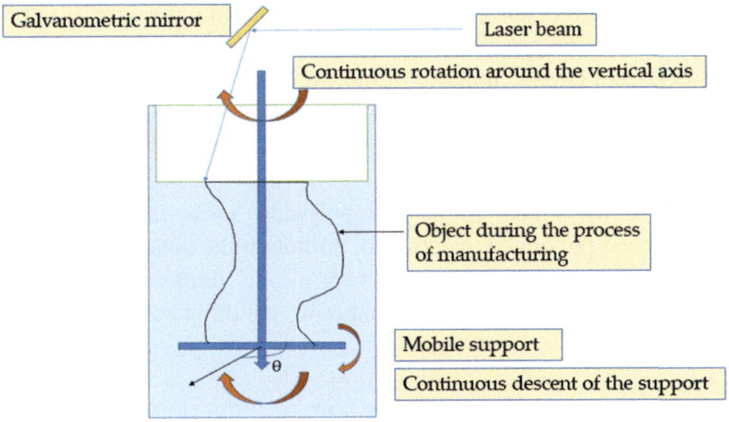

Figure 1.8. *"Screw" process*

With this patented system, André *et al.* [AND 91a, AND 91b] stated that it is no longer deemed necessary to use two rotating galvanometric mirrors positioned with two degrees of freedom if the angular speed of rotation of the falling object is constant. It was concluded that a single galvanometric mirror suffices. This system definitely avoids the waiting periods linked to the creation of layers; however, it is not sufficiently general in nature. In particular, although it allows for the manufacture of workpieces such as busts, it does not easily allow for the production of complex parts (see Figure 1.9).

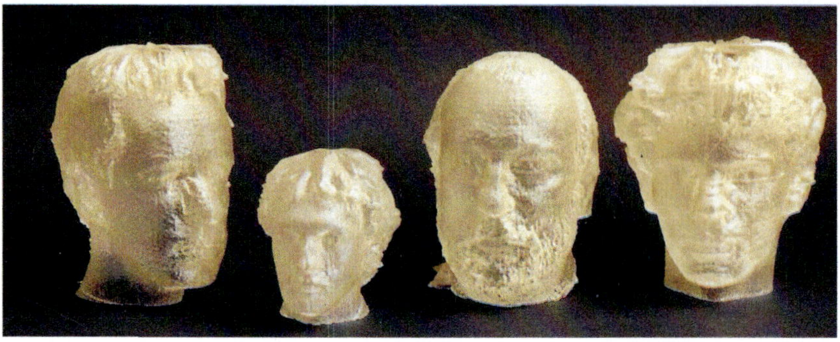

Figure 1.9. *Three-dimensional portraits of four of the inventors of the patent that is associated with this particular system (from left to right: Nonnenmacher, Schaeffer, André and Brullé)*

1.2.3.2. *The so-called "good idea"*

Recently, Tumbleston *et al.* [TUM 15] have proposed a very clever improvement of the stereolithography process. They exploited a constraint connected to the presence of oxygen, a classic inhibitor of radical chain polymerizations (see Volume 1 of this series). In their process, which is termed CLIP (continuous liquid interface production), by saturating the resin, this gaseous inhibitor in the vicinity of the light entry, the transparent surface which makes up the interface with the monomer, stays in contact with a liquid and not with a polymer. There is therefore no longer bonding, as has been previously mentioned. Although the concept of transforming a defect falls within divergent (and indeed very clever) thinking, all the same it has been necessary to produce a complex membrane, which is both transparent to light and porous to oxygen.

Figure 1.10 shows the process that has now been developed by Carbon 3D Inc. Irradiation happens through a continuous sequence of UV images, which are generated by a digital light projector. There are no layers to produce because it suffices to "pull" the object toward the top, to the extent that the interface between the entry window and the object is not, under normal use conditions, polymerizable. This opportunistic variable assumes the use of fairly fluid resins (the reaction of which is highly sensitive to oxygen) to allow both the transfer of oxygen within the reactant fluid and the limiting of mechanical constraints between the object under construction and the light entry window connected with a given resin viscosity. It is, however, highly attractive thanks to the significant time gain made possible. Fabulous [FAB 16] stated this to be seven times faster, as alluded to by André [AND 17] and Alex [ALE 16]. Scupteo [SCU 16] propounded that the resolution may be 75 µm. Also in this work, the author sets out a comparison between stereolithography and the CLIP process. An example part produced by this process is shown in Figure 1.11 [MAG 15].

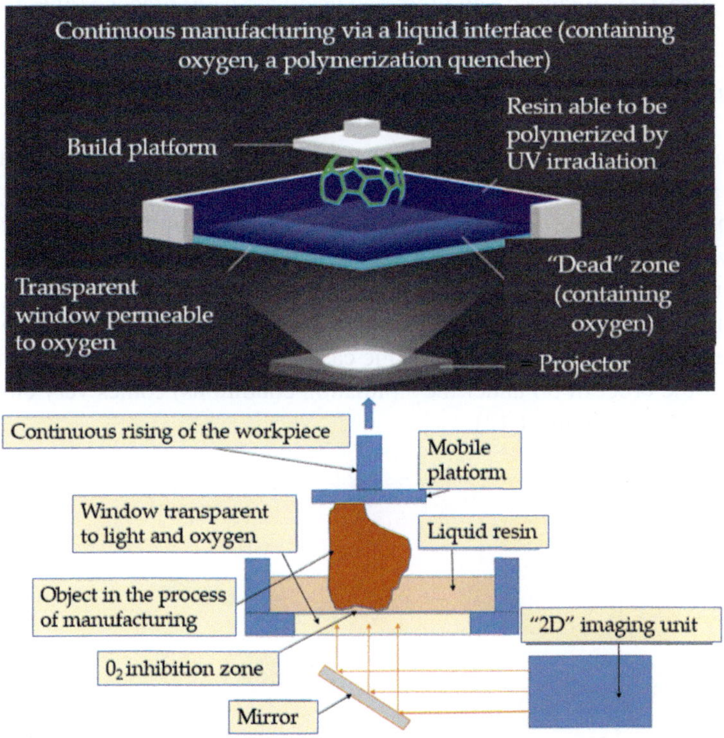

Figure 1.10. *Stereolithography according the 3D Carbon process ([TUM 15, WHE 15]; reproduced with kind permission from Carbon3D)*

Figure 1.11. *Example part produced using the CLIP process [ALE 15]*

1.2.4. *Simultaneous two-photon absorption*

Returning to the issue of irradiation, instead of using two parallel orthogonal beams, it appears conceivable to focus such beams. The author also thinks that it is conceivable to examine whether there is a possibility of having an acceptable contrast within the vicinity of their crossing point. To do this, the effect of light concentration was examined for focusing a beam at ± 45° (still working within a two-dimensional environment), leading to digital results that are more promising than those shown in Figure 1.4. This is the case because, under these conditions, the contrast (of the order of 50 under the simulation conditions) comes very close to the expected result (see Figure 1.12).

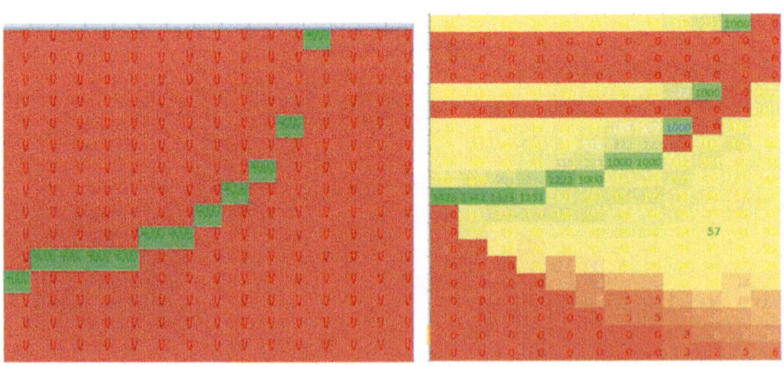

Figure 1.12. *Effect of focusing the beam (2D) (left, the control; right, simulation result). For a color version of the figure, see www.iste.co.uk/andre/printing2.zip*

When dealing with the focus of three-dimensional irradiation, the contrast is accentuated. It is greater than 100 between the green and orange zones. Yet, as was shown in Chapter 3 of Volume 1, photochemical polymerization is at the threshold, as Figure 1.13 shows, resuming this chapter's theme. Under these conditions, if we know how to absorb light at the focal point, it becomes possible to reach a state of liquid transformation being solid without undergoing implementation of the layering process. Areas with low levels of irradiation do not contribute to photo-polymerization, provided they are situated within zone I shown in this figure. However, the issue is then one of 3D printing, so we might say that the term "additive manufacturing" is therefore no longer required!

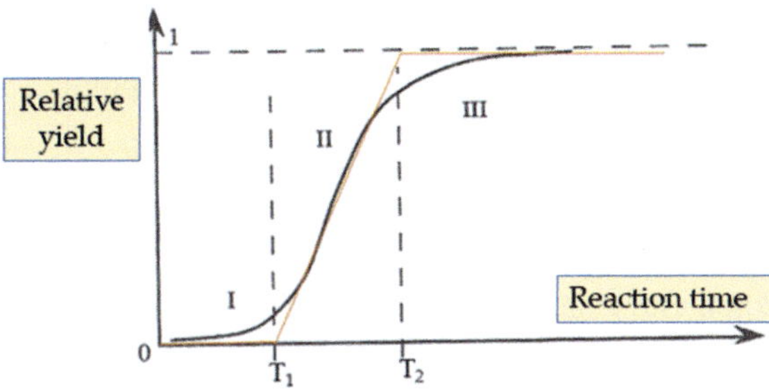

Figure 1.13. *Kinetic polymerization curve type (a point resumed from Chapter 3 of Volume 1) Relative yield*

1.2.4.1. *"Aides-memoires" on simultaneous two-photon absorption and its energizing potential*

The process of simultaneous absorption, shown in Figure 1.2, which is discussed here, was originally considered in 1931 by Gœppert-Mayer. Two-photon absorption (TPA) is the simultaneous absorption of two photons of either identical or different frequencies with the aim of exciting a molecule in a given electronic state (commonly the ground state) to reach a higher-energy excited electronic state. The energy difference between these two states is equal to the sum of the energies of the two photons. Simultaneous two-photon absorption is a third-order process, which is several degrees of magnitude weaker than the absorption of the so-called "linear" photon absorption defined in Volume 1. It differs from linear absorption because it is proportional to the square of the given light intensity, which undergoes a nonlinear process (see [ZHO 15]). This involves very high flow densities and hence the use of pulse light sources (for the most part, picosecond or femtosecond sources) in such instances.

It is worth repeating that single-photon absorption obviously follows, within non-diffused media, the Beer–Lambert law. Locally, for a given concentration (c), the absorbed light intensity (Ia) at the distance x is proportional to the product c. exp(−α.c.x), where x is the distance within the vat of the light input window, c is the concentration of the absorber and α is a coefficient called the molecular extinction coefficient. F(0) is the flow at the vat entry. It is expressed as:

$$Ia = \alpha.c.F(0).\ \exp(-\alpha.c.x)$$

For a given two-photon process, in making the reasonable hypothesis that the absorption is modest except within the vicinity of the focus point [SAL 07], the flow density is approximately equal to:

$$F(x,t) = F(0,t).\ R^2/(x+a)^2$$

where R is the photon entry beam in the vat, which has a parameter linked to both the light diffraction and the quality of the focus. Then, Ia is defined by:

$$Ia = dF(x)/dx = \beta.c.F(0,t)^2.R^4/(x+a)^4$$

where β is the equivalent of α for a given TPA. These use increased flux densities (>MW/cm^2) so that the responses of the various mediums are not proportional to the sources that produced them (see [TAO 09]). Several light sources may be used simultaneously to achieve this objective. However, it is also conceivable to focus a laser beam on a point in the space such as that which was applied in this simple calculation. We note that for a Gaussian pulse, only a very small part of the power emitted will be used for the actual absorption part of the process (much lower than 1%), which offers a decisive advantage, in terms of spatial resolution.

It can be observed that, if we take into consideration a two-photon process, the nonlinear system with which it is associated will translate into a better spatial resolution than that of the incident beam. This is markedly lower than the square of the radial power density, because it is necessary to have exceeded the consumption threshold of quenchers, as indicated in Figure 1.13. In order to be able to excite an initiator molecule by two-photon excitation, it is necessary that the photons are concentrated both spatially and in time. Indeed, in order for an electron with electronic orbitals lower than the initiator to move to an excited higher electronic state, it must receive a second photon quasi-simultaneously. It is from this excited state that unstable species, which are responsible for polymerization, are produced during a dissociative process. Consequently, we cannot normally use a continuous laser but one or more pulse laser(s). Indeed, by compressing the photons, both spatially and temporally, the likelihood of two-photon excitation is increased. The consequence of this type of illumination is that excitation is generally confined to the immediate vicinity of the given focal plane (which is the desired outcome). Figure 1.14 shows, in qualitative terms, that the gain in spatial resolution is a function of the square of flow density, as shown by the red curve.

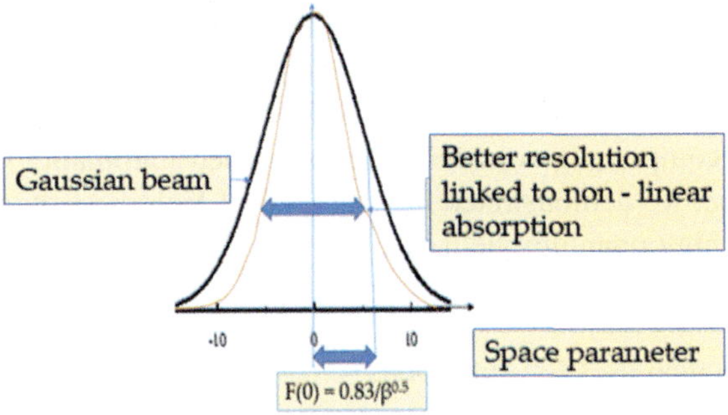

Figure 1.14. *Gain in spatial resolution induced by the TPA process. For a color version of the figure, see www.iste.co.uk/andre/printing2.zip*

With duration of a pulse ($\tau = 140 \times 10^{-15}$ s), repetition frequency (T = 80 MHz) and average power ($P_{average}$ = 3W), for peak power, we obtain $P_{peak} = 1.7 \times 10^6$ W, which corresponds to the magnitude at which two-photon absorption may lead to electronic excitation of the initiator. These approximate data define the framework of one of the constraints to be satisfied so that there is a possibility of two-photon absorption.

However, taking into account generally exothermic polymerization reactions and the non-radiative relaxation linked to the deactivation of the excited state created by TPA, there can be a local (and temporary) increase in temperature. This may locally induce a change in the refractive indices of the polymerizable liquid and solid polymer. Under these conditions, we should have object shapes to construct that are adapted to avoiding the problems mentioned above. In addition, it may be necessary to work within spatial and/or chemical conditions under which these phenomena are not preponderant and likely to be masked. This objective can be reached if we want to create, for example, small objects, a task which is well adapted to commercial picosecond lasers and microscopes.

Kannan *et al.* [KAN 01] and Belfield *et al.* [BEL 99, BEL 00] gave examples of values calculated for β from experimental data for two-photon absorption coefficients. Selimis *et al.* [SEL 15] suggested that molecules which have a "good" two-photon absorption output are often compounds from the benzophenone family (see, e.g. Woodward *et al.* [WOO 02]). Other substances have been discussed in several publications (Kim and Cho [KIM 09]; Devi *et al.* [DEV 15]; Liu *et al.*, [LIU 10, LIU 11]; Moura and Simas [MOU 10]; Terenziani *et al.* [TER 08]; Lee *et al.*, [LEE 04, LEE 08]; Chakrabarti and Ruud [CHA 09] and other authors). These are chiefly chemical compounds containing the so-called "aromatic" groups. It is in respect of these that we demonstrated in Chapter 3 of Volume 1 the potential for the formation, with a high quantum yield, of free radicals likely to initiate radical chain polymerization reactions.

COMMENT 1.– Practical uses for initiators:

Von Raumer *et al.* [VON 97] mentioned in the case of benzophenone the possibility of reactions between triplets leading to reactive species within this compound family. The production of these electronic states indeed falls within a one-photon process. However, the bi-molecular reaction between triplets induces a nonlinear process that may be exploited. Varadan *et al.* [VAR 01] moreover suggested using this type of initiation method in respect of the micro-stereolithography laser. In fact, it is actually possible for the benzophenone to take on its singlet state, and before decomposing into free radicals, it goes through the triplet state, which indeed absorbs a third photon, which then decomposes the molecule into free radicals. Thus, the process of initiation can be a little more complex than the author introduced in the previous sections. The development of the picosecond and/or femtosecond pulse laser sources might allow this concept to be demonstrated.

COMMENT 2.– Values of β:

Taouri [TAO 09] stated that the β coefficient values (β), which link a given power density (I) and absorption (dI/dx) through the relationship:

$$dI/dx = -\beta.I^2$$

are included between 0.04 and 0.4 10^{-8} cm.W^{-1}. This result means that for values in the range of Gigawatts per square centimeter, the probability of absorption is of the order of 0.004–0.04 for a layer crossing a distance of 0.1 mm. If, as for a sequential absorption, we want to produce 10^{-3} mole.l^{-1} of given reactive species (in the case of benzophenone) within a single voxel, we require energy availability of the order of some 2×10^{-7} J. For a pulse excitation duration of 0.5 ps, with an energy per pulse of 2 mJ and a repetition rate of 100 kHz, the firm Amplitude Systèmes® proposed an

"8 W" laser emitting at 1,030 nm. The power for the pulse duration is then 4 GW, which corresponds to an absorption of the order of 0.04 over a pulse duration (using a rectangular hypothesis), or at 10^{-3} J/pulse in the given voxel if the radiation is concentrated around 0.1×0.1 mm^2. This arrangement, as a consequence, leads to the potential transformation of 100–200 molecules of monomer per voxel following simultaneous two-photon absorption. From this approximate calculation, it is found that a commercial material is appropriate. This is chosen slightly by chance from the arsenal available on the market. However, according to its average increased power and the duration of its pulses, this leads to the possibility of inhabiting, by a single process, the excited electronic state, which is responsible for the polymerization sought using acrylic monomers. It is therefore conceivable, with a repetition frequency of the order of 10 kHz, to have a low average power available which is commercially accessible (but keeping the short duration of the pulses) to transform a liquid resin into a solid without any difficulty.

COMMENT 3.– Thermal effects:

Taking data from Lee *et al.* [LEE 08] by way of example, the following elements arise:

– an excitation at 780 nm; an average power of 100 MW and a frequency rate of 100 MHz, which correspond to a pulse of 10^{-9} J per pulse of 1 per second;

– absorption: 10^{-12} J per pulse (yield estimated at 1%);

– heat loss: 5×10^{-13} J per pulse (estimated at 5%).

With a heat capacity of 0.4 cal.g^{-1}, we are able to calculate the initial temperature rise, or $\Delta\theta = 10^2$ °C (with a very rapid temporal decrease around the voxel). Moreover, as Maruo and Ikuta [MAR 02] have already shown, using one-photon absorptions, initiations focusing upon a microscope lens could be photo-thermic. There is no reason for the same to apply with two-photon absorptions. Thus, this simplistic calculation shows that within micro-manufacturing, we have to take account of these temperature effects at least so as to avoid them.

We now examine, in a simplistic and approximate way, the typical average heat transfer time at the surface, for a distance of 1 mm, which is of the order of 10 s, according to the laws of Fick or Fourier (with a thermal diffusion coefficient of the order of 10^{-3} cm^{2}.s^{-1}). Alternatively, when expressed in another way, between two laser pulses, or 10 ns, heat transfer occurs over distances of the order of 15 nm. This result shows that with pulses every 10 ns, heat transfer at the surface has no time to take effect. We are therefore working under two-photon absorption conditions in which the local temperature may be superior to the average of that of the fluid situated a long distance from the nano-object in the course of construction. However, to the best of the author's knowledge, this phenomenon has not yet been commented upon by authors involved in the development of the process.

1.2.4.2. *Experimental manufacturing*

Taking into account the resolution potentialities, which are significantly lower than 1 μm, with the difficulty of producing stereolithographic layers, this principle of two-photon absorption (TPA) has particularly been used to produce objects with a very small size. At the point where the beam focuses, TPA is effective and is followed by polymerization. The movement in space of the given support (or in a reciprocal manner, the entire laser beam/lens device) allows, without the reactant fluid moving, little by little, the manufacture of a given object. Figure 1.15 illustrates the principle of this form of micro-fabrication.

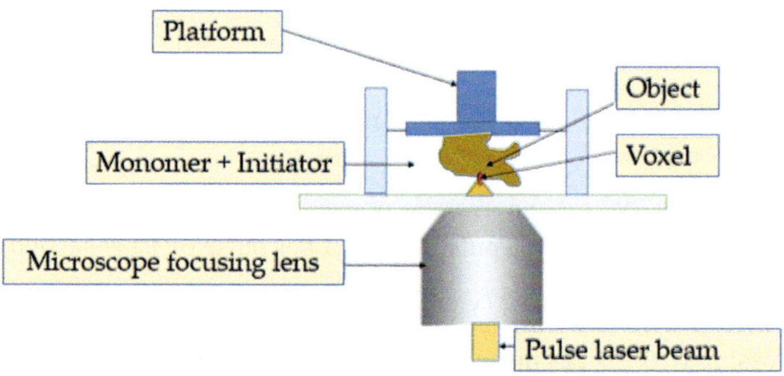

Figure 1.15. *Principle of two-photon stereolithography*

According to Park *et al.* [PAR 05, PAR 09a, PAR 09b]; Narayan *et al.* [NAR 10]; Chen *et al.* [CHE 12] and Ovsianikov *et al.* [OVS 11], through multi-photonic absorption, it is possible to have an excellent resolution available in this type of process. This is indicated in Figure 1.16. The number of other publications around this theme is still modest, but slowly the technology is developing (see, e.g. Lee *et al.* [LEE 04]; Al-Abaddi *et al.* [AL 12]; Jiang *et al.* [JIA 14, JIA 16]; Zhao *et al.* [ZHA 06]; Lim *et al.* [LIM 11]; Narayan *et al.* [NAR 10]; Xing *et al.* [XIN 15a, XIN 15b]; Jin *et al.* [JIN 14]; Engelhardt *et al.* [ENG 11]; Yokohama *et al.* [YOK 03]; Sun and Kawata [SUN 04]; Marder *et al.* [MAR 07]; Duan *et al.* [DUA 04]; Houbertz *et al.* [HOU 03]; Ovsianikov *et al.* [OVS 08]; Niesler and Werner [NIE 14]; Niesler and Hermatschweiler [NIE 15] and many others).

COMMENT 1.– Effects linked to initiation precision:

Waller and von Freymann [WAL 16] showed that two-photon-initiated precision can enable the production of micro-heterogeneous structures, allowing them to indirectly measure free-radical diffusion coefficients in the polymer material. This knowledge may be exploited to create spatially resolved refractive index structures.

COMMENT 2.– Induced birefringence:

Mendonca *et al.* [MEN 07] wrote that there is the possibility of inducing a given birefringence with the colorant, known as Disperse Red13, dissolved in PMMA (poly methyl methacrylate), which is linked to two-photon isomerization in polarized light to determine angular speed. After this treatment, through analysis in direct light, using a traditional one-photon polarized process, it is possible, within a volume of polymer, to observe a 3D image.

COMMENT 3.– Coupled effects:

By using two-photon absorption, Lim *et al.* [LIM 11] showed that we can exploit, simultaneously, 3D manufacturing and laser ablation that, taking into account the instantaneous forms of power, may occur. However, the shape of given objects should remain fairly simple to avoid the difficulties mentioned above. Other methods might be mentioned in exploiting the light-induced conversion of materials (insolubilization) (see [BRA 11]).

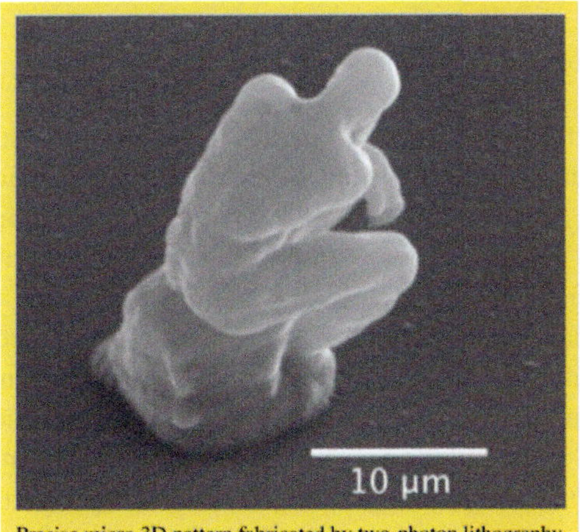

Precise micro 3D pattern fabricated by two-photon lithography.

Figure 1.16. *Multi-photon micro-stereolithography [PAR 09a, PAR 09b]*

COMMENT 4.– Holography and 3D printing:

Daqri, an American company, developed a 3D printing technology. This uses lasers to transform a photosensitive monomer into a polymer. However, instead of creating a shape layer by layer, this technology allows for the printing of a small object within a single stage, owing to holography [MEL 17, CON 17, GEO 17].

1.2.4.3. *Manufacturing time and voxel size*

We can devise the hypothesis of a voxel size 0.1 µm (such a voxel being assumed to be spherical, which is far from the case in reality) and then presume that each pulse (taking the data already used in the calculation produced in Comment 3) allows for the polymerization of the given voxel. As a consequence, we can then produce an object of volume 1 mm^3 by 10^{12} pulses, or ~10^4 s provided that the displacement system operates correctly. Alternatively, such a displacement system can last for a few hours. All of these conditions are considered acceptable for scientific works.

Yet what should we make of a complete part of 1 dm^3? The time would actually be, using this resolution of 0.1 µm, of the order of 10^{10} s or approximately 250 years. We put forward here, for the first time in this volume, a major issue, a type of "Heisenberg principle" applied to additive manufacturing. If we want to arrive at a short production time T, we must accept a level of precision linked to a modest voxel size V. Perhaps, it is thus necessary to propose a law (albeit very empirical) relating to shape:

Production time × Size of voxel =/= Constant

The issue is therefore no longer the quality in terms of resolution (100 µm often constitutes, for industrial parts, an acceptable compromise), which, in the preceding example, would give a production time of a few hours. Furthermore, 3D technology with TPA should be able to explore the possibility of a variable resolution taking advantage of the perspective of focus and the instantaneous power of the laser source. Thus, for what constitutes the internal part of the object (indeed the large part of it), there would be the possibility of using a poor resolution and producing the given object, step by step, by exploiting the potentialities of multi-photon absorption in terms of resolution. This aspect would, in principle, constitute significant added value when compared to current such production. At present, the majority of printing technologies have a fixed-value resolution.

To reach this objective, for a given distance z, there are solutions available to use Δz and Δzz, with, for example Δz = 4.Δzz. The perimeter of the object at the *area* z is polymerized several times with maximum precision (Δzz), followed by core polymerization with average precision Δz. Figure 1.17 illustrates this principle.

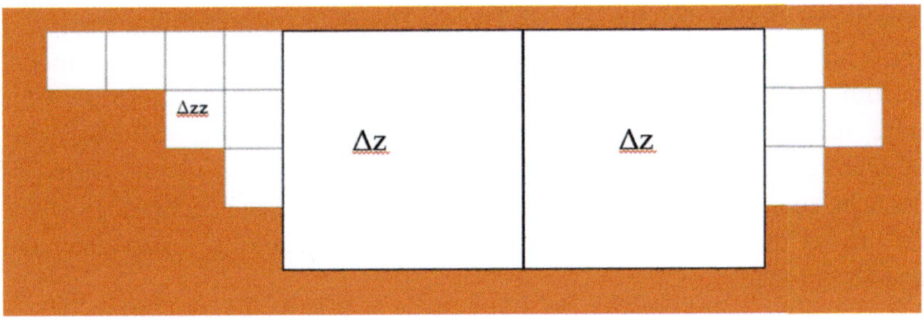

Figure 1.17. *Principle of the significance of resolution change*

Further on, the data used previously and by devising the hypothesis that we can rapidly change the laser beam focus system, if we want to produce, in one phase, the voxel with a size Δz, it is appropriate to have a pulse energy available that is 64 times that used previously (4^3). This would be an average power of 6.4 W, which is (still) commercially attainable. However, if we want to use a relationship Δz/Δzz of 10, we attain average power results, of the order of 100 W. This starts to pose a problem.

Nevertheless, this trend is worth bearing in mind. It should, as long as we find the "good idea", which is readily applicable, break down a highly significant barrier in additive manufacturing. This is manufacturing time.

1.2.4.4. *Analysis of innovations using photo-polymerization*

The partially historic approach described in section 1.2 starts from the acknowledgement that the time taken to create layers in the process of stereolithography may be prohibitive, despite numerous incremental innovations. It has already been technically possible to replace laser beams and their galvanometric mirrors with less expensive sources using irradiation mechanisms for an entire surface. However, the creation of layers to overcome the part being constructed sometimes remains a problem.

"Obvious" solutions have been both proposed and tested with mediocre success for various reasons explained in the paragraphs of this section. The transformation from a stereolithographical limit (oxygen) into an advantage in order for a highly original process to emerge, the aptitudes of which in terms of material usage and precision have to be explored further.

On the contrary, the exploitation of the nonlinear two-photon absorption process effectively avoids the production of layers. However, although the precision of objects is noteworthy, the manufacturing duration does not, in the end, compensate for the time taken to create layers. Even if we deal with "nanofabrication", it is therefore necessary to be content with compromise, the development of manufacturing mechanisms dedicated to a specific application. A relevance of this approach discussed here is indeed to show, for a particular technology, that the significance of the notion of additive generalist manufacturing apparatuses will be less and less limited.

1.3. Challenging the notion of layers

The Cartesian concept of orderly add-ons was a significant cultural element for the engineer in producing a given object using a rational approach. Moreover, the transition by way of cylindrical coordinates described above did not last long. Can we not rephrase the question in seeking to produce complex objects by other means? This paragraph approaches two possibilities, which are based on different principles, but which may require the fundamentals of additive manufacturing, within certain niche applications.

1.3.1. *Addition of prefabricated structures*

If we are able to assemble autonomous voxels of different sizes, as with a game of Lego® ([SOO 15]; see Figure 1.18), it is thus possible to produce complex three-dimensional structures. This is according to a principle that is intermediate between collaborative robots and "intelligent/smart" matter, which is mentioned in Volume 3 [CAM 14]. The objective of additive manufacturing without any loss of material (we recall the difference with traditional factory-based processes), with the concept of manufacturing with the so-called "right-material", should be retained. It is a matter of assembling components on the basis of "Robotic Building Construction by Contour Grafting" [KOS 04, KOS 14] or by doing so through techniques involving significant reductions. The "builder" is replaced by a robot, which uses the results of a CAD project. Koshnevis used a computer-controlled crane and a gantry to rapidly create desired buildings in the future at lower cost, requiring less material and using less time and workforce than those required for traditional construction methods. He considered that this technology will make it possible to construct large-scale architectural objects (see also Koshnevis *et al.* [KOS 06]).

Figure 1.18. *Interlocking using "Lego" constructions*

To reach this objective, the following are appropriate:

– from a given stock of voxels (see. Figure 1.19) to be able to move "a basic brick" (even several) to the part being constructed (e.g. with the aid of a robot system (see [MAN 13]). Perhaps, if we are capable of it, by morphogenesis [BIO 95, THÉ 95, RAM 02], traditional gripper systems may be envisaged or more sophisticated means [BOH 00];

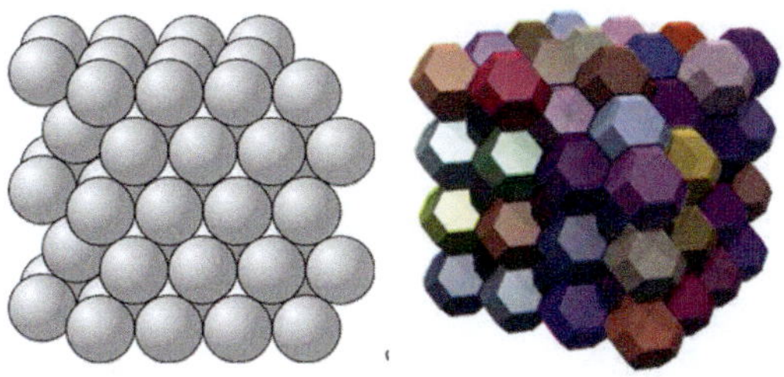

Figure 1.19. *Possible examples of voxels and their assemblies*

– to have isotropic materials (if we want to create optical experiments). The process described here allows us to avoid the issues of anisotropy produced by $2^1/_2$-D manufacturing and is henceforth considered conventional [SHA 14]. Moreover, it allows us to largely avoid the problems of aging and distortion, which intervene in traditional processes [MAN 07];

– these various voxels can be produced in one or indeed several materials with different properties (whether an insulator, conductor or numerous other properties);

– being able to link the voxels to the structure in the course of construction by different processes (electrostatic, thermal, photo or radiochemical bonding, magnetic attraction, mechanical fastening, "bio-inspired" assembly (LEMA project [LEM 13], and others);

– if necessary, to provide, whether locally or not, information within one or several voxels by different routes (optical, electromagnetic and other means) allowing for facilitated tracking or future tracking (see e.g. Meyer *et al.* [MEY 01]);

– equally, on the surface of each voxel it is possible to produce molecular grafting or a specific deposit allowing for surface reactivity, bonding or a form of adapted wettability (see, e.g. CEA [CEA 05]);

– Chapter 3 of Volume 1 showed the need to have available supports in respect of component parts, which are not yet supported, in the case of a process of transferring from a liquid to a solid or even from a powder to a solid. Within the process succinctly described here, there is no possibility to put in place such components. This leads production outside of the main three-dimensional structures, which can be added, within a future stage, onto the part under construction. This technique indeed therefore applies to complex structures, which themselves are partially hollow. Figure 1.20 shows the means to produce objects from sub-elements also using prefabricated components;

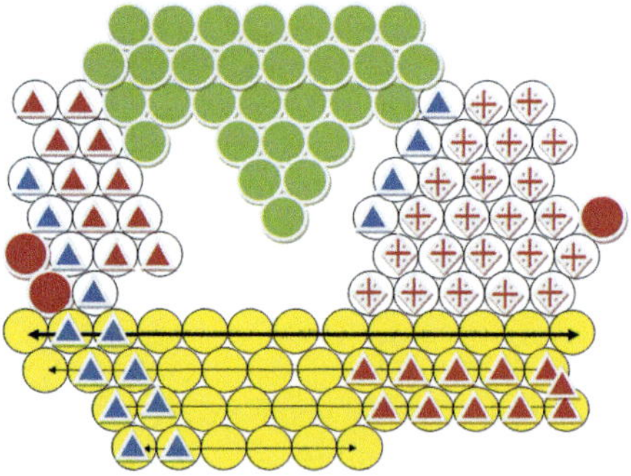

Figure 1.20. *Production (in 2D) of an object from sub-elements, some of which are pre-fabricated and stored in the so-called "library". This takes the form of a line of five circles each with a red triangle and its blue counterpart with four components. The yellow base can be advantageously produced from pre-fabricated components inside out. This is the same with the green component or that represented by the red crosses [SHI 03]. It is therefore by optimizing the procedure involving associated elements produced separately and by adding simple components (the red disks) that we can construct a given object. For a color version of the figure, see www.iste.co.uk/andre/printing2.zip*

– a methodology was developed by Vanek *et al.* [VAN 14] to define the zones of the given object being produced in the form of components, of variable and separate sizes (see the pictures in Figure 1.21). As the lower part of this figure shows, regarding the choice of shape (equilateral triangle or square), the basic components may induce different shapes to represent the same object for a given voxel size;

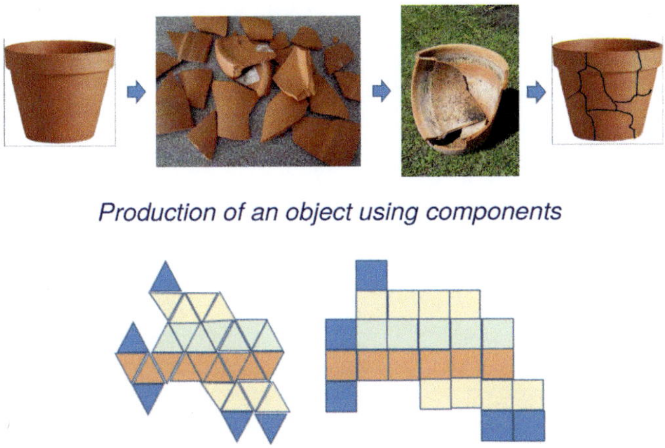

Figure 1.21. *Principle of segmentation and the production of an object using components and a given set of components (see. Vanek et al. [VAN 14]). For a color version of the figure, see www.iste.co.uk/andre/printing2.zip*

– exploiting a different path consists of either using solely appropriately shaped components or in combination with the previous item, which can be discarded at the end of production. This may be completed by means of either dissolution or meltdown (see Figure 1.22). It should be stressed that it is not necessary to fill all of the space, which can be emptied within a subsequent stage [STR 13]. The exception is where the researcher proceeds voxel by voxel within a simplified automated process;

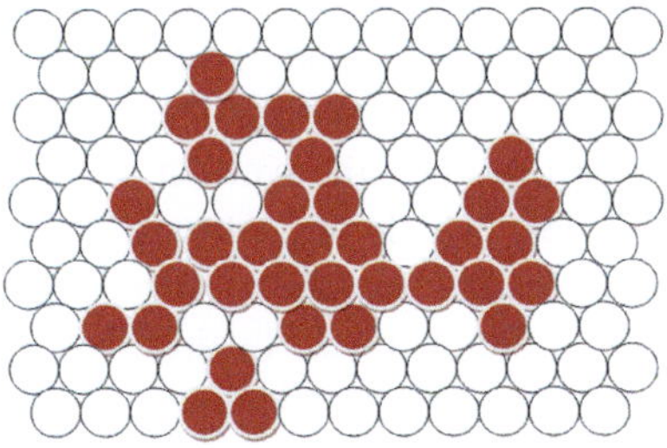

Figure 1.22. *Inclusion of solubilized support components (in red). For a color version of the figure, see www.iste.co.uk/andre/printing2.zip*

– on this basis, there is scope for the availability (even to save time during construction) of a library of sub-objects (including the choice of materials and shapes). Figure 1.23, which is more realistic, allows us to illustrate the concept [DEL 13];

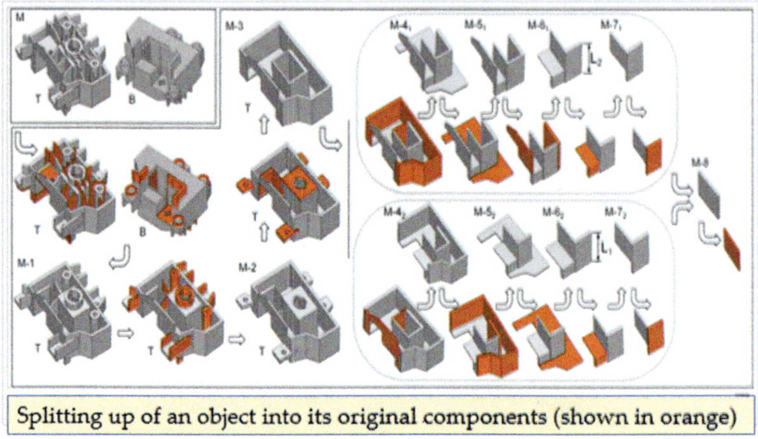

Figure 1.23. *Breaking up a given object into its core components (in this instance known as "original" components) under "creative commons". For a color version of the figure, see www.iste.co.uk/andre/printing2.zip*

– the bonding between components can usefully be produced with the help of a monomer of the same nature as that used previously to produce the elementary voxel. Thus, for example, by the illustration of the process through precise photochemical reticulation within a given area, it is possible to bring two or more voxels into contact and interlock them. Other techniques of the welding variety may be used with metals;

– within a prior stage, the component (which is not subject to volume withdrawals as generally happens in the field of stereolithography with the majority of uncharged monomers and with powder materials) may see its rigidity strengthened. This may happen through the wetting of the latter within a monomer bath, the moistening properties, which are adapted to the size of given voxels [BER 12]. A means envisaged for components made up of a spherical assembly consists of placing the object within a vat. In this vat, we can create a vacuum, allowing for, within a future phase, the introduction of the transformable fluid, from bottom to top, so that it can reach all of the accessible space within the given component. After discharge of the excess fluid, a reticulation/polymerization process can be carried out, allowing for the filling of the inter-voxel space (see Figure 1.24);

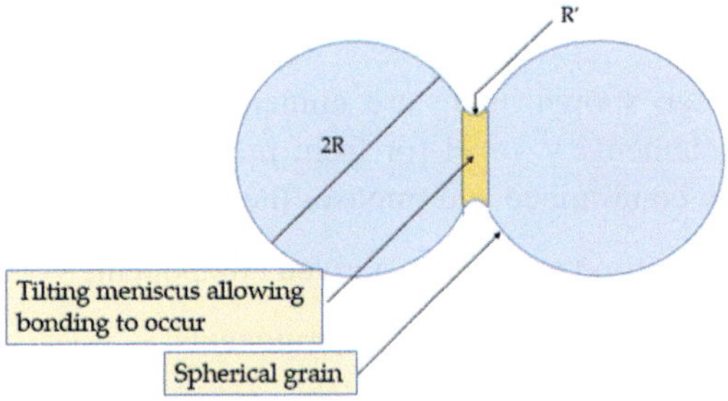

Figure 1.24. *Diagram of fluid distribution between two spheres [ANO 15]*

– it is possible to associate this process with that which is henceforth considered to be the more "standard" of additive manufacturing, particularly stereolithography (with necessary adaptations). This is able to generate structures which are finer than those of the voxels used in the process; and

– it is possible to envisage, as takes place with the Stratoconception® methods [CIR 15], subsequent substractive machining of the part so that its surface complies with the desired specification.

COMMENT 1.– Production time:

Although it is acknowledged that to have a time T to position a voxel at a given point in space, to produce a cube of a side containing n voxels, it is therefore necessary to have a time equal to T.n. Therefore, the higher the resolution sought, the longer the time to produce the component will be (with T = 1 s, a cube of 10 × 10 × 10 voxels will necessitate 1,000 s, whereas 100 × 100 × 100, about 250 h). Seeking the means to reduce T is therefore, as always, a necessity either by taking advantage of T itself (as within laser stereo-photolithography with reference times of the order of 10^{-3} s) or through more collaborative approaches (an example being computer-controlled mask irradiation). However, one of the major interests besides the possible choice of multi-materials is the possibility of producing objects with dimensions which comply with the instructions determined by the designer (e.g. avoiding the problem of withdrawals and tensions within parts) with a more limited number of object construction components (voxel systems of various shapes and sizes).

COMMENT 2.– Modeling:

As has already been shown, there are numerous ways to produce a complex component from an elementary voxel (or from prefabricated parts). The system is normally inadequately conditioned and imposes the following additional constraints:

– minimization of the time to produce a given component;

– knowledge of stocks of pre-fabricated components;

– accessible space for the production of sub-components;

– the selection (or non-selection) of given supports (voxels which may be non-reactive and, for example, solubilized);

– etc.

In such production (which is already at the stage of simulation), it may be necessary to detect un-cooperative situations operating between the various components (e.g. the impossibility of introducing a sub-object across the entire construction) and trying to delete them by re-organizing the programming [BON 94]. For example, it seems that it is possible to apply two approaches to lead the particular robot to place a component (whether unique or pre-fabricated) in a given position by avoiding collisions:

– analysis of trajectory planning (finding a trajectory to resolve the given problem) and monitoring (controls, which, in fact, produce the given trajectory). This approach may call upon CAD resources; and

– "behavioral" analysis, which should allow obstacles to be avoided while heading toward the given aim (an evasive strategy).

The overall movement of the robot is then obtained by means of structure cooperation between these two approaches (both anticipated and "reactionary"). The system resolution may also be influenced by other non-geometric factors: color, present information, material, texture, storage errors and predictable behaviors (the rigidity and fragility of a prefabricated component, e.g. linked with its prehension). It seems that taking into account all of these risk elements is particularly complicated. In a similar work, Ponche *et al.* [PON 12] showed the various stages involved in defining calculation procedures used in exploiting a layered process.

This may serve as a basis to produce the modeling of a construction process in which we would add the various components shown above.

1.3.2. *Proof of concept*

For a given "rustic" proof of concept (or "straw mattress corner"), 10 mm glass beads were bonded together using a bond of the "cyanolite®" variety to constitute three families of components of one, three and nine spheres. An example of these is shown in Figure 1.25.

Figure 1.25. *"Components of the library of sub-parts to produce a given object"*

These various components are bonded with each other as the two following figures (Figures 1.26 and 1.27) indicate:

Figure 1.26. *Partial assembly*

Figure 1.27. *Final assembly*

The use of a library of components therefore allows (and it is fairly obvious) for the production of three-dimensional objects by different means to standard additive manufacturing methods. Development may be necessary to allow for optimization between the number of components available and the main component to produce, by optimizing time-savings through moving construction components using a robotic transport device. One solution may be to combine additive manufacturing and machining determined within space, a little in the manner suggested by Maxey [MAX 15]. With the principle of carrying out preform production, it is possible to manufacture, either layer by layer or as a whole, with the necessary machining to achieve production as per the geometric instructions. This allows us to take into account the complexity of the object at the same time as optimizing the quantity of material used for production. Either way, this little working table area "tweaks" amount to a proof of concept, which it may prove judicious to explore.

1.3.3. *Synthesis*

Table 1.1 pulls together all of the elements to be mastered to develop this process (classed as + for "easy", = "achievable", – "existence of a barrier", – – "sticking point") with relevant commentaries.

Theme	Accessibility	Comments
Voxel	+	
Choice of materials	+	
Production of components taking the form of wires and sheets	+	
Robotization	-	Accessible space and principle of voxel prehension
Precision	--	The smaller the voxel size, the greater the chances of the construction time linked to physical movement in the manufacturing process becoming prohibitive
Optimized storage	-	Complex issue
Breaking up the object to create stored components	-	Ditto
The bonding of components	=	Various types of processes are accessible: welding, bonding and others
General principle	++	

Table 1.1. *Components to master to develop the process*

The issues to deal with stem from the author's skill set. Feasibility may even be produced by assembling identical volumes of given components in order to obtain a particular shape. However, it seems difficult to produce an instrumental mock-up of a device without the aid of automation experts (see the following chapter).

1.4. Optical-quality surface finish

It is common to produce an object with surface imperfections and then to treat its surface accordingly. The aim of this section is to show that it is possible to disregard this process by exploiting either the physicochemical properties of the material or indeed certain aspects of the process.

1.4.1. *Glasses lenses and contact lenses*

Several companies have embarked upon the niche of glasses printed in 3D, personalizing the shape of the glasses to make them perfectly tailored and comfortable as per individual requirements. This is unlike glasses manufactured by large-scale production. Although producing frames according to customer specifications depends on knowledge of additive manufacturing technologies, which is standard nowadays, the production of marketed optics is a recent innovation. The Dutch company, LuXeXcel [LUX 16], has proposed to print an entire pair of 3D glasses, including the lenses. It has also committed to print usable 3D contact lenses using this method.

According to Allard [ALL 15], this involves a stereolithography process by, means of depositing a resin material, at the end of the operation, which spreads out across the component under construction. After the process of polymerization, this allows us to achieve a sufficient level of transparency. The principle involved in such production is shown in Figure 1.28 (see also the first part of Volume 1 of this series). It is obvious that date issues around the treatment of the anti-UV surfaces and durability have not been studied within this field.

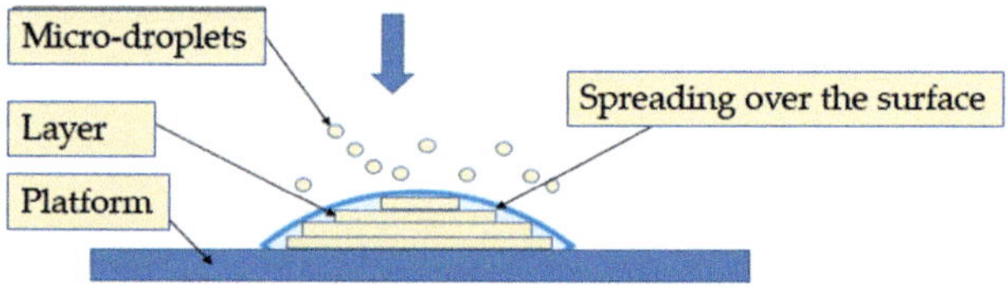

Figure 1.28. *Principle of producing a lens with optical features*

1.4.2. *Microlenses*

This is another interesting concept stemming from Lu's thesis work [LU 06]. It uses the influence of incident light flows across a given surface to modify the depth of polymerization. Knowing the influence of the threshold effect linked to the presence of inhibitors, complex shapes can be produced. By using the so-called "gray wedge", Figure 1.29 shows diagrammatically what may be obtained as a component made from a polymer material. It will be noted that the “gray wedge” uses a homogeneous, light-absorbing material, the transmission of which, independent of wavelength, is a function of the thickness of the material that it passes through. By adapting an additional parameter, the exposure time and the shape of microlenses may be fairly flexible.

In his work, Lu [LU 06] used this principle to produce sets of microlenses. This is an example of mass batch-processed manufacturing, made possible thanks to an increased number of “gray wedges” (see also [HUN 05]).

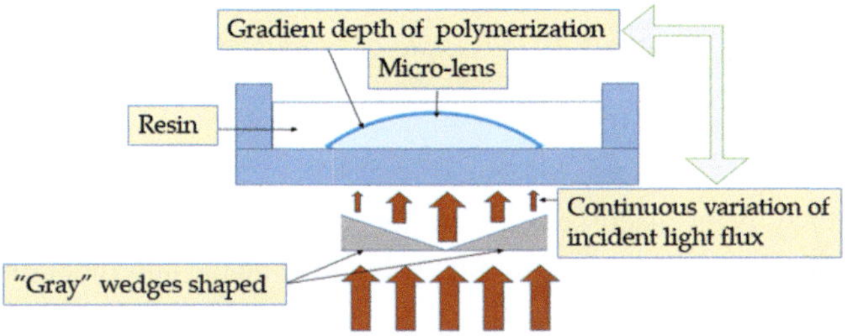

Figure 1.29. *Use of “gray wedges” to produce microlenses*

Other similar technologies could be used (see, e.g. [HUN 05]).

1.4.3. *Direct lens manufacture [AND 91a, AND 94]*

Rather than envisaging a production in several stages as was set out in Volume 1, it may be interesting to examine whether it is possible to calibrate a liquid voxel by giving it the form of an optic lens. It is then a matter of a principle of confinement in relation to the reactive part. The example shown in Figure 1.30 explains the concept. A cylindrical container is set to rotate at a constant angular speed, so that the liquid in it forms a vortex, which is normally a revolutionary paraboloid. Although this liquid is denser than the resin (immiscible) placed above, it constitutes the base of a type of mold. Thus, by irradiating the resin after hydrodynamic stabilization, we are able to produce an optical component.

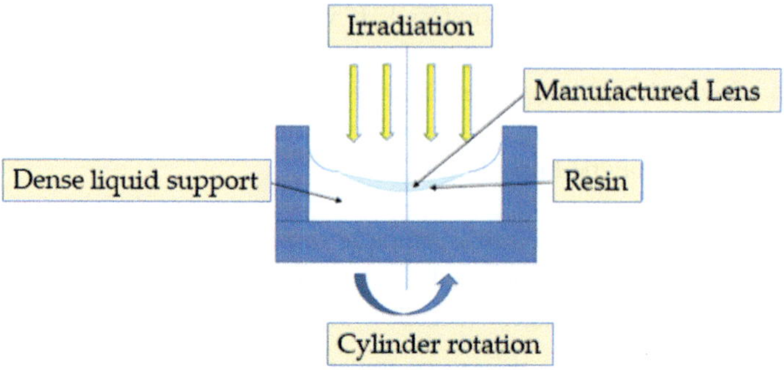

Figure 1.30. *Principle of production of a lens on a vortex with a single voxel*

As has already been discussed in Chapter 3 of Volume 1, the effects linked to interfacial tension may play a determining role in the shape and focal distance of the lens.

– There will be low interfacial tension in the face of inertia forces. The lens takes on a concave–convex shape (see Figure 1.30);

– when the tension forces across interfaces predominate and, in particular, at low rotational speeds, the shape taken on is biconvex; and

– by using two immiscible liquids, the resin voxel may be placed between the two fluids, which can allow the exploitation of the various parameters of influence. Such parameters are:

– the size of the rotating reactor,

– the volume of the voxel,

– the densities of the immiscible fluids,

– the interface tensions, and

– the rotational speed.

Figure 1.31, produced by André and Corbel [AND 94], illustrates the potentiality of this process.

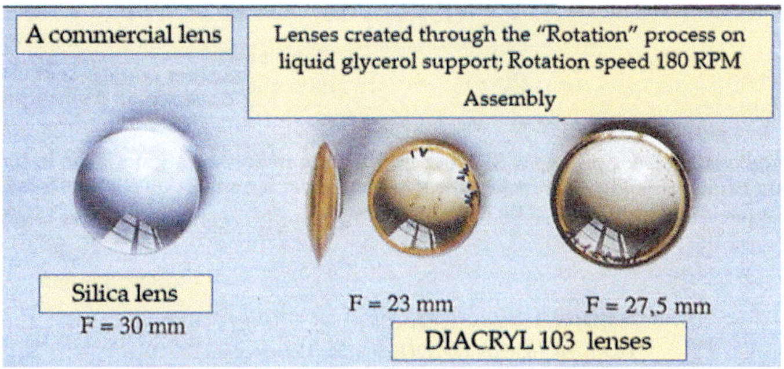

Figure 1.31. *Example of production of optical lenses using this process*

Brullé, in his thesis (completed in [BRU 92]), showed that it was possible to produce support fluids for resin voxels, by using phenomena other than rotation. He discussed the levitation of a given volume of resin, the molding on the surface of a ferromagnetic fluid subjected to a suitable magnetic field and other such cases. The limits of what is possible are therefore, even for a fairly specific application niche, fairly broad.

1.4.4. *Multi-mode optical fiber*

Yves Brullé showed that it was possible to resume Reynolds's experiment, as shown in Figure 1.32. The experiment relates to a liquid monomer containing a photochemical initiator flowing out at low velocity, surrounded by the flow from the same monomer at the same temperature, but not containing any initiator. When the two fluids meet, the light causes the polymerization of the central zone comprising the core of an optical fiber. At the crossing point of these two fluids, an initiator concentration gradient is established, leading to a polymerization gradient. Correspondingly, this translates optically, in practice, into an index gradient. Thus, in a single operation, within a continuous process, it is possible to produce the core and the sheath for a multi-mode optical fiber. The proof of this concept was demonstrated and patented [AND 91b].

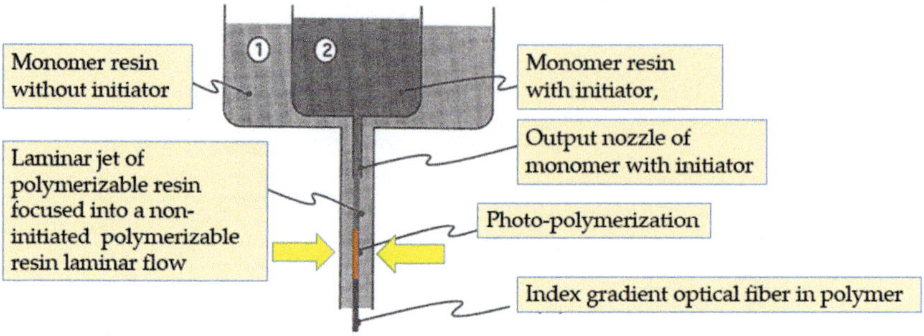

Figure 1.32. *Production of multi-mode optical fibers by photo-polymerization*

We note from these descriptions that it is the location of the voxel which defines the object and not the location of a form of energy (which is light in this situation, but it may also have had the ability to take on other energy forms).

1.5. Cold-cast metal 3D printing

In what has been shown throughout this work, there is indeed a need that is widely expressed by the manufacture of metal products. Certainly, several of the techniques shown in Chapter 3 of Volume 1 already allow for such production, but these go through several stages and a given number of works try to show that we may hope to achieve more with current "metal" processes, which are already broadly operational. These works are succinctly described below (see also [LUC 16]).

1.5.1. *Electrolytic deposition*

Research by LEPMI (*Electrochimie et Physicochimie des Matériaux et des Interfaces*, which translates as "Electrochemistry and physicochemistry of materials and Interfaces" laboratory [LEP 15]) has a particular emphasis. It deals with electro-crystallization of cold metals and their growth in relation to their respective electrochemical, chemical or physical properties. The objectives of such research aim to improve the superficial or structural properties of materials produced by electrochemistry (structural materials) or to confer new properties upon them (functional materials). The following diagram, shown in Figure 1.33, depicts the various modes of electrochemical synthesis for metallic phases from an electrolyte liquid, both globally and synthetically.

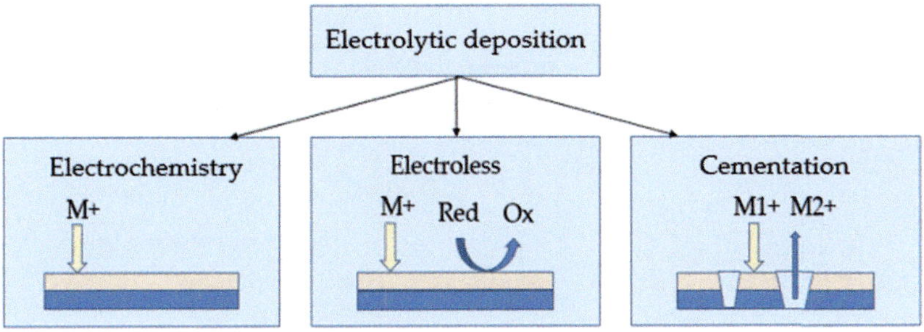

Figure 1.33. *Electrolytic deposit methods*

There are two major stages leading to the production of a metal phase from a liquid phase. As a first step, species of the liquid phase are adsorbed across the surface before their partial or total reduction. On the given surface, these adsorbed species may be diffused across the surface, before finding a given insertion site (for growth) or to create a new germ (nucleation). In both these cases, a phase change in species is seen with an electron exchange (the phenomenon known as electro-crystallization). According to LEPMI, there is a particular case of crystal growth given by deposits (UPD – Under Potential Deposition). The deposit obtained is generally of the thickness of the order of a few atomic monolayers. This situation is therefore not easily adaptable to additive manufacturing (perhaps apart from aspects of nanofabrication, which will be dealt with in this particular volume).

Other metal deposit methods also exist, such as "dynamic chemical plating" (DCP) developed to produce cold metal cast deposition (however with an absence of spatial resolution). This occurs by reacting in an aqueous solution of metallic salts with a reducer, such as KBH_4 (potassium borohydride). However, growth is very modest (of the order of 30 μm/h^{-1}, according to Stremsdoerfer [STR 03]). This type of process will not therefore be described in further detail in this book.

A more promising means was proposed by Seol *et al.* [SEO 15] aiming for metal deposition owing to droplets containing a metallic ion, the conversion into metal of which may be induced by an energy source, such as an electric current. This process, which uses pure metallic ions, is described in Figure 1.34. There remain some issues to be resolved. These include the positioning of the given drop in line with aspects of surface tension and wettability. There are also issues concerning the complex shape of the object and the speed of metal deposition.

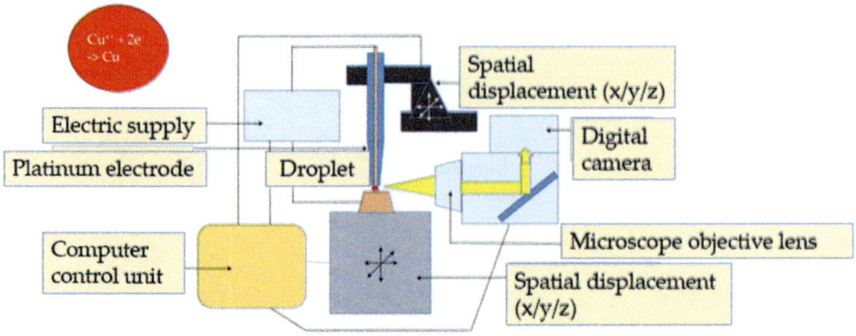

Figure 1.34. *Electroplating process by Seol et al. [SEO 15]*

In the conditions for the use of a given micropipette, the flow is perfectly controlled and regulated by the IT unit. Figure 1.35 illustrates the potential of the given technology (examples are bridge manufacture and micrometer resolution).

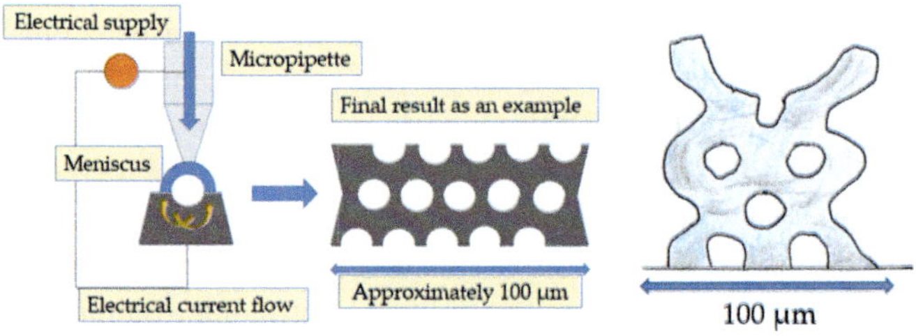

Figure 1.35. *Complex components likely to be produced by electroplating and design similar to an experiment by Seol et al. [SEO 15]*

Evidence suggests that this proof of concept has potential applications useful in micro-engineering.

NOTE.–

Concerning direct electrolytic deposits, on the micrometer or sub-micrometer scale, Brant and Sundaram [BRA 16] showed that it was possible to deposit metal through electrolysis, owing to the fine nature of a given electrode. Both of them, taking into account the nature of the technology, posed the issue of the shape of objects, which ought instead to have a shape that is close to convex. This should thus also have possible solutions to offset this disadvantage (also see [BRA 15, KAD 05, SUN 15] on this subject).

1.5.2. *Metallic ink*

In recent articles, [SHA 15, VIC 16, LUC 16, SHE 15], the creation of a start-up business in Israel called "Xjet" was referred to. This was intended to be capable of producing metal objects from metallic nanoparticles included within an "adequate" solution. By themselves settling upon the component being constructed, the nanoparticles would accumulate so as to make up a continuous metallic medium. The object being produced is constructed layer by layer, with a thickness of less than 2 µm followed by the evaporation of the support liquid (at 300 °C) so as to only leave metallic particles which are sintered at a later stage. The Xjet site is no more informative than this, but industrialization of the process seems to be near. If this was the case, a certain number of laser-based processes could most likely lose their appeal as long as the cost of the device is affordable to the consumer.

1.5.3. *Laser processes*

Stemming from the older works of Bohandy *et al.* [BOH 86], who had shown the possibility of depositing the metal in the form of thin sheets by excimer laser excitation, Zenou *et al.* [ZEN 15a, ZEN 15b] used pulse lasers to produce micro-machining for simple objects of a small size. The principle used by these authors is shown in Figure 1.36. The LIFT (laser-induced forward transfer) process, which serves as their basis [ALL 06, ALL 16, YUN 16] is based on irradiation and ejection, by a single pulse, from a layer of material laid beforehand upon a transparent substrate. The respective absorption coefficients of the layer and the substrate depend on the laser beam wavelength. According to these properties and the laser flow density, the process of various light–material interactions becomes relevant (whether thermal ablation of the material or mechanical ejection subject to high-intensity pressure waves). The process may be broken down into three phases: ablation, transfer and deposition. The LIFT process allows for local deposition and with a large spatial pixel resolution (using voxels) from a substrate on which the material was deposited beforehand as a thin film. This process may be recommended so as to exploit the potential of insoluble compounds, which cannot be deposited by more conventional processes such as ink jet printing [ALL 16, PAP 16, YU 15].

Zenou *et al.* [ZEN 15a, ZEN 15b] relied upon a variant of the LIFT process. The metallic material lies upon a transparent substrate within thin layers. The laser beam liquefies the metal or the alloy, and the molten zone is transferred to the receptor and solidifies. For the "LISAT" (laser-induced self-alloying transfer) part, which the article by Zenou *et al.* [ZEN 15a, ZEN 15b] involves, the process is identical, but they use several metallic layers in place of a single one. The resolution is of the order of a single micrometer, with modest growth speeds linked to the reduced voxel size.

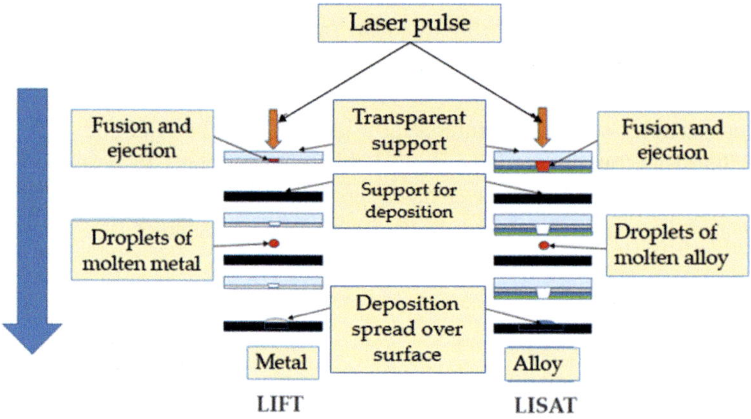

Figure 1.36. *LIFT and LISAT processes*

NOTE 1.–

In a spirit akin to this, Murr [MUR 16] set out a system for the deposition of metallic droplets, which are deflected owing to an electric potential of an oscilloscope type (in the manner of a "cell sorter" used in biology (see e.g. [HER 02]). The American company Vader Systems, according to Lucas [LUC 16], uses aluminum and its alloys to supply a machine in the form of a wire. This is heated in a chamber at a temperature of 750 °C. The liquefied metal is then directed via an electromagnetic field. "The projected drops, traveling at a rate of 1,000 μm/s, have a minimum size of 200 μm." Figure 1.37 shows the principle of the process. We should remember that this mechanism only allows for the manufacture of convex components and that its purpose is more micro-technology oriented. Designed with a print head capable of ejecting metallic drops at temperatures reaching 1,800 °C, caused by an electromagnetic force, this system may project silver, tin and/or copper. However, it does not quite fall under the section title, even if it is an innovative process.

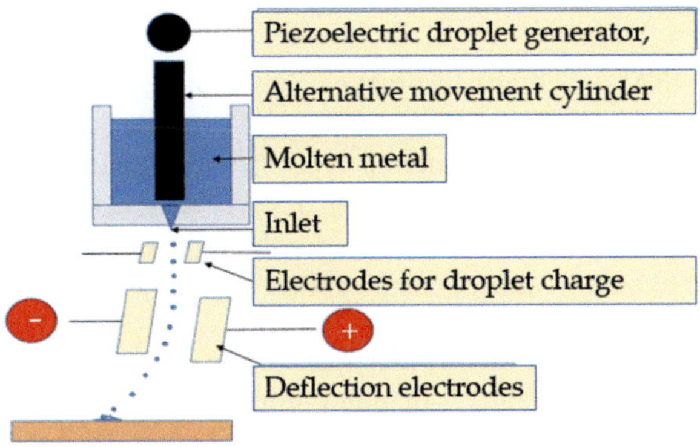

Figure 1.37. *Deposition of molten metal by electrical deflection determined on the given surface*

NOTE 2.–

On a similar basis to the method used in Note 1, there is a method which was invented in 1934 termed "electro-spinning". It consists of the deposition of a guided charged polymeric material through an intense electric field. By controlling the given field strength, it is possible to move the particular impact across a given surface and to, step by step, produce a given object [LEE 15, TEO 05].

1.5.4. *Photochemistry*

By giving "control" back to chemists, it is possible to find simple plating processes, as Ikeda *et al.* proposed in [IKE 01]: It is possible to use the oxidation–reduction (redox) properties of titanium oxide, irradiated by moderate levels of UV, to enable production of a copper-plated deposit. The copper-plated deposit produced has an electrical conductivity surface, which is sufficient for electrolytic deposition, allowing for an increase in the thickness of the metallic layer (see also Akamatzu *et al.* [AKA 05] with a slightly different process). Other metals can be deposited from this TiO_2 coupling as well as metallic ions: platinum, gold and palladium (see [MAI 11]).

1.5.5. *Silver metal*

Recently, the INM – Leibnitz Institute of New Materials [INM 16] published its works on the use of nanoparticles, containing silver oxide, which upon receipt of appropriate light energy, manufactures a type of latent image (see the traditional photography processes). This occurs using a silver salt bath that allows for the release of silver metal atoms. It also allows for the production of conductive layers, without any form of post-treatment being applied.

The article by Nie *et al.* [NIE 12] illustrates decomposition of silver citrate in the presence of a complexing agent 1,2-diamino-propane. This may be carried out with solvents such as methanol and isopropanol. These allow for the adjustment of viscosity and surface tension. They thus favor deposition at temperatures above 130 °C on the positron emission tomography (PET) surfaces. These authors showed an example of the application for the production of a circuit, which is usable within an RF (radiofrequency) antenna. The resolution may reach about 100 µm. Other compounds such as silver(I)-2-[2-(2-methoxyethoxy) ethoxy]-acetate were used by Jahn *et al.* [JAH 10]. It was thus possible to sequentially deposit from metallic silver. The latter is produced by thermal decomposition at approximately 250 °C.

With "inks" containing silver, the use of YAG lasers (488 and 515 nm) allows for the thermal conversion of the deposited material (continuous power densities are of the order of 10–150 kW.cm^{-2}) with resolutions that may reach 5 µm for silver (Aminuzzaman *et al.* [AMI 09]; see also [LEE 13]) and a few dozen nanometers for copper [LEE 14]. However it does not occur, strictly speaking, in photochemistry. Sakamato *et al.* [SAK 09], in an article targeting the photochemical production of metallic nanoparticles, set out two main mechanisms of metal atom production (see Figure 1.38).

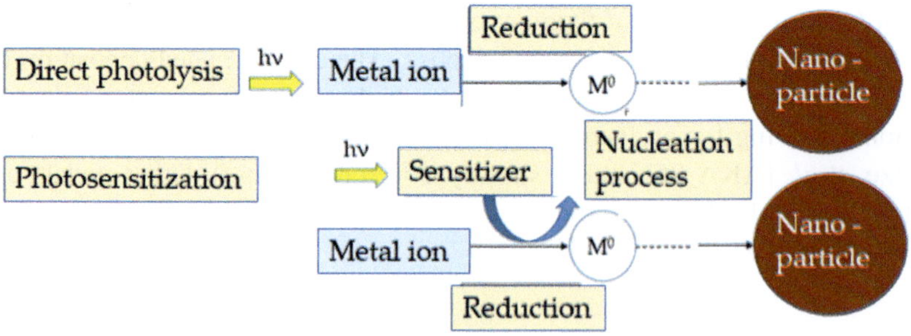

Figure 1.38. *Metallic atom photochemical production*

The first mechanism corresponds to photolysis of silver chlorate in an alcoholic medium. This is illustrated by Figure 1.39 (see also [AKA 03]). Reactions of the same nature have also been published for other metals such as nickel [PLY 92], platinum [GRI 92, GRI 91] and gold [MAL 01] using the principle of metallic deposition shown in Figure 1.39. The reduction mechanisms for the respective elements are then as follows:

– For silver:

$Ag^+ + RH$ --- $h\nu$ -→ $Ag^0 + H^+ + R.$

$Ag^+ + R.$ ----------→ $Ag^0 + H^+ + R'$

Ag^0xn ----------→ $(Ag^0)n$ (Nucleation)

– For gold:

$AuCl_4 - h\ \nu \rightarrow 2AuCl_3- + 2Cl_0$

$2AuCl_3 - \rightarrow AuCl_4- + AuCl_2-$

$AuCl_2- \rightarrow Au + Cl_0 + Cl^-$

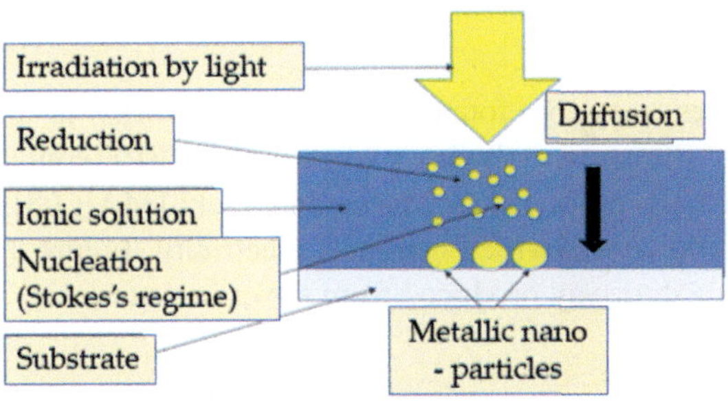

Figure 1.39. *Metal photochemical deposition*

The second mechanism uses, for example, chromophore absorption from the benzophenone family, which will react with a metallic ion, by means of reduction of the given ion leading to metal. Other chromophores may be used in this process, such as pyrene, thionine, titanium oxide and other such chromophores. In $2^{1/}{}_2$-D printing, the resolution is a few dozen micrometers.

It is a question of surface depositions, but with possibilities for the spatial resolution of convex objects. Furthermore, as nucleation occurs with Stokes' flow, the process occurs slowly, which, for standard industrial applications, may not be very worthwhile.

1.5.5.1. *Two-photon systems*

The general principle of this type of conversion induced by pulsed lasers is shown in Figure 1.40 [SAK 08, SAK 09]. This can apply to structures which are $2^{1}/_{2}$-D or 3D as long as the substrate is transparent to radiation types (see also [LIU 16]).

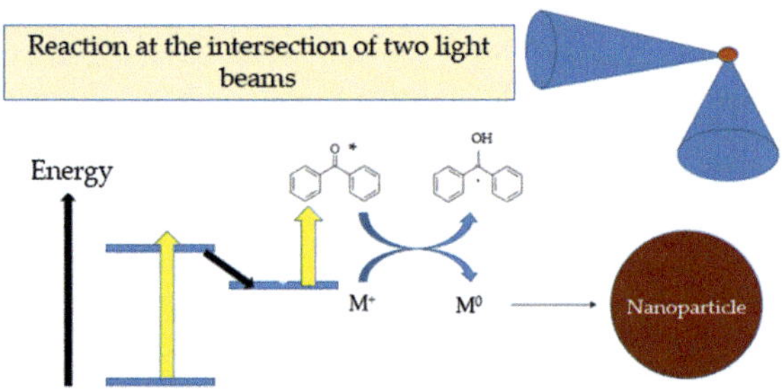

Figure 1.40. *Two-photon photo-reduction located within a given space*

Various examples may be found in specialist works in the field. With a femtosecond laser, films of polyvinylpyrrolidone (PVP) containing silver ions (silver nitrate) have been proposed by Maruo and Saeki [MAR 08]. This leads to the production of metallic silver, which conducts electricity. Figure 1.41 by Tanaka *et al.* [TAN 06] shows a doorway produced from silver in water.

Figure 1.41. *Approximate design of a conducting micrometer "doorway"*

Wu *et al.* [WU 00] produced this type of conversion by using a porous silica gel into which silver nitrate in an aqueous solution is introduced (through pores of 0.2 nm). With the help of a pulsed laser (120 fs, 800 nm) producing 5 TW.cm^{-2}.pulse^{-1}, metallic silver is produced within a 3D structure in a spectroscopy cell. Other examples can be found within field resources (see e.g. Soukoulis and Wegener [SOU 11]; Baldacchini *et al.* [BAL 05]; Park *et al.* [PAR 09a, PAR 09b]; Tosa *et al.* [TOS 08]; Terzaki *et al.* [TER 11]; Shukla *et al.* [SHU 10] and other works).

COMMENT.–

In the work published by Lenhart *et al.* [LEN 11], gold is the metal considered to have the most complex chemistry for preparation of a deposit with a usable fluid, having an "inject-printing" system. In order to reach the same objective, other compounds may be used, such as $[AuO_2CCH_2(OCH_2CH_2)_2OCH_3\text{-}(^nBu_3P)]$, containing 34.2% gold by mass [SCH 13].

Similarly, it is possible to use lasers such as excimer lasers (e.g. KrF) to convert a polyimide. The process involves initial thermal degradation and the formation of carbon atoms followed by a re-arrangement on a supra-molecular scale. This leads to the production of a graphitic polycrystalline electrical conductor network [QIN 01].

1.5.6. *Conducting polymers*

There are very few works related to the deposition of conducting polymers. However, we can cite the use of poly-pyrrole within some applications [ZHA 14, DES 97, AMB 16, LE 15, SEO 08, MUR 15, KIM 11]. A deposit conductor production principle is shown in Figure 1.42, taken from Kim *et al.* [KIM 11].

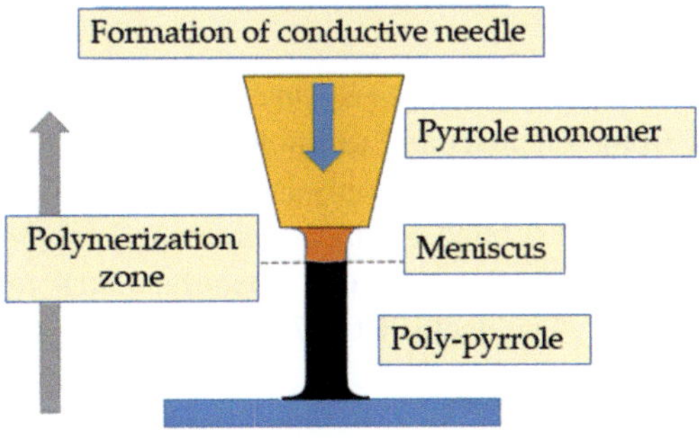

Figure 1.42. *Production of polypyrrole needles*

It is worth repeating that conductive polymers have appropriate electrical properties, as indicated in Table 1.2 [DES 97]. Research works could therefore be undertaken to test the relevance of the producing 3D objects containing non-metallic conductor components.

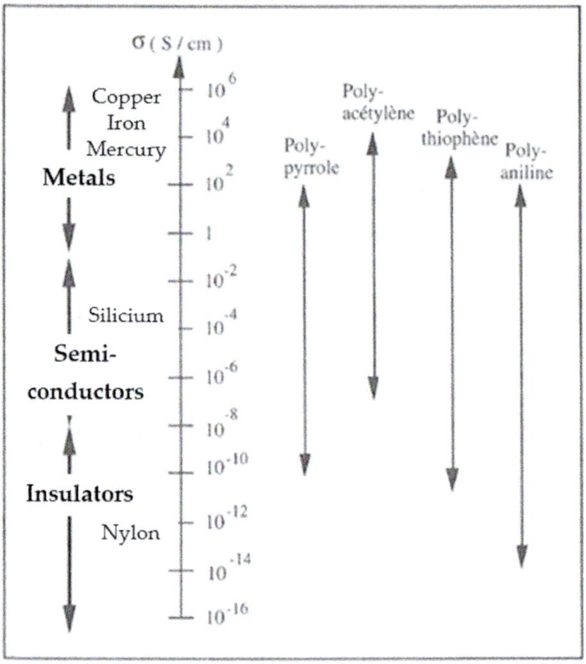

Table 1.2. *Comparison of the conductivities of various components including polymer conductors*

COMMENT.–

It is detrimental to the field that these metallic deposition processes receive insufficient attention from scientists and technologists. This is because it is already possible to produce connectors, which simply lack the conductor component. This is the subject of what might be termed a minor incremental research exercise carried out by Prodways (a French 3D machine manufacturer) to give a level of functionality to a particular 3D object [PRO 17]. This is shown in Figure 1.43.

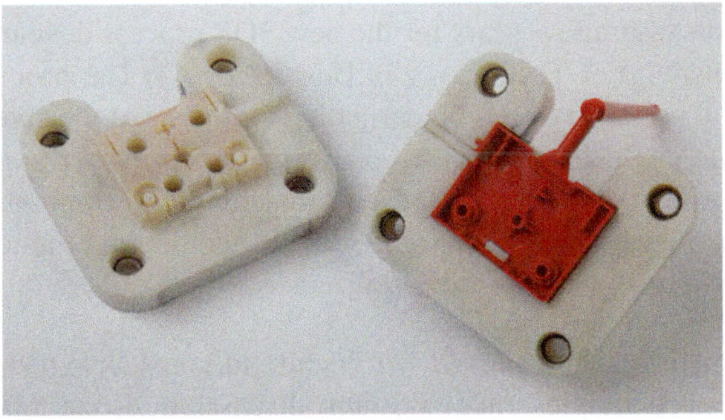

Figure 1.43. *Connector support [PRO 17] ©Prodways and ©Hamilton de Oliveira)*

1.6. Colored objects

By using several vats containing photo-polymerizable resins, associated with organic colorants of varying colors, it had already been possible to produce colored objects using stereolithography as early as 1992 [AND 92]. The system was complicated by the presence, between each given vat, of a solvent bath to eliminate the resin located around the object during construction, followed by a drying process before immersion within the vat containing the next color. Each layer might necessitate several colors, the position of which was defined by the computer used. Figure 1.44 shows some examples produced which are the precursors to 3D color photographs. Although, the proof of concept had been demonstrated, the transition from vat to vat was fairly tedious with the necessity to check the resin level within each vat and other factors. That having been said, apart from aspects relating to actual color changes, the mechanism worked. Despite this, there was, however, a considerable increase in the production time of a component by a factor of the order of between 5 and 10.

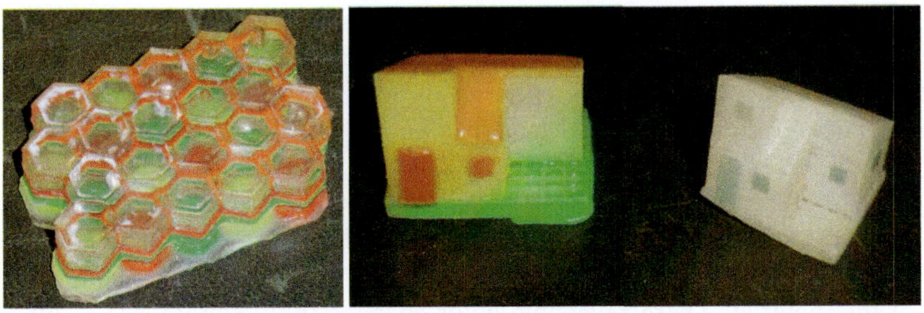

Figure 1.44. *Colored objects produced by the process of stereolithography. For a color version of the figure, see www.iste.co.uk/andre/printing2.zip*

We have had to wait a few years for the new 3DP process developed at the MIT (Massachusetts Institute of Technology in Boston, USA). The process, called 3DP (three-dimensional printing) or BJ (binder jetting), consists of the spreading of a fine layer of composite powder over a given platform. The print head will then deposit adhesive drops, which may or may not contain color. By combining drops when they contain color, we then obtain the desired color to produce a 3D color reproduction. This process is now industrialized.

Within this chapter, it has proved worthwhile to consider whether or not there exist alternative technologies, allowing for discussion of an object during its construction. The first example is fairly standard, corresponding to light irradiation by a pulsed source (the YAG laser type) powered by a given semiconductor material, such as titanium oxide (TiO_2). The light flow allows for a change in the stoichiometry of the given oxide leading to a corresponding change from white to gray or even black according to the following process (see Figure 1.45).

$$TiO_2 \text{ (white) + Nonlinear light absorption -> } TiO_{2\text{-}x} \text{ (black)}$$

Figure 1.45. *Pulsed irradiation stemming from YAG laser across a polymer containing titanium oxide (the right-hand side image is from Desprez [DES 97])*

On the basis of this principle, it is possible, either during manufacture or at the end of this, to irradiate, using a pulsed laser, the surface of the given object. This allows, in this example, for the provision of information around components, which contain semiconductor materials such as titanium oxide.

Crespo-Monteiro *et al.* [CRE 13], in their production, made this simple process more sophisticated. The photo-catalytic properties of this semiconductor allow us to modify the absorption spectrum of metallic nanoparticles and, as a result, alter the color of the material. These reversible color changes (photochromism) result from variations in the distribution of nanoparticle sizes, due to oxidation mechanisms within the material. The reduction of Ag(I) ions in this process is thermodynamically more favorable across particles that are already formed than with those formed from isolated sub-particles. With the mobility of ionic silver being large within these porous matrices, the use of lower light intensities (typically of the order of a few to few hundred milliwatts per square centimeter ($mW.cm^{-2}$)) tends to favor the growth of individual nanoparticles to the detriment of the actual quantity of such particles. They give rise to fairly large distributions ranging from several nanometers to a few dozen nanometers. According to the distribution of the given sizes obtained, the film has an absorption band, which is more or less extensive in terms of visibility, being orange, brownish or grayish.

Works of the same type have already been published on other systems. Ocana and Calatayud [OCA 16] experimented with titanium metallic surfaces, Penide *et al.* [PEN 15, PEN 16], aluminum, Mangaiyarkarsai *et al.* [MAN 05] over lenses containing silver ions. Depending upon the conditions, it is possible to go from black and white through to color.

With organic aromatic materials from the diarylethene family, Mori *et al.* [MOR 11] used a principle of multi-photon absorption with a femtosecond laser (35 fs; 1.28 μm) with energies lower than 1 nJ/pulse. The result obtained is the green colorization of the dye. They also showed that it is possible to erase the color by returning to the original color. This is due to a two-photon process, which brings back the initial material. A diagram of the reaction is shown in Figure 1.46.

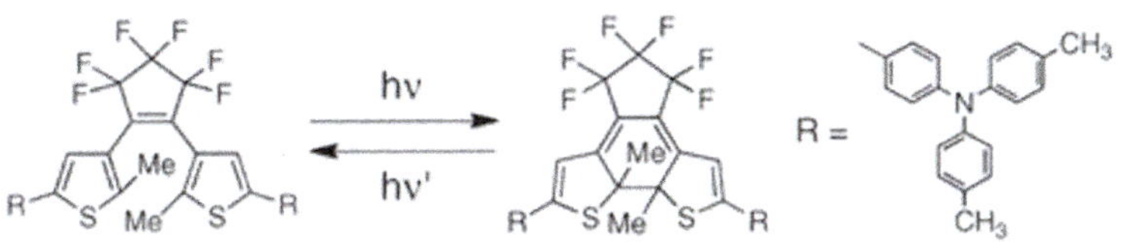

Figure 1.46. *Multi-photon process of reversible colorization from 1,2-bis(5-{N,N-bis(4-methylphenyl)aminophenyl}-2-methylthiophenyl-3)-3,3,4,4,5,5-hexafluorocyclopentene (APT)*

COMMENT.–

The so-called super-resolution may come from the inhibition of radical production induced by a dual excitation, allowing us to go beyond the limits of diffraction [FOR 13]. There is therefore, as shown in Chapter 3 of Volume 1, the possibility of gaining resolution, owing to a link between chemistry and light absorption. These authors have applied this principle by using photo-chromatic compounds not, in fact, allowing the production of objects but color images within a transparent matrix, consistent with Weis *et al.* [WEI 16].

Here, we have used a non-exhaustive approach to examine the possibility of adding colors to a 3D object produced by additive manufacturing. This overview analysis shows the possibilities that it would probably be necessary to look into in greater depth. These include the implementation of multi-color information for a given object in an even broader way to produce colored parts.

1.7. Conclusion

In this chapter, we have tried to show that there are numerous conceivable means to develop new 3D processes. These are increasingly specific and dedicated to particular, applications:

– from a localized transformation energy in the given space with specific physicochemical effects;

– linked to the form or the nature of this energy;

– by locating a particular voxel or by giving it a certain shape;

– or all of the above.

Within the list of proposals shown in this chapter, (which are probably not exhaustive), some processes are at a stage of actual industrialization. We might cite the case of 3D Carbon with its generic CLIP process. Others are aiming for high-added-value niches. Examples of these are optical lenses, microelectronics and other spheres. Works other than those revisiting stereolithography and a robotics system indeed focus upon microtechnological or nanotechnological aspects. We are still working within the standard paradigm of adding original material with energy source or material aspects, and/or physicochemical conversion of the given material. The essential barrier, linked to the manufacturing time, for a given quality in terms of precision and finding the "right material" remains an obstacle, which only the CLIP process can partially overcome. This particular research axis therefore seems to favor laying down the need to send the entire additive manufacturing concept back to the drawing board. However, no doubt it was necessary to travel this entire route to reach this conclusion.

Highly upstream spheres are not, however, legion, within this chapter. Moreover, we will not be interested, within the next chapter, in analyzing the reasons leading to this particular situation. We may suppose that, on the one hand, knowledge of reality and possibility, and, on the other hand, mastering ignorance of certain aspects, both serve to stimulate innovation, particularly on an incremental basis. It is from the proximity to known aspects that validation arises. It is because we know reasonably well what we must obtain. To invent within a framework of thought, to conceptualize the incremental concept, we must not move too far from justifiable knowledge. We simply take a form of distance concerning the aspects that we already know. When the issue is somewhat up in the air, academic research, which has, in theory, considerable freedom should help us to envisage the appropriate response. Without wishing to be too verbose(!), it is what proves, to a degree, that which we are trying to prove.

In addition, within the context of the strong competition which European industry is currently experiencing, the purpose of this chapter may have been to prove our capacity to develop an innovative technological offer. This would allow manufacturers to address the challenges of competitiveness and differentiation. Moreover, by this means, we could enter, via additive manufacturing, into the era of the so-called FoF "factory of the future". Yet, probably the ambition of given socioeconomic backgrounds focusing upon additive manufacturing is still too reliant upon cost rationalization, deadlines and the product lifecycle for new products, the growth of diversity and the quality of conceivable products (e.g. aspects such as high-end, production speed, customization, size and high accuracy involved in such production).

Briefly, the suggestions here are probably too upstream relative to a more immediate demand. Does that prove that creative researchers, without any network and motivation, actually accomplish very little? These observations may largely reduce the significance that we must attach to single individual creativeness. According to Chanut-Guieu and Guieu, "If this is not managed within a given motivating organizational mechanism it risks in the end not being very productive" [CHA 14]. In this section, from available works in the field (manufacturers' driving forces rarely form the subject of such publications), the opportunity comes uniquely from the mind of the given researcher. Save for his/her imagination, he/she starts with nothing or virtually nothing within his/her disciplinary competence. He/she transforms a scientific opportunity on a "bottom–up" basis into an applicable action [SHA 61]. This carries with it the risk of undertaking "inapplicable applied research". This forms, at least in part, the subject of risk-taking, which should be sought within the research sphere.

Indeed, we always look from a given hypothesis, with a return, which unfortunately cannot achieve a form of unity. We should remember that there are two forms of thought: deductive thinking, which starts from a hypothesis to reach a result. The author has attempted to use this approach in this chapter. There is also inductive thinking, which makes given assumptions. The latter approach will be further explored in Volume 3. This thinking comes from forms of intuition, divergent thinking and a willingness to simplify the world to formulate original promising concepts, which are breakthrough innovations [TEE 97, KEA 09, SAL 12]. It is then that we can explore other forms of uncertainty, those which do not rest upon a known issue. This is when nobody knows to ask the question because it is not (yet) being bandied about.

The initiatives described within this chapter already illustrate the possibility of going beyond the so-called "statuary" technologies set out in the previous volume. However, it must be noted that there are probably more concepts than actual economic successes to be expected in this branch of the field. Many of those set out are still at the proof-of-concept stage. The road is yet to be traveled before knowing whether the processes proposed can one day grapple with economic reality. However, for this to happen, the players in innovation must re-assert their will to try to explore the sphere. This is because they must understand that the hopes for long-term growth will be fulfilled. This is so even if the psychological warning shot may be harsh when a given process concept amounts to nothing. The El Dorado, which the researcher initially thought existed in this field, is not yet available, and the research group, or R&D department, must go through a phase of disappointment (which should be purely temporary).

1.8. Bibliography

[ADA 68] ADAMSON A.W., Method and apparatus for generating 3 dimensional patterns, US Patent 3609706 A, 6 December 1968.

[ADE 71] ADELMAN A.H., LEWIS J.D., Method and apparatus for generating 3 dimensional patterns, US Patent 3609707 A, filed December 9, 1968, patented September 28, 1971.

[AKA 03] AKAMATSU K., IKEDA S., NAWAFUNE H., "Site-selective direct silver metallization on surface-modified polyimide layers", *Langmuir*, vol. 19, pp. 10366–10371, 2003.

[AKA 05] AKAMATSU K., KIMURA A., MATSUBARA H. *et al.*, "Site-selective direct photochemical deposition of copper on glass substrates using TiO_2 nanocrystals", *Langmuir*, vol. 21, pp. 8099–8102, 2005.

[AL 12] AL-ABADDI M., SASSO L., DIMAKI M., "Fabrication of 3D nano/microelectrodes via two photon polymerization", *Microelectronic Engineering*, vol. 38, pp. 378–381, 2012.

[ALE 15] ALEX, "La technologie CLIP devrait révolutionner l'impression 3D!", available at: http://www.3dnatives.com/clip-revolutionner-impression-3d/, 2015.

[ALE 16] ALEX, "Carbon3D dévoile sa première imprimante 3D M1", available at: http://www.3dnatives.com/carbon3d-imprimante-3d-m1-04042016/, 2016.

[ALL 15] ALLARD F., "Les lunettes passent à l'impression 3D", available at: http://www.zesmallfactory.com/news/les-lunettes-passent-limpression-3d/, 2015.

[ALL 06] ALLONCLE A.P.B., THOMAS B., GROJO D. *et al.*, "Ablation laser pour la microélectronique plastique", *Journal de Physique IV*, vol. 138, pp. 213–221, 2006.

[ALL 16] ALLONCLE A.P.B., DELAPORTE P.C.M., "Procédé d'impression au moyen de deux lasers", Patent PCT/IB2016/052370, 3 November 2016.

[ALM 67] ALMGREN M., "The natural phosphorescence lifetime of biacetyl and benzil in fluid solution", *Journal of Photochemistry and Photobiology*, vol. 6, pp. 829–840, 1967.

[ALT 10] ALTER N., *L'innovation ordinaire*, PUF, Paris, 2010.

[AMB 16] AMBROSI A., PUMERA M., "3D-printing technologies for electrochemical applications", *Chemical Society Reviews*, vol. 45, pp. 2740–2755, 2016.

[AMI 09] AMINUZZAMAN M., WATANABE A., MIYASHITA T., "Fabrication of conductive silver micro-patterns on an organic-inorganic hybrid film by laser direct writing", *Thin Solid Films,* vol. 517, pp. 5935–5939, 2009.

[AND 91a] ANDRE J.C., BRULLE Y., CORBEL S. *et al.*, Procédé et dispositif de fabrication d'objets tridimensionnels par polymérisation d'une résine, French patent 91 07 983, 27 June 1991.

[AND 91b] ANDRE J.C., BRULLE Y., FAUGERAS P. *et al.*, Procédé de fabrication en continu de fibres et matériaux polymérisables, notamment de fibres optiques polymères, French patent 91 13 367, 25 October 1991.

[AND 92] ANDRE J.C., BRULLE Y., BARALDI U. *et al.*, Fabrication d'objets monochromes ou polychromes par stéréo-photo-lithographie à partir de poudres ou de granulés, French patent 92-08316, 06 July 1992.

[AND 94] ANDRE J.C., CORBEL S., *Stéréo-photo-lithographie laser*, Polytechnica, Paris, 1994.

[AND 17] ANDRE J.C., "Stéréo-lithographie par photo-polymérisation", *Techniques de l'Ingénieur*, ref. BM7910 V1, 2017.

[ANO 15] ANONYMOUS, "Frittage des céramiques", available at: http://flooose.free.fr/res1152/ceramiques.html, 2015.

[ANZ 88] ANZIEU D., *Le corps de l'œuvre: essai psychanalytique sur le travail créateur*, Gallimard, Paris, 1988.

[BAG 72/73] BAGDASARYAN K.S., KIRJUKHIN Y.I., SINITSINA Z.A., "Quantitative studies of biphotonic reactions", *Journal of Photochemistry*, vol. 1, pp. 225–240, 1972/1973.

[BAL 05] BALDACCHINI T., PONS A.C., PONS J. *et al.*, "Multiphoton laser direct writing of two-dimensional silver structures", *Optics Express*, vol. 13, pp. 1275–1280, 2005.

[BAR 16] BARRAU A., *De la vérité dans les sciences*, Dunod, Paris, 2016.

[BEL 99] BELFIELD K.D., HAGAN J., VAN STRYLAND E.W. *et al.*, "New two-photon absorbing fluorene derivatives: synthesis and nonlinear optical characterization", *Organic Letters*, vol. 1, pp. 1575–1577, 1999.

[BEL 00] BELFIELD K.D., SCHAFER K.J., LIU Y. *et al.*, "Multiphoton-absorbing organic materials for microfabrication, emerging optical applications and non-destructive three-dimensional imaging", *Journal of Physical Organic Chemistry*, vol. 13, pp. 837–849, 2000.

[BEN 71] BENSASSON R., LAND E.J., "Triplet-triplet extinction coefficients via energy transfer", *Transactions of the Faraday Society*, vol. 67, pp. 1904–1915, 1971.

[BER 12] BERENSCHOT E.J.W., BUROUNI N., SCHURINK B. *et al.*, "3D nanofabrication of fluidic components by corner lithography", *Small*, vol. 8, pp. 3823–3831, 2012.

[BOH 86] BOHANDY J., KIM B.F., ADRIAN F.J., "Metal deposition from a supported metal film using an excimer laser", *Journal of Applied Physics*, vol. 60, pp. 1538–1539, 1986.

[BIO 95] BIONDI J., MICHEL O., "Evolution de structures neuronales; application à un robot mobile autonome", *Journées de Rochebrune*, pp. 23–28, March 1995.

[BOH 00] BOHRINGER K.F., DONALD B.R., KAVRAKI L.E. *et al.*, "Part orientation with one or two stable equilibria using programmable force fields", *IEEE Transaction on Robotics and Automatics*, vol. 16, pp. 157–170, 2000.

[BON 94] BONABEAU E., THERAULAZ G., *Intelligence collective*, Hermès, Paris, 1994.

[BRA 11] BRANDI F., ANJUN F., CESERACCIN L. *et al.*, "Rigid biodegradable photopolymer structures of high resolution using deep-UV laser photo-curing", *Journal of Micromechanics and Microengineering*, vol. 21, doi:10.1088/0960-1317/21/5/054007, 2011.

[BRA 15] BRANT A.M., SUNDARAM M.M., KAMARAJ A.B., "Finite element simulation of localized electrochemical deposition for maskless electrochemical additive manufacturing", *Journal of Manufacturing Science and Engineering*, vol. 137, p. 011018, 2015.

[BRA 16] BRANT A.M., SUNDARAM M., "A novel electrochemical micro additive manufacturing method of overhanging metal parts without reliance on support structures", *Procedia Manufacturing, 44th Proceedings of the North American Manufacturing Research Institution of SME*, vol. 5, pp. 928–943, available at: http://www.sme.org/namrc, 2016.

[BRI 72] BRINEN J.S., INNES I.E., KAZAN J., "T-T absorption of molecules with extremely high extinction coefficients", *Chemical Physics Letters*, vol. 15, pp. 69–72, 1972.

[BRU 92] BRULLE Y., Procédés photoniques pour l'optique: lentilles et fibres, PhD Thesis INPL, Nancy, France, 1992.

[CAM 14a] CAMPBELL T.A., TIBBITS S., GARRETT B., "The next wave: 4D printing programming the material world", available at: http://www.atlanticcouncil.org/images/publications/The_Next_Wave_4D_Printing_Programming_the_Material_World.pdf, 2014.

[CAM 14b] CAMPBELL T.A., TIBBITS S., GARRETT B., "The programmable world", *Scientific American*, vol. 311, pp. 60–65, 2014.

[CAR 98] CARACOSTAS P., MULDUR U., Society, the endless frontier – a European vision of research and innovation policies for the 21st century, Office for Official Publications of the European Communities, Luxembourg, EUR 17655, 1998.

[CEA 05] CEA, Doter les nano-objets de propriétés moléculaires par greffage moléculaire, Clefs CEA, pp. 67–83, 2005.

[CHA 09] CHAKRABARTI S., RUUD K., "Large two photon absorption cross section: molecular tweezer as a new promising class of compounds for nonlinear optics", *Physical Chemistry Chemical Physics*, vol. 11, pp. 2592–2596, 2009.

[CHA 14] CHANUT-GUIEU C., GUIEU G., "Créativité, PME et entrepreneuriat: des zones d'ombre et de lumière", *Revue internationale P.M.E.: économie et gestion de la petite et moyenne entreprise*, vol. 27, pp. 35–63, 2014.

[CHA 11] CHAVEL S., *Se mettre à la place d'autrui; l'imagination morale*, Presses Universitaires de Rennes, Rennes, 2011.

[CHE 12] CHEN Z., CHONG T.C., MINGHUI H., "Laser nano-fabrication: a route toward next generation mass production", *Industrial Laser Production*, vol. 27, pp. 18–20, 2012.

[CIR 15] CIRTES, "Stratoconception", available at: http://www.cirtes.com/presentation/ressources/stratoconception, 2015.

[CNR 15] CNRS/BIOLOGIE, "Débat sur la Stratégie Nationale de la Recherche", available at: http://sciences.blogs.liberation.fr/home/2014/06/d%C3%A9bat-sur-la-strat%C3%A9gie-nationale-de-la-recherche.html, 2015.

[CON 17] CONDLIFFE J., "This super-fast 3D printer is powered by holograms", available at: https://www.technologyreview.com/s/603605/this-super-fast-3-d-printer-is-powered-by-holograms/?set=603690, 2017.

[CRE 13] CRESPO-MONTEIRO N., NADAR L., DESTOUCHES N. *et al.*, "Contrôle optique de la croissance et de la déformation de nanoparticules métalliques au sein de matrices méso-poreuses de TiO_2", *UVX 2012*, 01011, doi: 10.1051/uvx/201301011, 2013.

[DEL 13] DELANOY J., "Etude intuitive de forms d'objets 3D: Analyse et interactions", available at: https://ensiwiki.ensimag.fr/index.php/Johanna_Delanoy_:_Edition_intuitive_de_formes_d%E2%80%99objets_3D_:_Analyse_des_interactions, 2013.

[DES 92] DESCARTES R., *Discours de la méthode*, Flammarion, Paris, 1992.

[DES 97] DESPREZ P., Métallisation d'objets stéréo-lithographiques, PhD Thesis INPL, Nancy, France, 1997.

[DEV 15] DEVI C.L., YESUDAS K., MAKAROV N.S. *et al.*, "Fluororenylethynperylene derivatives with strong two photon absorption: influence of substituents on optical properties", *Journal of Materials Chemistry C*, vol. 3, pp. 3730–3744, 2015.

[DUA 04] DUAN Y.M., SUN M.B., KAWATA S., "Microfabrication of two and three dimensional structures by two-photon polymerization", *Journal of Photopolymer Science and Technology*, vol. 17, pp. 393–396, 2004.

[ECR 05] ECRIN, *Technologies du futur; enjeux de société*, OMNI Science, Paris, 2005.

[ENG 11] ENGELHARDT S., HU Y., SEILER N. *et al.*, "3D microfabrication of polymer-protein hybrid structure with a Q-switched micro-laser", *Journal of Laser Micro/Nanoengineering*, vol. 6, pp. 54–58, 2011.

[FAB 16] FABULOUS, "L'impression 3D: son histoire pour une révolution en marche", available at: http://www.fabulous.com.co/guide-impression-3d/en-bref/histoire/, 2016.

[FEY 79] FEYERABEND P., *Contre la méthode*, Le Seuil, Paris, 1979.

[FOR 02] FORAY D., "Ce que l'économie néglige ou ignore en matière d'analyse de l'innovation" in ALTER N. (ed.), *Les logiques de l'innovation*, La découverte, Paris, 2002.

[FOR 13] FORMAN D.L., COLE M.C., MCLEOD R., "Radical diffusion limits to photo-inhibited super-resolution lithography", *Physical Chemistry Chemical Physics*, vol. 15, pp. 14862–14867, 2013.

[GAG 11a] GAGLIO G., *Sociologie de l'innovation*, PUF, Paris, 2011.

[GAG 11b] GAGLIO G., LAURIOL J., DU TERTRE C., *L'économie de la fonctionnalité: une voie nouvelle vers un développement durable?*, Octarès, Paris, 2011.

[GEO 17] GEORGE B., "Un hologramme pour accélérer l'impression 3D", available at: http://www.lesechos.fr/idees-debats/sciences-prospective/0211812864347-un-hologramme-pour-accelerer-limpression-3d-2066551.php#gLSw8u8BbeuWKavy.99, 2017.

[GRI 91] GRIVIN V.P., KHMELINSKI V.F, PLYUSNIN V.F., "Primary photochemical processes of the $PtCl_6^{2-}$ complex in alcohols", *Journal of Photochemistry and Photobiology A*, vol. 59, pp. 153–161, 1991.

[GRI 92] GRIVIN V.P., KHMELINSKI V.F, PLYUSNIN V.F. *et al.*, "Laser pulse photolysis study of the reaction of the intermediate platinium (III) complex $PtCl^{2-}{}_5$ and 4-(2-p-dimethylaminophenylethynil)-6,6-dimethyl-3-cyano-5,6-dihydro-2(1H)-pyridone", *Journal of Photochemistry and Photobiology A*, vol. 63, pp. 7–14, 1992.

[GUY 14] GUYOT J.L., BRUNET S. (eds), *Construire les futurs – Contribution épistémologique et méthodologiques de la démarche prospective*, Presses Universitaires de Namur, Namur, 2014.

[GUZ 95] GUZZETTI L., *A Brief History of European Research Policy*, Office for Official Publications of the European Communities, Brussels, 1995.

[HER 02] HERZENBERG L.A., PARKS D., SAHAF B. *et al.*, "The history and future of the fluorescence activated cell sorter and flow cytometry: a view from Stanford", *Clinical Chemistry*, vol. 48, pp. 1819–1827, 2002.

[HOU 03] HOUBERTZ R., FROHLICH L., POPALL M. *et al.*, "Inorganic-organic hybrid polymers for information technology: from planar technology to 3D structures", *Advanced Engineering Materials*, vol. 8, pp. 551–555, 2003.

[HUN 05] HUNG K.Y., TSENG F.G., CHOU H.P., "Application of 3D gray mask for the fabrication of curved SU-8 structures", *Microsystem Technologies*, vol. 11, pp. 365–369, 2005.

[IKE 01] IKEDA S., AKAMATSU K., NAWAFUNE H., "Direct photochemical formation of Cu patterns on surface modified polyimide resin", *Journal of Materials Chemistry*, vol. 11, pp. 2919–2921, 2001.

[INM 16] INM, "Hannover Messe: Photochemical metallization allows the manufacture of touchscreens in a single step", available at: http://www.leibniz-inm.de/en/2016/04/hannover-messe-photochemical-metallization-allows-the-manufacture-of-touchscreens-in-a-single-step/, 2016.

[JAH 10] JAHN S.F., JAKOB A., BLAUDECK T. *et al.*, "Inkjet printing of conductive patterns with an aqueous solution of [$AgO_2C(CH_2OCH_2)$] without any additional stabilizing ligands", *Thin Solid Films*, vol. 518, pp. 3218–3222, 2010.

[JIA 14] JIANG L.J., ZHOU Y.S., XIANG W. *et al.*, "Two photon polymerization: investigation of chemical and mechanical properties of resin using Raman micro-spectroscopy", *Optics Letters*, vol. 39, pp. 3034–3037, 2014.

[JIA 16] JIANG L.J., CAMPBEL J.H., LU Y.F. *et al.*, "Direct writing target structures by two-photon polymerization", *Fusion Science and Technology*, vol. 70, pp. 295–309, 2016.

[JIN 08] JIN M., MALVAL J.P., VERSACE D.L. *et al.*, "Two-photon absorption and polymerization ability of intramolecular energy transfer based photo-initiating systems", *Chemical Communications*, vol. 48, pp. 6540–6542, 2008.

[JIN 14] JIN M., XIE J., MALVAL J.P. *et al.*, "Two photon lithography in visible and NIR ranges using multi-branched- based sensitizers for efficient acid generation", *Journal of Materials Chemistry C*, vol. 2, pp. 7201–7215, 2014.

[KAD 05] KADEKAR V., FANG W., LIOU F., "Deposition technologies for micro-manufacturing: a review", *Journal of Manufacturing Science and Engineering*, vol. 126, pp. 787–795, 2005.

[KAN 01] KANNAN R., HE G.S., YUAN I. *et al.*, "Diphenylaminofluorene-based two-photon-absorbing chromophores with various π-electron acceptors", *Chemistry of Materials*, vol. 13, pp. 1896–1904, 2001.

[KEA 09] KEARNEY E., GEBERT D., VOELPEL S.C., "When and how diversity between teams: The importance of team members' need for cognition", *Academy of Management Journal*, vol. 52, pp. 581–598, 2009.

[KEL 82] KELLMANN A., TFIBEL F., "Radical produced from the laser induced photoionization of acridine in solution", *Journal of Photochemistry*, vol. 18, pp. 81–88, 1982.

[KHE 10] KHERKHOVE S., *La dictature de l'immédiateté; sortir du présentalisme*, Y. Michel, Gap, 2010.

[KIM 09] KIM H.M., CHO B.R., "Two photon materials with large two-photon cross section: structure property relationship", *Chemical Communications*, pp. 153–164, 2009.

[KIM 11] KIM J.T., SEOL S.K., PYO J. *et al.*, "Three-dimensional writing of conducting polymer nanowire arrays by meniscus-guided polymerization", *Advanced Materials*, vol. 23, pp. 1968–1970, 2011.

[KLE 15] KLEIN E., "Comment faire progresser l'idée de progrès?", available at: http://www.up-magazine.info/index.php?option=com_content&view=article&id=5305:comment-faire-progresser-l-idee-de-pro-gres&catid=103:analyses&Itemid=542&utm_source=newsletter_175&utm_medium=email&utm_campaign=up-nl-vnew, 2015.

[KOS 04] KOSHNEVIS B., "Automated construction by contour grafting – related robotic and information technologies", *Journal of Automation Construction*, vol. 13, pp. 5–19, 2004.

[KOS 06] KOSHNEVIS B., HWANG D., YAO K.T. *et al.*, "Mega-scale fabrication by contour grafting", *International Journal of Industrial and Systems Engineering*, vol. 1, pp. 301–320, 2006.

[KOS 14] KOSHNEVIS B., "Robotic building construction by contour grafting", available at: http://contest.techbriefs.com/2014/entries/machinery-automation-robotics/4737, 2014.

[KUH 01] KUHLMANN S., "Future governance of innovation policy in Europe – three scenarios", *Research Policy*, vol. 30, pp. 953–976, 2001.

[LAN 68] LAND E.J., "Extinction coefficients of triplet-triplet transitions", *Proceedings of the Royal Society of London A*, vol. 305, pp. 457–471, 1968.

[LAR 03] LAREDO P., MUSTAR P., "Politiques publiques de recherche et d'innovation", in MUSTAR P., PENAN E. (eds), *Encyclopédie de l'innovation*, Economica, Paris, 2003.

[LE 15] LE T., SONG B., LIU Q. *et al.*, "A novel strain sensor based on 3D printing technology and 3D antenna design", *IEEE – Electronic Components & Technology Conference*, pp. 983–986, 2015.

[LEE 04] LEE K.S., KIM S.O., YANG H.K. *et al.*, "Lithographic microfabrication by using two photon absorbing phenylene-vinylene derivatives", *Molecular Crystals and Liquid Crystals*, vol. 424, pp. 35–41, 2004.

[LEE 07] LEE K.S., KIM R.H., PRABHAKARAN P. *et al.*, "Two photon stereolithography", *Journal of Nonlinear Optical Physics & Materials*, vol. 16, pp. 59–72, 2007.

[LEE 08] LEE K.S., KIM R.H., YANG D.Y. *et al.*, "Advances in 3D nano/microfabrication using two photons induced polymerization", *Progress in Polymer Science*, vol. 33, pp. 631–681, 2008.

[LEE 13] LEE D.G., KIM D.K., MOON Y.J. *et al.*, "Effect of laser-induced temperature field on the characteristics of laser-sintered silver nanoparticle ink", *Nanotechnology*, vol. 24, 265702 (9 pp.), 2013.

[LEE 14] LEE J., LEE B., JEONG S. *et al.*, "Microstructure and electrical property of laser-sintered Cu complex ink", *Applied Surface Science*, vol. 307, pp. 42–45, 2014.

[LEE 15] LEE J.W., "3D nano-printing technologies for tissue engineering applications", *Journal of Nanomaterials*, Article ID 213521, available at: http://dx.doi.org/10.1155/2015/213521, 2015.

[LEM 13] LEMA PROJECT, "Assemblage de microcomposants en milieu liquide", available at: http://lema.isir.upmc.fr/index.php?perma=tache_3, 2013.

[LEN 11] LENHART N., CROWLEY K., KILLARD A.J. *et al.*, "Inkjet printable polyaniline-gold dispersions", *Thin Solid Films*, vol. 519, pp. 4351–4356, 2011.

[LEP 15] LEPMI, "Electro-cristallisation – dépôts métalliques", available at: http://lepmi.grenoble-inp.fr/ei/electrocristallisation-depots-metalliques-125597.kjsp, 2015.

[LES 04] LESOURNE J., BRAVO A., RANDET D., *Avenirs de la recherche et de l'innovation en France*, La Documentation Française, Paris, 2004.

[LEW 12] LEWANDOWSKA-ANDRALOJC A., HUG G.L., HÖRNER G. *et al.*, "Unusual photobehavior of benzophenone triplets in hexafluoroisopropanol. Inversion of the triplet character of benzophenone", *Journal of Photochemistry and Photobiology A*, vol. 244, pp. 1–8, 2012.

[LIM 11] LIM T.W., SON Y., YANG D.Y. *et al.*, "Selective ablation-assisted two photons stereolithography for effective nano- and micro-fabrication", *Applied Physics A*, vol. 103, pp. 1111–1116, 2011.

[LIU 10] LIU Y., NOLTE D.D., PYRAK-NOLTE J., "Large format fabrication by two-photon polymerization", *Applied Physics A*, vol. 100, pp. 181–191, 2010.

[LIU 11] LIU X.T., REU A.M., GUO J.F. *et al.*, "Theoretical investigation of one- and two-photon spectra of pyrazabole chromophores", *Theoretical Chemistry Accounts*, vol. 130, pp. 37–50, 2011.

[LIU 16] LIU Y., HU Q., ZHANG F. *et al.*, "Additive manufacture of three dimensional nanocomposite based objects through multiphoton fabrication", *Polymers*, vol. 8, p. 325, doi:10.3390/polym8090325, 2016.

[LOU 17] LOUISOT J.M., "Optimiser la prise de risques", *Risk-Assur*, vol. 474, pp. 28–37, available at: https://www.riskassur-hebdo.com, 2017.

[LU 06] LU Y., Micro/nano fabrication of polymeric materials by DMD-based micro-stereolithography and photothermal imprinting, PhD Thesis, Austin, TX, USA, 2006.

[LUC 16] LUCAS T., "Trois technos pour imprimer en 3D des métaux sans poudre", available at: http://www.usinenouvelle.com/editorial/trois-technos-pour-imprimer-en-3d-des-metaux-sans-poudre.N465383, 2016.

[LUX 16] LUXEXCEL, "LUXeXcel, The Netherlands-Proprietary Printoptical Technology Set to Disrupt Markets", available at: http://additivemanufacturing.com/2015/05/12/luxexcel-the-netherlands-proprietary-printoptical-technology-set-to-disrupt-markets/, 2016.

[MAG 15] MAGEE C., "With $100M in funding, Carbon3D will make 3D manufacturing a reality", available at: https://techcrunch.com/2015/08/20/with-100m-in-funding-carbon3d-will-make-3d-manufacturing-a-reality/, 2015.

[MAI 11] MAICU M., HIDALGO M.C., COLON G. *et al.*, "Comparative study of the photodeposition of Pt, Au and Pd on pre-sulphated TiO_2 for the photocatalytic decomposition of phenol", *Journal of Photochemistry and Photobiology A*, vol. 217, pp. 275–283, 2011.

[MAL 01] MALLICK K., WANG Z.L., PAL T., "Seed-mediated successive growth of gold particles accomplished by UV irradiation: a photochemical approach for size-controlled synthesis", *Journal of Photochemistry and Photobiology A*, vol. 140, pp. 75–80, 2001.

[MAL 07] MALVAL J.P., DIETLIN C., ALLONAS X. *et al.*, "Sterically tuned photoreactivity of an aromatic α-diketones family", *Journal of Photochemistry and Photobiology A*, vol. A192, pp. 66–73, 2007.

[MAL 11] MALVAL J.P., JIN M., MORLET-SAVARY F. *et al.*, "Enhancement of the two-photon initiating efficiency of a thioxanthone derivative through a Chevron-shaped architecture", *Chemistry of Materials*, vol. 23, pp. 3411–3420, 2011.

[MAN 05] MANGAIYARKARSAI D., KAMADA K., SAITO N. *et al.*, "Large area multi-color changes induced by femtosecond laser pulses in soda-lime silicate glass embedded with Ag nanoparticles", *Journal of Non-Crystalline Solids*, vol. 351, pp. 3156–3159, 2005.

[MAN 13] MANNHEIM T., KURTENBACH S., BOUDOU K. *et al.*, "Augmenter la flexibilité des proceeds de manipulation automatisés par l'utilisation de la structure reconfigurable PARAGRIP", *21ème congrès de mécanique*, p. 6, available at: http://documents.irevues.inist.fr/bitstream/handle/2042/52509/a_727ZJN87.pdf?sequence=1, Bordeaux, 26–30 August 2013.

[MAN 07] MANSOUR S., GILBERT M., HAGUE R., "A study of the impact of short-term ageing on the mechanical properties of a stereolithographic resin", *Materials Science and Engineering A*, no. 447, pp. 277–284, 2007.

[MAR 07] MARDER S.R., BREDAS J.L., PERRY J.W., "Materials for multiphoton 3D microfabrication", *MRS Bulletin*, no. 32, pp. 561–565, 2007.

[MAR 02] MARUO S., IKUTA K., "Submicron stereolithography for the production of freely movable mechanisms by using single-photon polymerization", *Sensors and Actuators A*, vol. 100, pp. 70–76, 2002.

[MAR 08] MARUO S., SAEKI T., "Femtosecond laser direct writing of metallic microstructures by photo-reduction of silver nitrate in a polymer matrix", *Optics Express*, vol. 16, pp. 1174–1179, 2008.

[MAX 15] MAXEY K., "A profile of hybrid additive manufacturing", available at: http://advancedmanufacturinginsight.com/archived-articles/item/a-profile-of-hybrid-additive-manufacturing-technology, 2015.

[MCG 84] MCGINNISS V.D., SCHWERZEL R.E., Photo-polymerizable composition containing a photosensitizer donor and photo-initiator acceptor, US Patent 4571377 A, 23 February 1984.

[MEL 17] MELANIE, "Daqri, la startup qui imprime en 3D à partir d'hologrammes", available at: http://www.3dnatives.com/impression-3d-hologramme-daqri-27022017/, 2017.

[MEN 07] MENDONCA C.R., NEVES U.M., DE BONI L. *et al.*, "Two-photon induced anisotropy in PMMA film doped with Disperse Red 13", *Optics Communications*, vol. 273, pp. 435–440, 2007.

[MEY 01] MEYER J.R., RHOADS G.B., LÉVY K.L., Liaison d'objets audio et d'autres objets médias, PCT/US2001/002609, August 2 2001.

[MOR 11] MORI K., ISHIBASHI Y., MATSUDA H. *et al.*, "One-color reversible control of photochromic reactions in a diarylethene derivative: three-photon cyclization and two-photon cycloreversion by a near-infrared femtosecond laser pulse at 1.28 μm", *Journal of the American Chemical Society*, vol. 133, 2621–2625, 2011.

[MOU 10] MOURA G.L., SIMAS A.M., "Two photon absorption of fluorine derivatives: systematic molecular design", *Journal of Physical Chemistry*, vol. 114, pp. 6106–6116, 2010.

[MUR 15] MURR L.E., *Handbook of material structures, properties, processing and performance*, Springer, Switzerland, 2015.

[MUR 16] MURR L.E., "Frontiers of 3D printing/additive manufacturing: from human organs to aircraft fabrication", *Journal of Materials Science & Technology*, vol. 32, pp. 987–995, 2016.

[MUS 14] MUSSO P., *L'imaginaire industriel*, Manucius, Paris, 2014.

[NAR 10] NARAYAN R.J., DORAISWAMY A., CHRISEY D.B. *et al.*, "Medical protoptyping using two photon polymerization", *Material Today*, vol. 13, pp. 42–48, 2010.

[NIE 98] NIELSEN B.R., JORGENSEN K., SKIBSTED L.H., "Triplet–triplet extinction certiesefficients rate constants of triplet decay and rate constants of anthracene triplet sensitization by laser flash photolysis of astaxanthin, β-carotene, canthaxanthin and zeaxanthin in deaerated toluene at 298 °K", *Journal of Photochemistry and Photobiology A*, vol. 112, pp. 127–133, 1998.

[NIE 12] NIE X., WANG H., ZOU J., "Inkjet printing of silver citrate conductive ink on PET substrate", *Applied Surface Science*, vol. 261, pp. 554–560, 2012.

[NIE 14] NIESLER F., WERNER A., "3D laser lithography", available at: http://www.cytofluidix.com/3d-laser-lithography/, 2014.

[NIE 15] NIESLER F., HERMATSCHWEILERM., "Two-photon polymerization – a versatile microfabrication tool; from mask-less lithography to 3D printing", *Lasertechnik Journal*, vol. 3, doi:10.1002/latj.201500019, 2015.

[OCA 16] OCANA R., CALATAYUD A., "Laser induced coloring on small titanium surfaces", *Physics Procedia*, vol. 83, pp. 225–232, 2016.

[OCD 05] OCDE, "Manuel d'Oslo", available at: http://www.oecd.org/dataoecd/35/58/2367554.pdf, 2005.

[OVS 08] OVSIANIKOV A., VIERTI J., CHICHKOV B. *et al.*, "Ultra-low shrinkage hybrid photosensitive material for two photon polymerization microfabrication", *ACS Nano*, vol. 2, pp. 2257–2262, 2008.

[PAP 16] PAPAVLU A.P., MATTLE T., TEMMEL S. *et al.*, "Highly sensitive SnO_2 sensor via reactive laser-induced transfer", *Scientific Reports*, vol. 6, p. 25144, 2016.

[PAR 05] PARK S.H., LEE S.H., YANG D.Y., "Subregional slicing method to increase three-dimensional nanofabrication efficiency in two-photon polymerization", *Applied Physics Letters*, vol. 87, p. 154108, 2005.

[PAR 09a] PARK S.G., LEE S.K., MOON J.H. *et al.*, "Holographic fabrication of 3 dimensional nanostructures for microfluidic passive mixing", *Lab on a Chip*, vol. 9, pp. 3144–3150, 2009.

[PAR 09b] PARK S.H., YAN D.Y., LEE K.S., "Two photon stereolithography for realizing ultraprecise three dimensional nano-/micro- devices", *Laser & Photonics Reviews*, vol. 3, pp. 1–11, 2009.

[PAU 14] PAULING L. cited by BRYNJOFSSON E., MCAFEE A., *Le deuxième âge de la machine; travail et prospérité à l'heure de la révolution technologique*, Odile Jacob, Paris, 2014.

[PEN 15] PENIDE J., QUINTERO F., RIVEIRO A. *et al.*, "High contrast laser marking of alumina", *Applied Surface Science*, vol. 336, pp. 118–128, 2015.

[PEN 16] PENIDE J., QUINTERO F., RIVEIRO A. *et al.*, "Everlasting dark printing on alumina by laser", *Physics Procedia*, vol. 83, pp. 233–239, 2016.

[PLY 92] PLYUSNIN V.F., KUZNETZOVA E.P., KHMELINSKI V.F. *et al.*, "Photochemistry of bis[3-(hydroxylamino)-3-methyl-2-butanone-oximato-(2-)-N,N']nickel in CCl_4- containing solutions", *Journal of Photochemistry and Photobiology A*, vol. 63, pp. 269–301, 1992.

[PON 12] PONCHE R., HASCOET J.Y., KERBRAT O. *et al.*, "A new global approach to design for additive manufacturing", *Virtual and Physical Prototyping*, vol. 7, pp. 93–105, 2012.

[PRO 17] PRODWAYS, "Ensemble de photographies d'objets 3D réalisés par Prodways", personal communication from Gorgé C., Prodways Technologies (Secretary General), 2017.

[QIN 01] QIN Z.Y., DU B.Y., ZHANG J. *et al.*, "Thermal influence on excimer-laser-induced electrical conductivity on polyimide film surfaces", *Applied Physics A*, vol. 72, pp. 711–715, 2001.

[RAM 02] RAMOS V., "On the implicit and on the artificial", in MAUBANT J.L., MOURA L. (eds), *Art, Architecture and Science*, Ministerio da ciencia e technologia Ed., Lisbon, Portugal, 2002.

[SAK 08] SAKAMOTO M., TACHIKAWA T., FUJITSUKA M. *et al.*, "Three-dimensional writing of copper nanoparticles in a polymer matrix with two-color laser beams", *Chemistry of Materials*, vol. 20, no. 6, pp. 2060–2062, 2008.

[SAK 09] SAKAMATO M., FUJISTUKA M., MAJIMA T., "Light as a construction tool of metal nanoparticles: synthesis and mechanism", *Journal of Photochemistry and Photobiology C*, vol. 10, pp. 33–56, 2009.

[SAL 12] SALAZAR M.R., LANT T.K., FIORE S.M. *et al.*, "Facilitating innovation in diverse science teams through integrative capacity", *Small Group Research*, vol. 43, pp. 527–558, 2012.

[SAL 07] SALEH B.E.A., TEICH, M.C., *Semiconductor Photon Detectors, in Fundamentals of Photonics*, John Wiley & Sons Inc., New York, 2007.

[SCH 61] SCHACKLE G.L.S., *Decision, Order and Time in Human Affairs*, Cambridge University Press, Cambridge, 1961.

[SCH 13] SCHONER C., TUCHSCHERER A., BLAUDECK T. *et al.*, "Particle-free gold metal–organic decomposition ink for inkjet printing of gold structures", *Thin Solid Films*, vol. 531, pp. 147–151.

[SCH 11/99] SCHUMPETER J., *Théorie de l'évolution économique*, Dalloz, Paris, 1911/1999.

[SCH 04] SCHUMPETER J., *Histoire de l'analyse économique*, Gallimard, Paris, 2004.

[SCU 16] SCULPTEO, "SLS and Carbon's CLIP: our prized in-house technologies", available at: https://www.sculpteo.com/blog/2016/09/14/sls-vs-clip/, 2016.

[SEL 15] SELIMIS A., MIRONOV V., FARSANI M., "Direct laser writing: principles and materials for scaffold 3D printing", *Microelectronic Engineering*, vol. 132, pp. 83–89, 2015.

[SEO 08] SEOL S.K., KIM J.T., JE J.H. *et al.*, "Three-dimensional (3D) polypyrrole microstructures with high aspect ratios fabricated by localized electro-polymerization", *Macromolecules*, vol. 41, pp. 3071–3074, 2008.

[SEO 15] SEOL S.K., KIM D., LEE S. *et al.*, "Electrodeposition-based 3D printing of metallic microarchitectures with controlled internal structures", *Small*, vol. 32, pp. 3896–3902, 2015.

[SHA 14] SHAFFER S., YANG K., VARGAS J., *et al.*, "On reducing anisotropy in 3D printed polymers via ionizing radiation", *Polymer*, vol. 55, pp. 5969–5979, 2014.

[SHA 15] SHANAH D., "3D printing ready for its next big sprint-metal", available at: http://www.timesofisrael.com/3d-printing-ready-for-its-next-big-sprint-metal/?utm_source=Start-Up+Daily&utm_campaign=7eefb85158-2015_11_05_SUI11_5_2015&utm_medium=email&utm_term=0_fb879fad58-7eefb85158-55014345, 2015.

[SHE 15] SHER D., "Xjet to launch revolutionary inkjet metal 3D printing nanotechnology", available at: http://3dprintingindustry.com/2015/11/12/xjet-launch-revolutionary-inkjet-metal-3d-printing-nanotechnology/, 2015.

[SHI 03] SHIN B.S., YANG D.Y., CHOI D.S., *et al.*, "A new rapid manufacturing process for multi-face high-speed manufacturing", *International Journal of Advanced Manufacturing*, vol. 22, pp. 68–74, 2003.

[SHU 10] SHUKLA S., FURLANI E.P., VIDAL X. *et al.*, "Two-photon lithography of sub-wavelength metallic structures in a polymer matrix", *Advanced Materials*, vol. 22, pp. 3695–3699, 2010.

[SIN 09] SINGH-RACHFORD T.N., CASTELLANO F.N., "Low power visible to UV up-conversion", *Journal of Physical Chemistry A*, vol. 113, pp. 5912–5917, 2009.

[SOO 15] SOOCURIOUS, "Un passionné de Lego reproduit de façon stupéfiante toutes sortes d'objets du quotidien", available at: http://soocurious.com/fr/lego-construction/, 2015.

[SOU 11] SOUKOULIS C.M., WEGENER M., "Past achievement and future challenges in the development of 3D photonic multi-materials", *Nature Photonics*, vol. 5, pp. 523–530, 2011.

[SPI 94] SPINOZA B., *L'éthique*, Folio – Essais, Paris, 1994.

[STR 13] STRANO G., LIANG H., EVERSON R.M. *et al.*, "Surface roughness analysis, modelling and prediction in selective laser melting", *Journal of Materials Processing Technology*, vol. 213, pp. 589–597, 2013.

[STR 03] STREMSDOERFER G., GHANEM F. *et al.*, "Coper deposition by dynamic chemical plating", *Journal of Materials Science*, vol. 38, pp. 3285–3291, 2003.

[SUN 04] SUN H.B., KAWATA S., "Two-photon photopolymerisation and 3D lithographic fabrication", *APS*, vol. 170, pp. 169–273, available at: http://nathan.instras.com/MyDocsDB/doc-712.pdf, 2004.

[SUN 15] SUNDARAM M.M., KAMARAJ A.B., KUMAR V.S., "Mask-less electrochemical additive manufacturing: a feasibility study", *Journal of Manufacturing Science and Engineering*, vol. 137, p. 9, 2015.

[TAN 06] TANAKA T., ISHIKAWA A., KAWATA S., "Two-photon-induced reduction of metal ions for fabricating three-dimensional electrically conductive metallic microstructure", *Applied Physics Letters*, vol. 88, p. 081107, 2006.

[TAO 09] TAOURI A., Etude de l'absorption à deux photons dans des matériaux organiques à l'aide de la technique dite de la lentille thermique, PhD Thesis, Angers University, France, 2009.

[TEE 97] TEECE D., PISANO G., SHUEN A., "Dynamic capabilities and strategic management", *Strategic Management Journal*, vol. 18, pp. 509–533, 1997.

[TEO 05] TEO W.E., KOTAKI M., MO X.M. *et al.*, "Porous tubular structures with controlled fiber orientation using a modified electrospinning method", *Nanotechnology*, vol. 16, pp. 918–924, 2005.

[TER 08] TERENZIANI F., KATAN C., BADAEVA E. *et al.*, "Enhanced two-photon absorption of organic chromophores: theoretical and experimental assessments", *Advanced Materials*, vol. 20, pp. 4641–4687, available at: https://hal.archives-ouvertes.fr/hal-00490264/file/review_070208.pdf, 2008.

[TER 11] TERZAKI K., VASILANTONAKIS N., GAIDUKEVICIUTE A. *et al.*, "3D conducting nanostructures fabricated using direct laser writing", *Optical Materials Express*, vol. 1, pp. 586–597, 2011.

[THE 95] THÉRAULAZ G., BONNABEAU E., "Essaims sur réseau et modélisation des processus de construction collective chez les insectes sociaux", *Journées de Rochebrune*, pp. 165–175, 20–24 March 1995.

[TOS 08] TOSA M., VITRANT G., BALDECK P.L. *et al.*, "Fabrication of 3D metallic micro/nano-structures by two-photon absorption", *Journal of Optoelectronics and Advanced Materials*, vol. 10, pp. 2199–2204, 2008.

[TUM 15] TUMBLESTON J.R., SHIRVANYANTS D., ERMOSHKIN N. *et al.*, "Continuous liquid interface production of 3D objects", *Science – Research Reports*, vol. 347, pp. 1349–1352, 2015.

[VAN 14] VANEK J., GARCIA-GALICIA J.A., BENES B. *et al.*, "PackMerger: a 3D print volume optimizer", *Computer Graphics Forum*, vol. 33, pp. 322–332, 2014.

[VAR 01] VARADAN V.K., JIANG X., VARADAN V.V., *Microstereolithography and Other Fabrication Techniques for 3D MEMS*, John Wiley & Sons, Chichester, 2001.

[VIC 16] VICTOR, "Un procédé de jet d'encre métallique dévoilé lors du salon Formnext", available at: http://www.3dnatives.com/xjet-impression-3d-07112016/, 2016.

[VON 97] VON RAUMER M., SUPPAN P., JACQUES P., "Photoinduced charge transfer processes of triplet benzophenone in acetonitrile", *Journal of Photochemistry and Photobiology A*, vol. 105, pp. 21–28, 1997.

[WAL 16] WALLER E.H., VON FREYMANN G.,"Spatio-temporal proximity characteristics in 3D μ-printing via multi-photon absorption", *Polymers*, vol. 8, p. 297, doi:10.3390/polym8080297, 2016.

[WEI 16] WEIS P., WANG D., WU S. "Visible-light-responsive azopolymers with inhibited π–π stacking enable fully reversible photo-patterning", *Macromolecules*, vol. 49, pp. 6368–6373, 2016.

[WHE 15] WHEELER A., "Legacy effects uses Carbon 3D printer for progressive Ad & Terminator: Genisys", available at: http://3dprintingindustry.com/2015/06/22/legacy-effects-using-carbon-3d-printer-for-progressive-commercial-and-new-terminator-movie/, 2015.

[WOO 02] WOODWARD J.R., LIN T.S., SAKAGUCHI Y. *et al.*, "Biphotonic photochemistry of benzophenone in DMSO: a flash photolysis EPR study", *Molecular Physics*, vol. 100, pp. 1235–1244, 2002.

[WU 00] WU P.W., CHENG W., MARTINI I.B. *et al.*, "Two-photon photographic production of 3D metallic structures within a dielectric matrix", *Advanced Materials*, vol. 12, pp. 14398–1441, 2000.

[XIN 15a] XING J.F., ZHENG M.L., DUAN X.M., "Two photon polymerization microfabrication of hydrogels and advance 3D printing technology for tissue engineering and drug delivery", *Chemical Society Reviews*, vol. 44, pp. 5031–5039, doi: 10.1033/C5CS00278H, 2015.

[XIN 15b] XING J.F., LIU L., SONG X. *et al.*, "3D hydrogels with high resolution fabricated by two-photon polymerization with sensitive water soluble initiators", *Journal of Materials Chemistry B*, vol. 3, pp. 8488–8491, 2015.

[YOK 03] YOKOYAMA S., NAKAHAMA T. MIKI H. *et al.*, "Fabrication of 3D microstructures in optical gain medium using two-photon-induced photopolymerization technique", *Thin Films*, vol. 30, pp. 438–439, 452–456, 2003.

[YU 15] YU C.C., CHEN H.L., "Nanoimprint technology for patterning functional materials and its applications", *Microelectronic Eng.*, vol. 132, pp. 98–119, 2015.

[YUN 16] YUNG W.K.C., SUN B., MENG Z. *et al.*, "Additive and photochemical manufacturing of copper", *Scientific Reports*, vol. 6, Article number: 39584, 2016.

[ZEN 15a] ZENOU M., SA'AR A., KOTLER Z., "Laser transfer of metals and metal alloys for digital microfabrication of 3D objects", *Small*, vol. 11, pp. 4082–4089, 2015.

[ZEN 15b] ZENOU M., SA'AR A., KOTLER Z., "Laser jetting of femto-liter metal droplets for high resolution 3D printed structures", *Scientific Reports*, vol. 5, 2015, Article number 17265, doi: 10.1038/srep17265, 2015.

[ZHA 06] ZHAO Y., LI X., WU F. *et al.*, "Novel multi-branched two-photon-polymerization initiators of keto-coumarin derivatives", *Journal of Photochemistry and Photobiology A*, vol. 177, pp. 12–16, 2006.

[ZHA 14] ZHAO C., WANG C., GORKIN III R. *et al.*, "Three dimensional (3D) printed electrodes for interdigitated supercapacitors", *Electrochemistry Communications*, vol. 41, pp. 20–23, 2014.

[ZHO 15] ZHOU J., LIU Q., FANG W. *et al.*, "Up-conversion luminescent materials: advances and applications", *Chemical Reviews*, vol. 115, pp. 395–465, 2015.

PART 2

Additive Manufacturing Pushed to its Limits

Large-sized construction components – photograph reproduced with the kind permission of XtreeE

Introduction to Part 2

"We do not have time to make technological advancements, even though we derive no benefit by not doing so". [ROS 13]

"What would be your impression of a football team that, for reasons of efficiency, divided the pitch into eleven rectangles with each of its players covering one such space, each being prohibited from encroaching upon that of his/her teammates?". [HER 11]

"Indeed, a group only exists insofar as it achieves a certain amount of agreement around beliefs, values (which are often implicit), what it stands for and customs. Political or economic structures are indeed insufficient to constitute the group, and values and morals amount to structural modifications. At the same time, they put in motion what the group stands for". [ELL 13]

"So as to completely release a portion of available energy to decide to explore radically different areas, certain requirements should be met. We must stimulate our capacity for surprise, for shifting our point of view, and for creative reflection. We must update new perspectives and new levers which both make sense and, in an indiscernible moment, can put processes into gear". [LAG 13]

"If you wish to build a boat, you must generate a yearning for the sea in your men's hearts". [SAI 00]

"Technology has immense power to erode distance... To begin with every technological change is achieved in remarkable locations. When a State can control the effects of such activity, it is in a hegemonic position". [FOU 14]

"All sufficiently-advanced technology is indiscernible from magic". (Clark, quoted by Sadin [SAD 13])

"The concept of an unmodified repetition of the past, and that of a radical rupture with all of the past, are two symmetrical results of a single time conception". [LAT 97]

"What is a machine? We take a machine to be a device that converts input messages into output messages. An engineer would say, using his/her particular jargon, that a machine is a transducer of multiple inputs and outputs". [WIE 00]

"We will come to understand that scientific intellect actually consists of banishing all dreams and musings. It is even within its nature to relentlessly rule them out. However, it is thanks to our imagination feeding our subconscious, and our musings, that scientific intellect may deploy its initial theoretical creativity". [BAC 77]

"Henceforth, we make scientific advances through successive coherences. Within this grid of interacting phenomena, within this kind of quagmire where we are seeking to find our way, the explanatory hypotheses that we form do themselves take a complex twist. We must therefore agree to a conceptual outlay around experimental action". [LEG 97]

"With time, events shape the given subject and the course of its development. To truly know the subject, above all you have to uncover the secrets of its history, and the 'lineage' of which it is both the single result and a future projection". [BRU 76]

"Who has the greatest need for epistemology? The answer to this is engineers, those who have the most urgent need of a theory of particular knowledge, and the best probability to create it". ([PAP 68]; quoted by Le Moigne [LE 94])

"There are numerous recent examples, or indeed those yet to come. These illustrate both the scale of what is at stake and the difficulties

encountered by governments to determine a given line of response. Such a solution allows all businesses, both those providing the innovation as much as those who are affected by it, to anticipate market evolutions and to invest within a sufficiently predictable framework". [FRA 17]

"We live in an era in which humanity, in its truest sense, is unsettled. Civilisation takes precedence over existence. Greed is an accepted notion everywhere". [CIT 17]

I.1. Introduction

The engineer's initial viewpoint (applying the KIS "Keep It Simple" system) is to produce a design as "simple as possible", using the tools or materials available to him/her for a particular purpose, for a given economic market. Using processes which remain "rustic" in their operative principles, Volume 1 proved that the engineer is capable of manufacturing parts from inert matter within highly varied spheres. The process using light, patented in 1984, rested upon knowing the coordinates for the object being manufactured, memorized within a computer operating mirror galvanometers and configuring the light beam displacements. The aim of such a configuration is to change a resin into a solid by the polymerization of a single layer, sequentially "voxel" after "voxel" (using the 3D analogy with the pixel concept). Adding a second layer, then a third, and so on allows us to create the component, just as we have been doing for a long time, for example, when constructing a wall. This basis has always served and continues to serve as a foundation concept for all additive technologies. However, although the purpose of Volume 2 is to show that this basis is currently cracking, this chapter explores this "old" concept further, but using new niche applications.

It is hard to believe that a high-level technocrat may define social needs and the functions in response to these needs as well as the principles for separating the populations intended to occupy these functions. It is therefore difficult to tack a stereotypical strategy on to these emerging spheres (hence a certain weakness in the current evaluation methods). However, at the same time, it is important not to make too many errors in choices, allowing for the progress enabled by science. How can we take research risks around the decisions we make? There thus exists a form of "squaring the circle", closely aligned to the foundations of the second principle of thermodynamics – Heisenberg's uncertainty principle. Indeed, the concept of widespread appropriate meritocracy comes up against difficulties linked to the complexity of research systems and the absence of "reliable" economic forecasting. This is so, particularly within fields where initiatives can be very large, and lastly, around talent scarcity (and, doubtless, the difficulty of bringing about the creation of such talent). All of this takes place in a world of vast mass research involving acknowledged disciplinary "castes" or "silos". The issue is indeed one of producing integrated high-risk research within the activity of the overall body of research. This takes place more particularly within the sphere of additive manufacturing.

With cross-community debates around the subject of research into the risks of additive manufacturing, consideration should be given to the following. It appears to be necessary that after reflecting upon the significance of proofs of concept, only a limited number of courses of action are chosen. However, it is also necessary to keep traceability in mind. We must thus consider "how" the selection takes place, upon which rational principles it is based on or is not made [LEP 13], and how it is able to support mass "imaginary run-off" over time [STE 95]. The basis of this reflection will be the subject matter at the end of this work, particularly in raising the "old" issue of interdisciplinarity, and the difficulty of convergence of disciplines which are too isolated. However, these are examples of "run-off", which often come to nothing.

The particular targets set out here essentially concern a number of aspects. These include microfluidics, nanomanufacturing, electronics and robotics. These are initiated for development with particular applications in mind. In contrast, they may use scientific and technological principles operated within these niche areas, such as for example the use of robots to operate 3D machines!

I.2. Overall framework

General comments concerned with risk-taking and interdisciplinarity apply each time research involves several disciplines. These disciplines need to work together in terms of research (and not "simply" by providing scientific technical information). We will now consider cultural differences between the various relevant spheres in this chapter on additive manufacturing stretched to its limits. We will not consider here the image provided by XtreeE found in the introduction to this section, linked to the construction industry and civil engineering. The issues in this section essentially deal with microscales and nanoscales, which are collated in summary form in Table I.1. It is possible to reveal forms of facilitated knowledge-sharing at the same time as severe tensions, indeed difficulties, which may contribute to explain why important spheres have not yet exploited 3D printing to gain its rightful position. These spheres already have well-managed technologies and 3D printing methods should demonstrate their added value relative to widely proven robust processes.

Field	"Standard" scientific and technological culture	"Standard" 3D printing
Concept	Material removal process	Additivity
Manufacturing	Mass-production of a single component	One-off components
Platform	Clean rooms	"Industrial" environment
Object size	Small (from a few μm to a few cm) Larger size for robotics	Either micro-metric objects (two-photon polymerization), or larger objects (> 1 cm)
Surface quality	Excellent (< a few dozen nm)	Average (> 100 nm)
Homogeneity	Often necessary	Often delicate to achieve, owing to the nature of the processes involved
Materials	Specific materials for electronics (purity) Corrosion resistance (intensification) Multi-materials	Standard 3D printing materials Mainly monomaterial

Table I.1. *Disparities between aspects of "standard" methods compared to those of 3D printing*

Apart from robotics, for which it is easy to produce usable components as robotic parts, a sphere which can easily rely upon additive manufacturing during the research and R&D stage of its development, this summary table (which is probably a little simplistic) highlights the main axes to explore if we wish additive manufacturing to compete (at least at the prototype stage) with manufacturing technologies, concepts of which have widely been proven. The following may be advantageous to end users:

– developing mass manufacturing methods: the example of stereo lithography techniques shown in the introduction (see the μCOP – microscale continuous optical printing process – devised by Zhu *et al.* [ZHU 15]. This could serve as a starting point to reach this objective which could, moreover, be of interest to other spheres);

– surface quality: the use of more refined manufacturing methods with variable voxel sizes may allow for optimization of manufacturing time while at the same time enabling surface quality to achieve the required specification;

– materials: it is possible to develop these 3D technologies so that they are able to work with materials in the sphere so as to manufacture electrical conductors. One example was published by Yamada *et al.* [YAM 09a]: They polymerized a standard resin using a 2½D process. As a part of this process, they used a two-photon process to transform (with the assistance of a femtosecond laser emitting at 800 nm, a transparency zone of the single-photon initiator) the material containing pyrroles. The polypyrrole conducts electricity (see also Kim *et al.* [KIM 01]);

– intelligent/smart material: developing materials, the properties of which are modifiable (see 4D printing in Part 2);

– multi-materials: the 3D technologies may envisage sequential or simultaneous use of either various materials or a single material, the properties of which are specifically modified during manufacturing;

– processes: envisaging the integration of additive manufacturing in all of the other manufacturing processes in the sphere;

– and other similar factors.

These various development steps assume the existence of partnerships between various disciplines, being above all the industry and research users with their respective spheres concerned. However, there are also partnerships between specialists in materials associated with those that, with a little creativity, may develop machines dedicated to connected technologies within the productive sphere and additive manufacturing. This further assumes a greater will to engage in an operation which encourages innovation (this is less risky than bio-printing, shown in Part 3). Thus, to conclude, we must continue to move forward and always, to quote Montaigne [MON 84], "rub and polish our brains against other people's brains".

I.3. Bibliography

[BAC 77] BACHELARD G., *La formation de l'esprit scientifique*, Vrin, Paris, 1977.

[BRU 76] BRUTER C.P., *Topologie et perception*, Doin & Maloine, Paris, 1976.

[CIT 17] CITOUR C.E., "Qu'est-ce que l'Humain? La science brouille les frontières", available at: http://www.agoravox.fr/tribune-libre/article/qu-est-ce-que-l-humain-la-science-188249, 2017.

[ELL 13] ELLUL J., *Déviances et déviants dans notre société intolérante*, ERES, Toulouse, 2013.

[FOU 14] FOUCHER M., "Science et géopolitique" in WIEVOSKA M. (ed.), *La science en question(s)*, Les Entretiens d'Auxerre, France, 2014.

[FRA 17] FRANCE STRATÉGIE, "2017/2027 – Répondre à l'innovation disruptive – actions critiques", available at: http://www.strategie.gouv.fr/publications/20172027-repondre-linnovation-disruptive-actions-critiques, 2017.

[HER 11] HERY J.F., BESNIER J.M., HIRSCH E., *Ethique et recherché, un dialogue à construire*, Quae, Paris, 2011.

[KIM 01] KIM Y., FUKAY S., KOBAYASHI N., "Photo-polymerization of aniline derivatives in solid state and its application", *Synthetic Metals*, vol. 119, pp. 208–212, 2001.

[LAG 13] LAGADEC P., *Piloter en univers inconnu*, Préventique, Bordeaux, 2013.

[LAT 97] LATOUR B., *Nous n'avons jamais été modernes; essai d'anthropologie symétrique*, La Découverte, Paris, 1997.

[LE 94] LE MOIGNE J.F., *La théorie du système général ; théorie de la modélisation*, PUF, Paris, 1994.

[LE 94] LE MOIGNE J.F., "La théorie du système général; théorie de la modélisation", PUF, Paris, available at: http://www.mcxapc.org/inserts/ouvrages/0609tsgtm.pdf, 1994.

[LEG 97] LEGAY J.M., *L'expérience et le modèle; un discours sur la méthode*, INRA, Paris, 1997.

[LEP 13] LEPELTIER T., *Histoire et philosophie des sciences*, Sciences Humaines, Auxerre, 2013.

[MON 84] MONTAIGNE (de) M., *Essais*, Gallimard, Paris, 1984.

[PAP 68] PAPERT S., VOYAT G., *Cybernétique et épistémologie*, PUF, Paris, 1968.

[ROS 13] ROSA H., *Accélération, une critique sociale du temps*, La Découverte, Paris, 2013.

[SAD 13] SADIN E., *L'Humanité augmentée; l'administration numérique du monde*, De l'Echappée, Montreuil, 2013.

[SAI 00] SAINT EXUPÉRY A., *Citadelle*, Folio, Paris, 2000.

[STE 95] STENGERS I., *L'invention des sciences modernes*, Champs Sciences, Paris, 1995.

[WIE 00] WIENER N., *Sur quelques points de collision entre cybernétique et religion*, Editions de l'éclat, Paris, 2000.

[YAM 09a] YAMADA K., SONE J., CHEN J., "Three-dimensional photochemical microfabrication of conductive polymers in transparent polymer sheet", *Optical Review*, vol. 16, pp. 208–212, 2009.

[YAM 09b] YAMADA M., KONDO M., MIYASATO R. *et al.*, "Photomobile polymer material – various three dimensional movements", *Journal of Materials Chemistry*, vol. 19, pp. 60–62, 2009.

[ZHU 15] ZHU W., LI J., LEONG Y.J. *et al.*, "3D-printed artificial micro-fish", *Advanced Materials*, vol. 27, pp. 4411–4417, 2015.

2

µ-Fluidics (or Microfluidics)

2.1. Introduction

Microfabrication is related to the field of microfluidics and involves the production of microchannels that have diameters of less than a millimeter, therefore necessitating techniques adapted to such sizes. When the process first started, and even nowadays, it has always placed enormous reliance upon microelectronic silicon technologies. The selection of materials used has expanded to other materials such as glass, steel and polymers. Microsystems intended for industrial use (such as micromixers) are made of stainless steel. Carrier [CAR 12] has said that two main technologies known as "hard or soft" are subject to consideration. "Hard technologies (glass, silicon and other materials) are based upon the processes of printing, lithography and matter deposition. Soft technologies (involving elastomers or plastics) are based upon a so-called direct method or replication method." Microfabrication processes depend, for the most part, on lithography, and in particular photo-lithography [TAB 03] and molding [HAM 08]. An important element is that of the surface finish of ducts in different projects which 3D technology users focus upon [VIL 04, REY 02, BAR 04]. The purpose of this chapter is to analyze competing works targeting the exploitation of additive manufacturing technologies in microfluidic applications.

Figure 2.1 (proven by the works of Ho *et al.* [HO 15] and of O'Neill *et al.* [ONE 14]) corresponds to the development of the number of cataloged publications in inverted commas. These are namely "microfluidics" and "3D printing" with a quantitative significance of around 0.5‰. This is when compared to the field of additive manufacturing taken globally. However, there has been a pronounced over-exponential development in these fields over the last two years. This compares to the field in general, which is tending to become under-exponential. The appeal for biological applications appears to be the source of this particular marked development. It is worth bearing in mind the comments of Fouet [FOU 16] that there may possibly have been around 25,000 publications per year since 2012 on the subject of microfluidics. This is another form of positioning for additive manufacturing within the microfluidics sphere (which is around 4.10^{-3}).

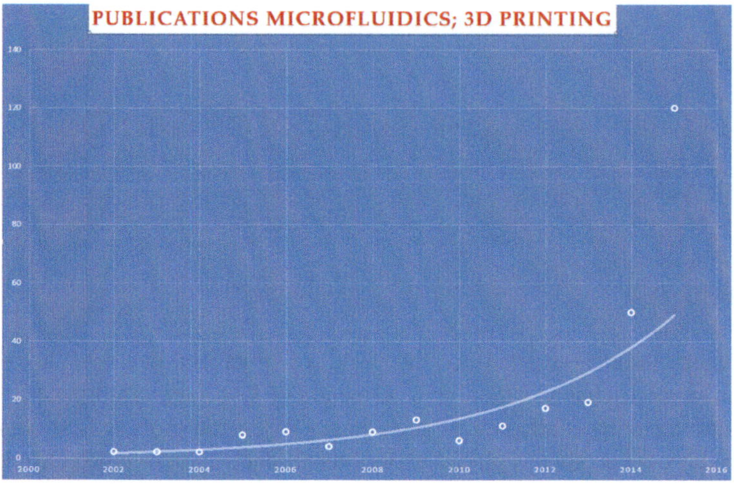

Figure 2.1. *Development of the number of publications in the two linked fields of microfluidics and 3D printing*

2.2. Review of microfluidics

Returning to technology and its applications, microfluidics is both the science and technology of systems manipulating fluids, of which at least one of the dimension specifications is of the order of the µm (one micrometer). Wikipedia [WIK 15] states that Whitesides [WHI 06] defined microfluidics as "the science and technology of systems manipulating small volumes of given fluids (from 10^{-9} to 10^{-18} liters), by using channels which are a few dozen micrometers in dimension". Tabeling said [TAB 03] that microfluidics is a science revolving around fluid flows which are either simple or complex, mono-phase or multi-phase, within artificial microsystems. Thus they are made using new techniques. This particular author states that by "new techniques" he means those of micro-manufacturing inherited from the microelectronics sphere. The first definition indeed reveals the duality of microfluidics. As a scientific discipline, it encompasses the study of phenomena and fluid mechanics, both on a nanoscale. As a technological discipline, it also includes an applicative dimension. However, the Whitesides definition [WHI 06] places the "micro" prefix upon the given duct size. This prefix is not concerned with fluid volumes within which its definition ranges from the nanoliter to the attoliter.

A topological criterion having µ-fluidic effects encompasses two types of duct porosities [CAR 06]:

– open or connected porosity, which has possible access from the outside of the given material and participates in carrying the given material;

– closed pores, which do not participate in material transfer and combine resistance, lightness and sealing. These characteristics are particularly adapted to be used as a structural material with greater or less functionality (depending on whether or not it combines additional properties, such as sound and/or heat insulation).

In the case of open pores, the winding nature of the paths through the pore space (that is to say straight line divergence) and/or the presence of areas which are barely connected to the rest of the network (leading to dead ends) are as much factors influencing the material transfer processes. Consequently they may have a role in both reactivity and selectivity of µ-fluidic reactions (refer to Figure 2.2).

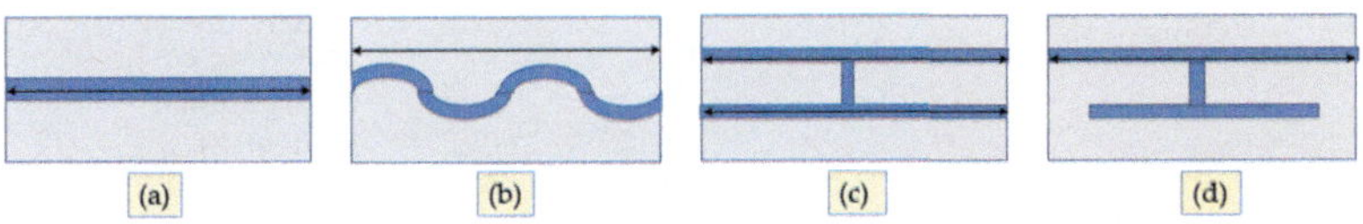

Figure 2.2. *Various types of ducts intervening in µ-fluidics processes (microfluidics) (a: straight ducts; b: winding; c: open interconnections and d: partially-open interconnections)*

However, we see that some microfluidic systems, for example dielectric electro-dampening mechanisms function by means of drops and without ducts. Tabeling [TAB 03] provides a definition of an engineer which, as he himself stresses, excludes all natural microfluidic systems such as blood capillaries or the transport of sap in plants.

For around 10 years, man has imitated nature by producing objects through micrometric flows. There are numerous applications for this. Generally, manipulating on the micrometer scale enables work to be done faster, within a cleaner and safer environment, but possibly at a higher cost. As real "micro-processors" for biology and energy and matter conversion, microfluidic systems enable the substitution of instruments or cumbersome and highly costly equipment. In addition, conversions are produced which are inaccessible to standard systems in the field. Nowadays, the volume of economic microfluidic technologies is estimated to be valued at tens of billions of euros [INS 15]. The number of potential industrial applications is considerable. These include those within medicine, energy, green chemistry, cosmetics, the food processing industry and biology [ONE 14]. This is all the more the case since it may be to our advantage in terms of resolution by using photo polymerization manufacturing having shorter wavelengths or using accelerated protons [VAN 11]. Figure 2.3 (from Gonzalez and Hervas [GON 09]) indicates spheres corresponding to microfluidics and several applications.

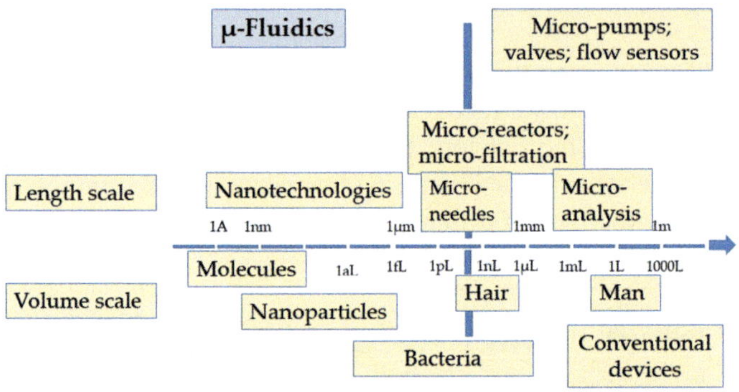

Figure 2.3. *Spatial spheres and µ-fluidic applications*

One of the advantages of changing scales is that we can apply the scaling laws. We thus enable reactions which would not necessarily be present within other spatial spheres with the majority of laminar flow cases having very low Reynolds numbers (Re) as indicated in Table 2.1 [ABK 11].

Scaling laws	Standard reactive system	µ-fluidic system
Exothermic reaction	Energy: Average cube length – L	Ditto, but having L small; this has increased controllability
	Conduction: Proportional to L	Ditto
Flow system	Generally turbulent	Creeping flow (reversible)

Table 2.1. *Microfluidic potential so as to explore the effects of scale changes*

The Reynolds number is the relationship between inertial forces and the relevant viscous forces. It is defined by Re = ρVL/μ where ρ is the fluid volume mass ($kg.m^{-3}$), V the fluid flow rate ($m.s^{-1}$), L a specific length (m) and μ the dynamic viscosity (Pa.s). The various types of flow are included in Table 2.2.

Reynolds number (Re)	Flow system
< 1	Viscous creeping flow (reversible); deterministic diffusion-dependent Brownian motion
$1 < Re < 100$	Laminar flow
Around 10^3	Transition to turbulence
$> 3.10^3$	Turbulent flow; low sensitivity to the given Re value

Table 2.2. *Reminder of flow system*

By selecting a fluid such as water ($\mu = 10^{-3}$ Pa.s) with a velocity of 100 $\mu m.s^{-1}$, a specific length of the order of 100 µm the Reynolds number value is 10^{-2}. This is far from the values used in "traditional" process engineering.

Some microfluidic examples are given below:

– ink-jet printhead;

– micro-pumps;

– the "lab-on-a-chip" for systems enabling an actual heart attack diagnosis [REN 13];

– genotyping chip;

– electronic paper;

– 3D touch screen;

– urine pregnancy test;

– blood test for AIDS detection;

– morphogenesis and cellular growth [JOH 14, CHU 12];

– synthetic biology [HUA 14].

2.3. Applications

Dragone *et al.* [DRA 15] set out spheres of application for microfluidic methods in the conversion of matter. Nowadays, there are numerous applications which are subject to intensive scientific work, in particular within the sphere of process intensification (see, for example Falk *et al.* [FAL 10]) and Figure 2.4 based upon the functional reactor which is fairly easily achieved in additive manufacturing. Please also see the scientific journal illustrating this sphere's development and its relevance for the scientific community. This is entitled *Chemical Engineering and Processing: Process Intensification* published by Elsevier:

https://www.journals.elsevier.com/chemical-engineering-and-processing-process-intensification/.

Within the context of particular overall economic collapse, process intensification may become an asset for the chemical industry in an environment desperately searching for competitiveness: "Process intensification consists of the development of methods, techniques and adapted devices to design more compact and economic devices. The production capacity of such devices is several times greater than a conventional process" [FAL 15]. Figure 2.5 (inspired by DOD – Drop-on-Demand [DOD 15]) explains the taxonomies of intensified processes, illustrating the richness in this particular field. We will examine, in the following section, how additive manufacturing technologies may provide a whole new dimension in this field's development.

Figure 2.4. *The principle of a blender using divided and intertwined ducts – the so-called "baker's problem" in fluid mechanics (reproduced with the kind permission of Laurent Falk)*

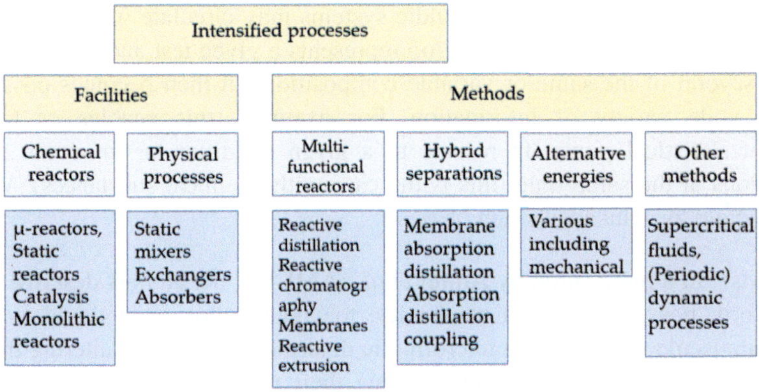

Figure 2.5. *Taxonomy of intensified processes*

Outside this knowledge production aspect within the sphere of conversion of matter and energy, studies on toxicology are developing. We now have the REACH regulations – a European Union regulation aimed at improving the protection of human health and the environment from chemical risks. Microfluidics are being used in this process (for example with GDR – French Research Network – "Micro and Nano Fludique", which specializes in nanofluidics and microfluidics [GDR 16]). We are aware that the regulatory context is evolving rapidly as much at the European level as it is at the international level. Methods of relevant evaluation are necessary for the swift acquisition of possible toxicological determinants for substances undergoing industrialization (with a large number of chemicals for which the physicochemical, toxicological and eco-toxicological properties must be assessed). This will thus meet the demands of this regulation.

This must be accompanied by development and adjustment of the current methods used. These often resort to animal experiments, which should be limited as much as possible. Trials should take place during the development phase for new molecules with a view of running experiments and favoring molecules with an advantageous profile (Safe by design; [LEG 15]). The advancement of new hazard analysis may take place by developing tests within the sphere of miniaturized fluidic tools so as to benefit from the capacities of screening these tools [OHN 08]. The GDR Micro-Fluidique team proposes using a microfluidic approach. In this approach the microorganism being studied is confined within a single drop. This drop contains the chemical substance that is carried within a continuous immiscible phase. Drops generated within the fluidic systems may circulate within duct flows, but equally may be stored. As each drop represents a given test and it is possible to generate several of the same or variable compositions, it then becomes possible to access a wide variety of information. For example, this enables us to have significant statistical power in respect to a given condition or to screen several compositions at the same time (this is the case with chemical substances). We can thus accelerate the data acquisition phases.

Likewise, in a publication by Amini *et al.* [AMI 13], the authors describe how it is henceforth possible, in fluid mechanics, to create tailor-made liquid flows by placing minuscule pillars within microfluidic determined ducts. By altering the flow velocity, multiplying the pillars and varying their respective thicknesses, position and direction, it is possible to produce a large number of flows which are fully controlled [HPC 13, SCH 02]. This research can enable the separation of leukocytes (white blood cells) from other cells in a blood sample to improve mixtures in industrial applications or to produce "lab-on-a-chip" processes more quickly. These include DNA sequencing or chemical sensing. These activity sectors are, in the long-term, highly important economic opportunities. "Together, they are revolutionizing microfluidics" (this emerges from GDR Micro and Nano-Fluidique [GDR 16]).

Each barrier has its own signature. Thus by accumulating them, it is possible to create distortions, which may be combined and adjusted for a given use. These types of tools enable the exploitation of liquid flows. It is thus possible to create a sequence of pillars capable of diverting leukocytes towards the edges of a given duct to isolate them and then to redirect towards the center, to capture them (a process involving separation by given barriers). More generally, the potential capacity of pillars to improve fluid mixing may find other applications such as heat evacuation during microprocessor fabrication or very fine control of manufacturing processes at micrometric and nanometric scales. However, this assumes good control over the surface condition, dependent upon the technology chosen for the given microfluidic application [WAL 15, OCO 15].

According to Fouet [FOU 16], "Historically, one of the important themes of fluidic system manufacture is photolithography (hard, using materials such as silicon and glass; or soft, using polymers), coming as it did from the microelectronics industry, which comprised the basics of microfluidics manufacture. Other micro-manufacturing techniques were developed (for example, lamination, roll-embossing,

3D printing) with new materials that appeared (such as microfluidic paper, Teflon and biodegradable lacquers). Within commercialized systems, plastics (Polystyrene – PS, Polymethyl methacrylate – PMMA) are used most often. Known and robust processes such as injection molding and ultrasonic welding..." Within this particular work, only the effective or potential contribution of additive manufacturing methods will be stated.

2.4. Return to additive manufacturing

By analyzing these examples, the technological origin of this scientific action framework still demonstrates much of the microelectronics know-how and additive manufacturing technologies, which are also generally 2½D [MOS 10]. However, the 3D methodologies have significant room in which to operate and grow so as to enable a forthcoming emergence in the sphere. There are various developments which are described below using recent examples.

In the 1990s, the team from Nancy had developed the concept of micro-stereolithography by pushing the technology to its limits. Figure 2.6 shows a demonstration example of microtechnic "stemmed glasses", the height of which is of the order of millimeters with a resolution of around several dozen µm [AND 94]. The volume of a given voxel was around 10^{-11} liters, which was still a little further from the generally stated requirement.

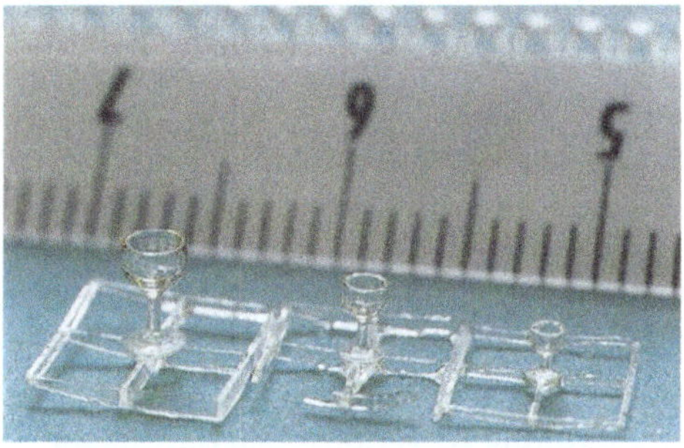

Figure 2.6. *Micro-stereolithography*

From 1999, by using a dynamic mask system it has been possible to produce parts with a resolution of some dozen µm using a one-photon process (please see Figure 2.7 by Monneret *et al.* [MON 99], and also Monneret [MON 04]).

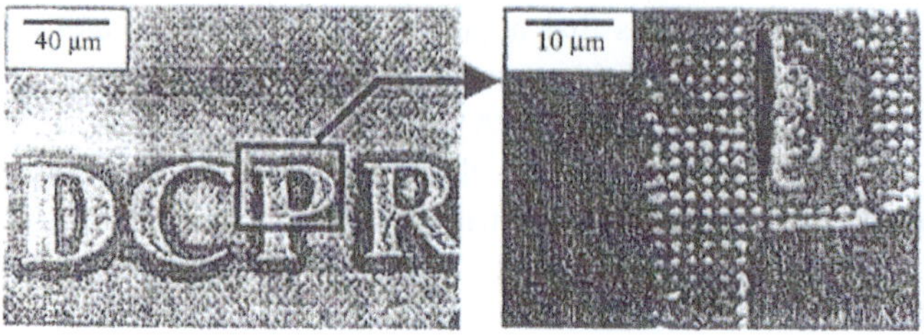

Figure 2.7. *Resolution obtained using a one-photon process and a dynamic mask system*

With regard to the surface conditions, Figure 2.8 (left) derives from nanoscribe® and Figure 2.8 (right) is an approximate design by Ovsianikov *et al.* [OVS 11]. The latter illustrates the possibility of achieving (very) good surface conditions with two-photon photochemical polymerizations (see also Xing *et al.* [XIN 15]).

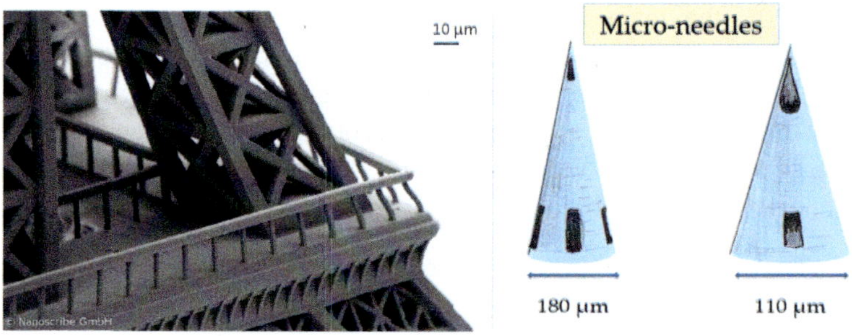

Figure 2.8. *Quality of surface conditions using two-photon stereolithography*

However, apart from generally transparent polymers, despite significant progress in the manufacture of 3D printing in recent years, major challenges remain. This is within the manufacture of structures on a microscopic and nanoscopic scale. In addition, it applies to the printing of various materials having desired electrical and mechanical properties to develop microreactors with more free structures than those envisaged by microelectronic technologies, valves and micropumps, and other technologies. Microstructures, in which it is possible to invest for scientific purposes, may be used for various applications including manufacturing Micro-Electromechanical Systems, micro-antennae and "lab-on-a-chip" devices. Direct manufacture from new types of practical conductor materials [ROT 14] for 3D printing and applicable microstructures within the sphere of microelectronics and sensors also reverts back to the given approach [BER 11]. It also returns to the use of additive manufacturing for the production of molds [BON 10].

However, as Figure 2.9 indicates, there is a particular problem that 3D technologies emerging within the microfluidics sphere have to surmount: being able to compete with existing technologies. This in particular concerns the surface finish [CHA 13]. Currently, apart from two-photon stereolithography and the LIFT process set out below, the best resolutions are in the order of 10 µm. This covers, with significant deficiencies, the field of µ-fluidics [AU 14]. Taking bibliographical approaches produced nowadays into account, only the multi-photon stereo lithography technologies appear capable of providing the desired surface finish in all situations, but they do so using polymeric materials.

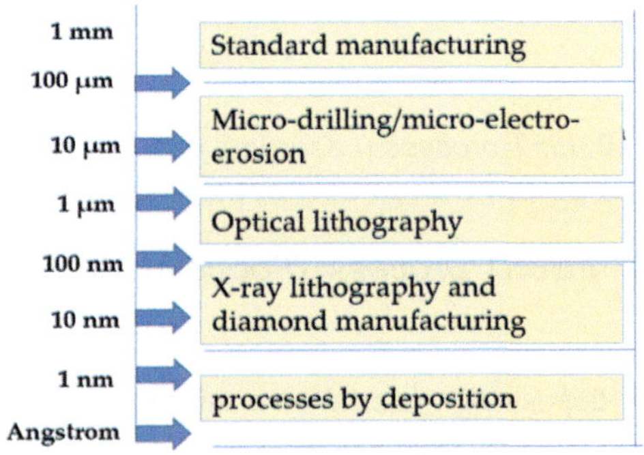

Figure 2.9. *Surface quality obtained by techniques (apart from 3D printing) and applied in the microfluidics sphere*

2.4.1. *Comment 1: LIFT process (Laser-Induced Forward Transfer)*

This is a digital printing technology which enables the production of micrometric structures (between 10 and 50 µm) with a resolution dependent on the applications targeted but in the order of a few micrometers. The principle is shown in Figure 2.10. A substrate is deposited on a surface, which is transparent to laser light. This absorbs the light pulse which will be caused by local vaporization towards a given support upon which the object is being constructed. Repetition frequencies may be of the order of Mhz and sections of the surfaces of the order of 20 µm (for thicknesses which may be low, dependent upon the given deposit). It is however necessary to anticipate using production time for a "layer" of the order of

4 seconds/cm^2. This could be satisfactory for surface treatment. Technology experts do not write about the possibility of producing objects which take complex forms but are non-convex [DEL 16, HUI 15]). In the latter paper, the attempt at collective manufacturing is laid out using a set of coupled galvanometric mirrors.

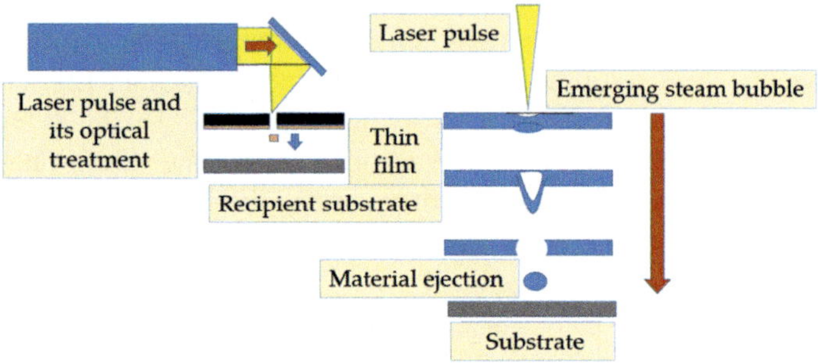

Figure 2.10. *LIFT process (Laser-Induced Forward Transfer)*

2.4.2. *Comment 2: FEBID process (Focused Electron Beam Induced Deposition)*

Another means for depositing material on a microscopic scale was suggested by Huth *et al.* [HUT 12] using a focused electron beam such as is shown in Figure 2.11 (also see Fowlkes *et al.* [FOW 16]; Utke *et al.* [UTK 08]; Randolph *et al.* [RAN 06]). A precursor gas is split up by the electronic flow leading to a deposition which may be metallic. The resolutions provided may be on a micron level. However, the deposition rate is less than that of the LIFT process.

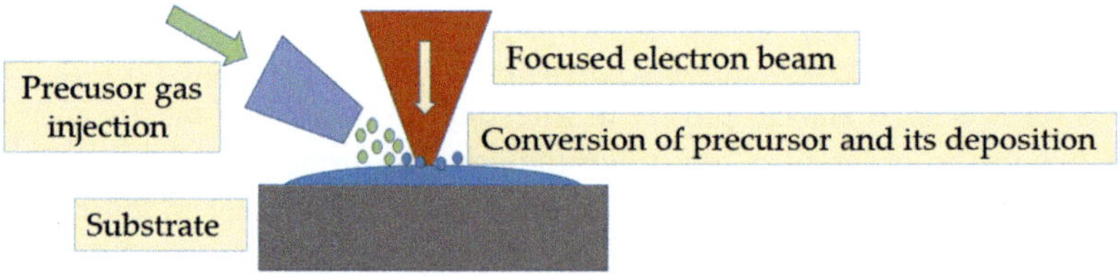

Figure 2.11. *FEBID process*

In these two examples, the use of two different processes may be useful with regard to surface treatments insofar as the flow of matter and energy are able to achieve the desired objective (there is a potential issue with the interiors of micron pores used in μ-fluidics).

2.4.3. *Other methods*

Table 2.3 stemming from the works of Ho *et al.* [HO 15] sets out the additive manufacturing technologies used within the microfluidic sphere, their appeal and their limits, together with their linked applications within biology. We may note upon perusing this table that it is the main one-photon and two-photon stereo lithography methods that are of interest to scientists owing to the resolution potentialities (two-photon stereo lithography) which they enable. Although originally applications aiming to produce molds were targeted, we can now produce parts directly, insofar as the given geometries are simple to produce. In the future, it is envisaged that it will be possible to produce single parts with complex internal structures including the set of elements traditionally used within microfluidics.

Process	Energy	Materials	Comments
Stereolithography (SLA)	UV Laser	Polymers	Finish (> 10 µm), difficulty to withdraw the µ-supports; mold production, use of "lab-on-a-chip".
Fluid-jet printing	UV	Polymers	Quick manufacturing rate, mechanical durability sometimes mediocre; production of molds, use of "lab-on-a-chip".
2-Photon SLA	Pico/femto second laser	Polymers	Very high resolution; slow process; basic research studies.
FDM	Heat	Polymers and elastomers	Average resolution, quick manufacturing rate, non-transparent materials, average mechanical durability; application in biological analysis.
Ink-jet	UV	Polymers	Quick manufacturing rate, multi-materials, difficulty to withdraw the µ supports; application within biological analysis.
Bio-printing	UV/laser	Polymers, hydrogels and other related substances	Slow process; multi-materials and production of µ-fluidics systems for vascular artificial cells and systems.

Table 2.3. *3D printing applications in microfluidics*

According to O'Neill *et al.* [ONE 14] who produced a synthesis of 3D printing methods applied to μ-fluidics, it is possible, independently of two-photon stereo lithography methods, to produce parts with resolutions which often achieve satisfactorily that which is demanded by "professionals" in the sphere. Such objects have surface resolutions x and y ranging from 16 to 300 μm and a depth, z, comprising between 1 and 50 μm.

It is important to return to the voxel notion before approaching this technological aspect. This elementary volume is as it were a good approximation of the solution which we wish to achieve. Within the various additive manufacturing processes that were set out in Volume 1, we revealed technologies which might be of significance for their microfluidic aspects. This is because the size of voxels which is conceivable is of the order of a μm or less. However, although the conversion time induced by all forms of energy of a volume of matter within a given single voxel is (as an initial approximation) independent of time, or τ, it would be necessary to produce a volume of 1 dm^3, with a resolution of 1 mm a time of $10^9.\tau$, and for a resolution of 1 μm a time of $10^{18}.\tau$. It is therefore necessary to take into account the surface roughness that you wish to achieve and the manufacturing time of a given object. One possibility in the sphere is offered by the bibliography, allowing for the coupling of a process of additive manufacture with a so-called "laser direct-to-plate" technology which uses photo-ablation. Although the purpose of this experiment was to increase the porosity of the material, this type of coupling connected with a measurement of surface quality may enable the linking of two types of resolution and, as a result, save in terms of the production time of a given object made using microfluidics [MAL 14]. Indeed, this association of manufacturing processes enables us to progress from a 3D problem (with varying production time of the cube of size ratio from the given object size of the voxel) to a 2D problem (with the time no longer varying, other than with the square of the relationship of sizes). It works likewise for another method, which exploits grain deposition determined on the surface linked to surface machining, enabling the production of the top layer with a better resolution ([SPA 14] and Figure 2.12).

In a recent paper, Au *et al.* [AU 14] carried out a comparison between the traditional molding method and additive manufacturing. These authors suggest that 3D printing has obvious advantages in terms of cost and flexibility for low volume quantities. In producing instrumental models, Liao *et al.* [LIA 12] propose the use of femtosecond lasers which may produce very high-precision polymer ablation as indicated in Figure 2.13. In these conditions, the microfluidics mechanism will be produced as illustrated using a stacked-layer process.

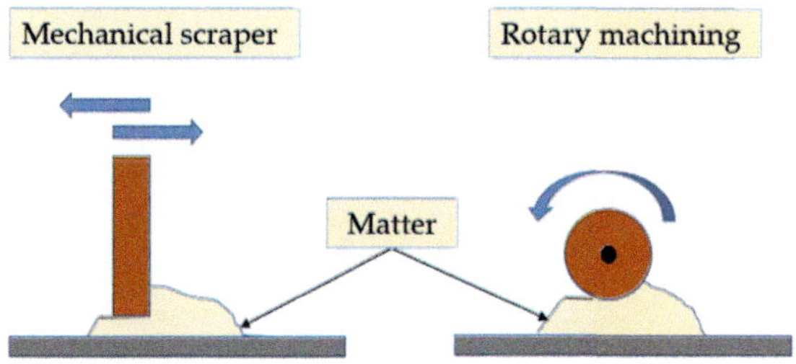

Figure 2.12. *Re-machining the surface*

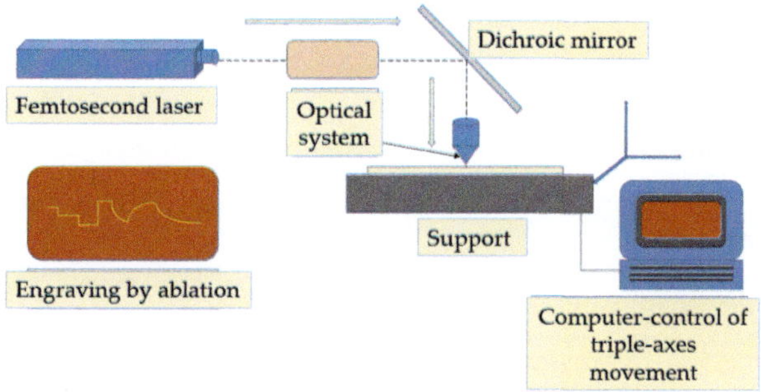

Figure 2.13. *Laser ablation process to create microfluidics systems*

Other examples come from China with a recent article by Xu and Sun [XU 15] using femtosecond lasers for additive manufacturing as well as ablations. Integrated systems are thus produced using resolutions of a few nm. Lastly, the University of Twente in the Netherlands [GAM 15], by optimizing the laser flow within a standard SLS process, was able to obtain resolutions of 100 nm from metallic powders. On this basis, any structure with a small size is conceivable.

Moreover, as will be shown later in this analysis of different processes, there are various levels of precision amongst the standard technologies. These stem from different physical aspects (solubility, wavelength and development of liquid layers, powder-particle size, energy coupling material and other factors). Figure 2.14 produced by Vaezi *et al.* [VAE 13] shows this comparison.

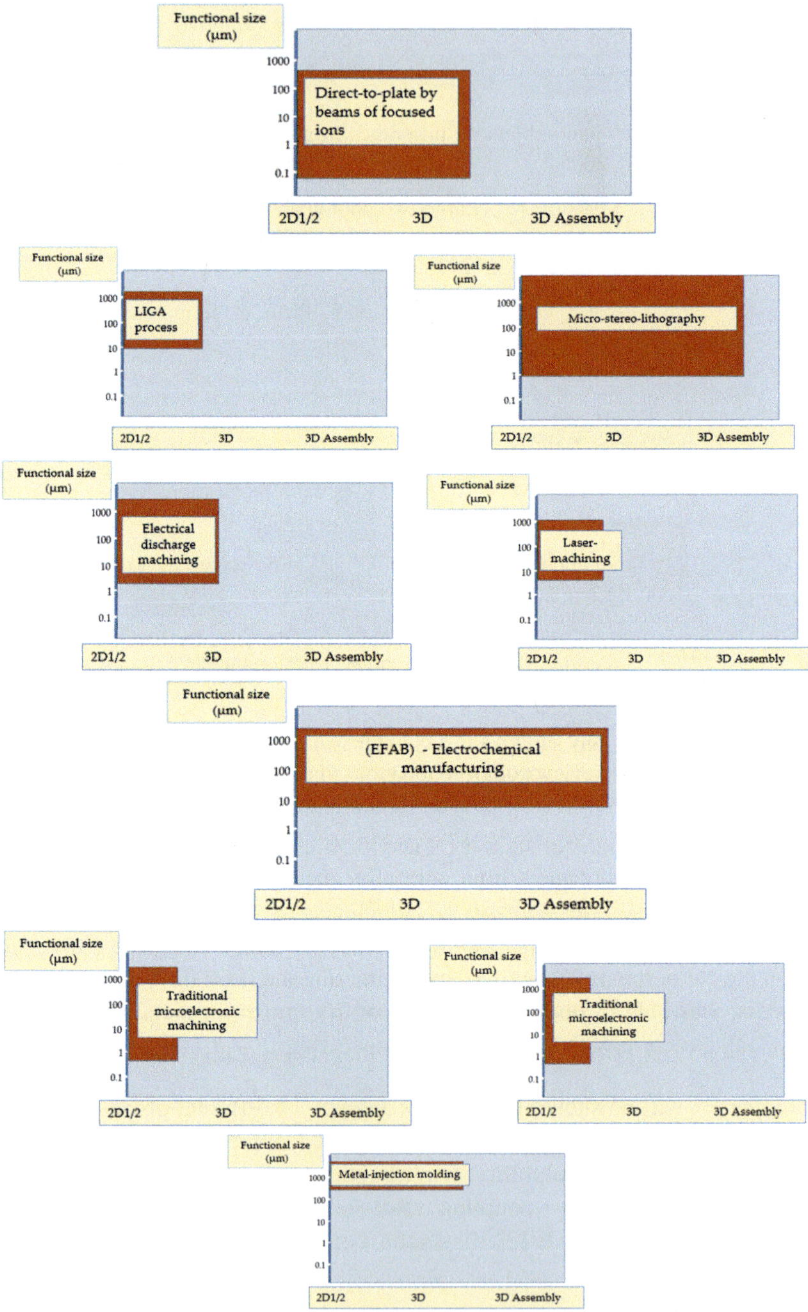

Figure 2.14. *Precision of 3D microfluidics technologies*

However, the use of more sophisticated technologies such as those used in multi-photon processes makes it possible to go beyond standard resolutions which are of the order of 100 nm or even below this value [EAT 12]. Figure 2.15 enables visualization of the process (as it traditionally produces a better resolution in a preferred given direction and a manufacturing duration which may be higher). Likewise, there is the possibility of generating molten metal droplets which solidify to create three-dimensional structures. The resolution is then linked to the size of given microspheres of a few µm [LAD 13].

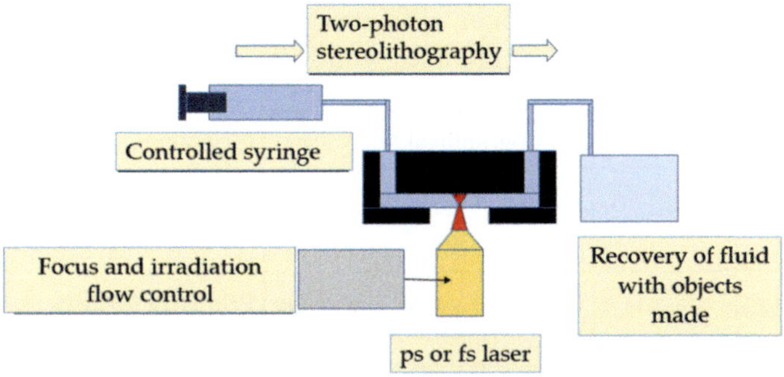

Figure 2.15. *Two-photon photopolymerization – sub-micron resolution within a µ-fluidic duct [HON 11]*

As part of the technological breakthroughs, Figure 2.16 by Madden and Hunter [MAD 96] is an original process consistent with producing spatially-determined electrolysis by electro-deposition (using a layer-by-layer process), enabling the production of metallic parts.

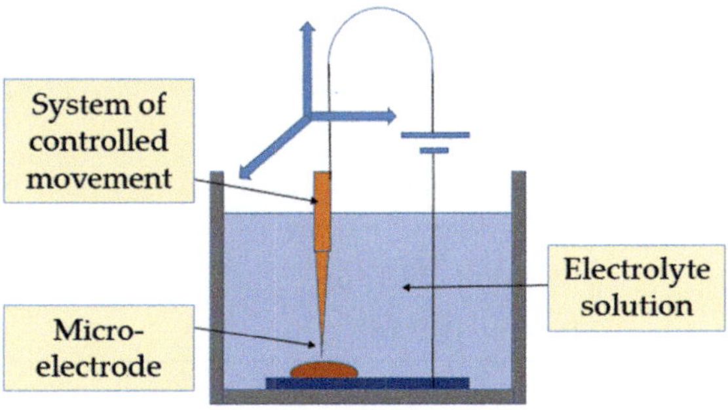

Figure 2.16. *Spatially-determined electroplating*

An original idea was set out in 2015 by He *et al.* [HE 15] (see also Gelber and Bhargava [GEL 15]) using a sugar deposition technology at the outflow to a micro-injector which enables the production of 3D structures. These are so-called "foundry processes with lost wax" comprising "counter-mold" in PDMS (Plant Design Management System) thus enabling, by dissolving sugar, the creation of microfluidic ducts.

Thus there exist methods which enable the production of objects using applications including bioprinting which may necessitate the presence of micro-ducts [JOH 12]. Table 2.4 compiles a number of applications and methods with recent bibliographical references combined with some comments.

Process/Application	Bibliographical references	Comments
Two-photon process	Honegger *et al.* [HON 11]	Resolution of a few nm
LIFT process	Delaporte and Alloncle [DEL 16]	Primarily surfaces
FEBID process	Huth *et al.* [HUT 12]	Ditto
Automation	Au *et al.* [AU 15]	
Chemistry/Processes	Johnson [JOH 12] Kitson *et al.* [KIT 12] Gross *et al.* [GRO 14] He *et al.* [HE 16]	Chemistry + biotechnology
Biology	O'Neill *et al.* [ONE 15] Gupta *et al.* [GUP 15] Moulton and Wallace [MOU 14] Bettinger and Borenstein [BET 10] Lind *et al.* [LIN 16] He *et al.* [HE 16]	Delivery of therapeutic drugs Creation of biological tissues
Micro-channels	Kamei *et al.* [KAM 15]	
Molds	Hwang *et al.* [HWA 15] He *et al.* [HE 14]	Studies of transport within given mechanisms

Valves and micro-pumps	Moore *et al.* [MOO 11] Au *et al.* [AU 11] Lai and Folch [LAI 11] Begolo *et al.* [BEG 14] Habhab *et al.* [HAB 16]	
Electrodes	Erkal *et al.* [ERK 14]	
Conductor components	Czyzewski *et al.* [CZY 09]	
T-shaped mechanism	Donvito *et al.* [DON 15]	Validation of droplet production
Physical sensor mechanisms	Vaezi *et al.* [VAE 13]	Dual-material 3D printing including one able to be disposed of
"Paper" sensors	Xia *et al.* [XIA 13]	
Micro-batteries	Ferrari *et al.* [FER 15]	Lithium ion

Table 2.4. *A number of 3D-printing microfluidics applications*

In particular, the biological sphere is developing owing to the possibility of producing "lab-on-a-chip" products (a concept which comes out of the present work). We should, however, remember that the situation stems from the fact that there is scope to precisely control the biological environment which is being determined. We can produce mechanisms by automated means (collective approaches), which enables both an increase in durability and a reduction in analytical costs.

2.4.4. *Hybrid methods*

Sugioka *et al.* [SUG 14] have recently put forward a hybrid method for the production of microfluidic systems shown in Figure 2.17. The figure shows that it is possible to use a subtractive manufacturing method involving microfluidic ducts with femtosecond lasers. This destroys the glass material by "fracturing" it in specific places. After heat and chemical treatments, this enables us to create channels and, in the second stage, having filled the microducts with a resin, to carry out a two-photon photochemical high-resolution polymerization within the area of interest. By using such technologies, it is therefore possible to save manufacturing time (see Volume 1). Various microelectronic operations used may then be exploited (for example metallic deposition and other methods). Techniques where the operation starts by using polymerization/de-polymerization methods through light connected within a second stage by high-resolution printing were put forward by Hengsbach and Díaz Lantada [HEN 14].

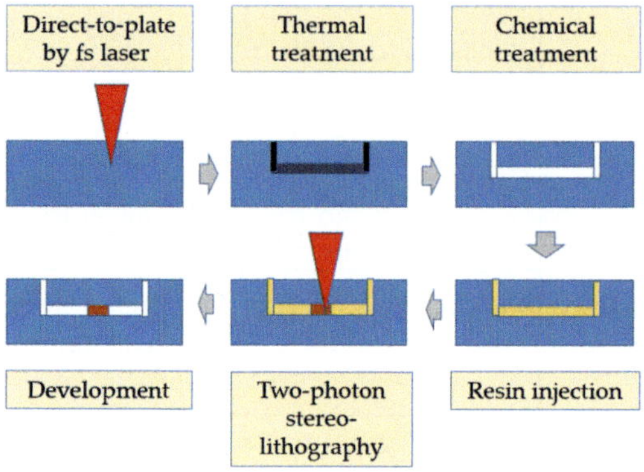

Figure 2.17. *Subtractive-additive hybrid system in μ-fluidics*

2.5. Conclusive outcomes

Nowadays, microelectronics technologies have enabled the promotion of applications within the microfluidics sphere. However, these processes, even when well-controlled, do not have all of the "adaptability" of the manufacturing desired. This leads to research in alternative methods such as 3D printing. However, on reading this section, the reader will be able to notice the relative weakness of the proposed ideas. These are limited, above all, to two-photon stereo lithography. However, although the μm is desirable within given situations, the majority of 3D technologies may satisfy many needs immediately and the acceptable resolution comes to the order of 50 μm or slightly above this value. Taking into account the increasingly high demand, research for new methods still appears to be required to access a robust and applicable technology.

In particular, it appears judicious, as Sugioka *et al.* undertook [SUG 14], to "exploit" additive technologies coupled with subtractive methods. Within such processes each technology can provide its own added value to manufacturing couplings.

2.6. The converse problem: a potential µ-fluidics application to additive manufacturing

2.6.1. *3D sintering*

In a recent paper, Hirt *et al.* [HIR 16] proposed the use of a microfluidics system to provide metallic powders connected to an AFM laser. The original mechanism for this is shown in Figure 2.18. It enables the exploitation of the resolution linked to injection-system size and the capability to produce metallic parts for the watchmaking industry (but not this industry alone).

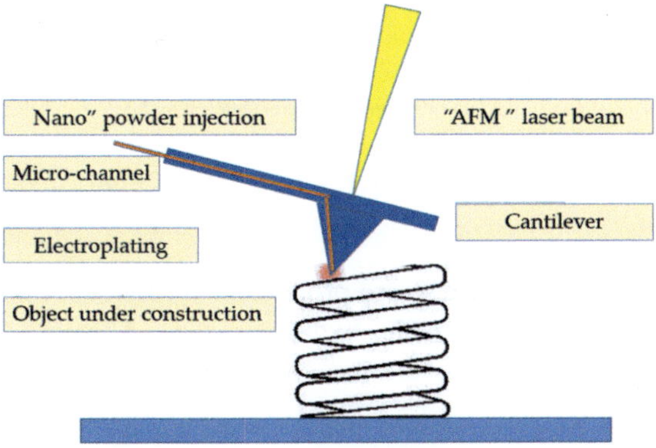

Figure 2.18. *Micro-printing connecting microfluidic powder flow and electroplating [HIR 16]*

2.6.2. *Deposition of polymerized particles*

Hashimoto *et al.* [HAS 15] state that it is possible to produce porous media from various types of biocompatible materials enabling us to envisage regeneration processes for biological tissues on scaffolds (see also Oh *et al.* [OH 07]). Indeed, the open porous structure enables cellular growth, vascularization, nutrients to be carried and the disposal of "waste" [MUR 10]. In the paper by Hashimoto *et al.* [HAS 15], the production of a porous media is made possible by a Michael addition chemical reaction of modified acryloyl groups (CHPOA) around the penta-erythritol-tetra-(mercaptoethyl)-polyoxyethylene (PEGSH) leading to the stacking of micrometric spheres. Other similar techniques also exist (see Lu [LU 15]).

Ratner [RAT 09] proposed another method where a fluid material is introduced into a cluster of sintered spheres which then solidifies. In the last stage, the initial spheres are dissolved which enables the creation of a porous media, in a negative manner, that has an appearance similar to that shown in Figure 2.19.

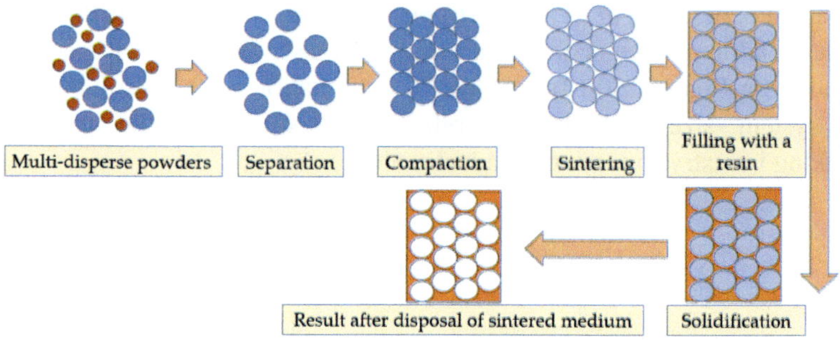

Figure 2.19. *Production process for a porous media per Ratner [RAT 09]*

Other sub-millimeter structures were produced by Griffin *et al.* [GRI 15] around other principles using biodegradable hydrogels (see Ahmed [AHM 15] for a synthesis with hydrogels). These were molded by microfluidic techniques (see also Rezende *et al.* [REZ 13]).

The purpose of this paragraph is to examine whether it is possible to envisage the production of 3D porous structures with spheres of known sizes to, for example, use in bio-printing applications. These applications have pore sizes which are compatible with cellular sizes or other applications (for example µ-batteries). Cox *et al.* [COX 15], in producing bio-printing "scaffolds", produced porous structures with highly variable pore size diameters and local structure diameters. This may be of interest for some applications, but slightly less so if we wish to control the geometric parameters, for example in tissue regeneration. Typically, it is necessary to achieve resolutions which should be equal to or less than 100 µm if we wish to have a process which will be of potential significance for this type of subject.

2.6.2.1. *Basic concept*

The general concept is to examine whether microfluidics methods are likely to enable the manufacture of porous media. As Figure 2.20 indicates, there is the possibility of dropping drops of fluid which are potentially reactive on a given surface [SHU 10] or to carry them via a mechanism of aerosolization (see Figure 2.21). Ink-jet printheads, which appeared in the 1990s, incorporate an ink reservoir, a heating element for activating fluid movement, and a micronozzle. Nowadays, tens of millions of ink-jet printers use microfluidic technologies and billions of documents are both written and read owing to this technique.

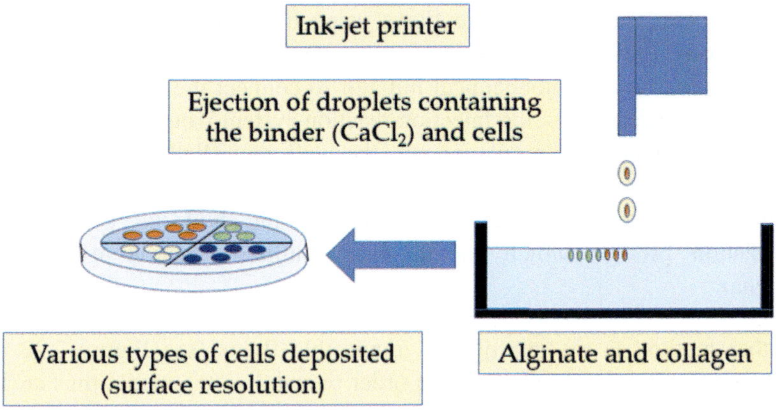

Figure 2.20. *Ink-jet bio-printing*

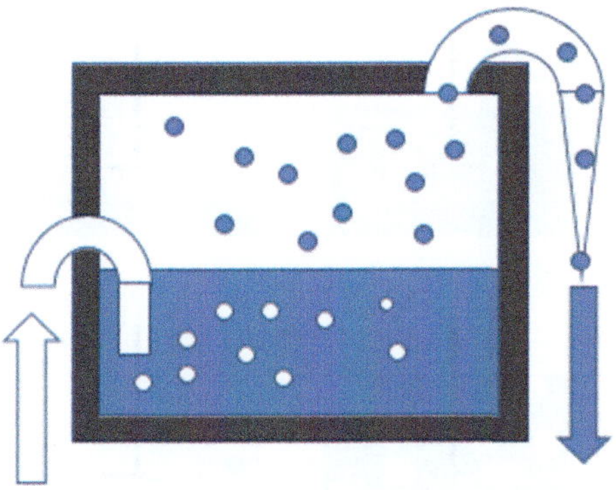

Figure 2.21. *Extract from Small Lab [SMA 15], the fluid moved in aerosol form and projected upon the surface being covered in the process*

In the event that we have to introduce a form of post-treatment (polymerization), the localization of spheres produced may pose an issue. In the first instance, the resolution may be of good quality when exiting the micronozzle. However, there is no space to undertake the process of polymerization. This polymerization cannot take place within the microduct if it is photochemical. This is due to the risks of sticking within the ejection device. Within the second case, we should avoid particles sticking on the ejector surface. This is the case even if there is the possibility of concentrating particle flows using the Corona effect [PAR 15]. In these conditions, only a monomer heat treatment [KRU 14] containing a thermal initiator and possibly a photochemical initiator to connect the spheres together at a later phase is required. Thus, the injection/reaction system would comprise of a dual mechanism (for which the regulation will be well-controlled) in the event that a heated particle, which is not totally polymerized is ejected. Between ejection and reaching the applicable surface, the sphere should both be cooled using a polymerization process which should then continue (this involves opposing phenomena).

Another method is the use of two immiscible fluids (see Figure 2.22 by He [HE 09]) within an agitated medium. In order to stabilize the media thus created, it is possible to use lipophilic surfactants which are organized around monomers that could be polymerized in the given batch before nebulizing and transmiting them towards the application surface.

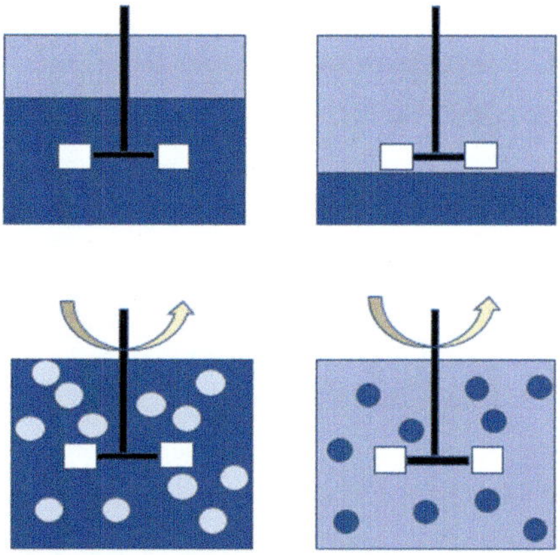

Figure 2.22. *Mechanical formation of micro-droplets (left, immersion of agitator within the heavy phase; right: the reverse)*

In these situations which have now become standard, the fluid, whether charged or not, spreads out across the surface and will react to produce, voxel after voxel, the desired object. In such conditions, there is therefore no possibility of easily creating a porous media. The concept is as follows. It consists of choosing, using a monomer or oligomer material, to polymerize it (for example by photochemical means, without a spatial resolution being sought) at least partially. The principles of such a polymerization have been described to a great extent in Volume 1 and it is not necessary to return to this aspect of the process. Keeping in mind, as Figure 2.23 shows [AND 94], that there are more or less three zones within polymerization kinetics:

– Zone (I), with very slow kinetics where free radicals formed within photochemical initiation consume the inhibitors present (including the oxygen in the air);

– Zone (II) the actual polymerization; and

– The final stage, connected to the confinement of free radicals which can no longer achieve reactive bonds of the remaining monomers. This is due to the passage of a fluid material to a highly polymerized entity in which the diffusion of reactive species is severely hampered.

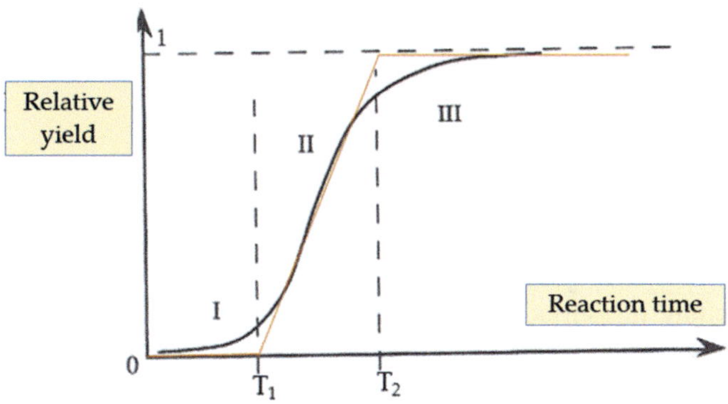

Figure 2.23. *Kinetic polymerization curve*

Within Zone II of Figure 2.23 if we are able to irradiate an aerosol containing a monomer and a photochemical initiator (within the example to illustrate the idea), it is in principle possible:

– to enable a potential solvent to evaporate out of the droplet; and

– to pre-polymerize the droplet.

As the oxygen in the air is likely to spread within the spheres reacting to the surface, we may achieve two zones within the material. There is one which is internal where molecular transport has not had time to take effect, Zone II, which must be reached and the other on the surface involving transport coupling and reactivity. In this case, the material is virtually non-polymerized and certainly has a small thickness.

In formulating for example, the hypothesis of a core polymerization in 10 seconds having an oxygen diffusion coefficient within the viscous monomer in the order of 10^{-8} $cm^2.s^{-1}$, by using Fick's law to describe the transport element, we then get:

e = 10 µm approximately (it would be 1 µm for a time of 0.1 second).

In these conditions, a mini sphere of 0.1 mm of a beam would be defined by a polymerized core and a fine "skin" with low thickness (see Figure 2.24).

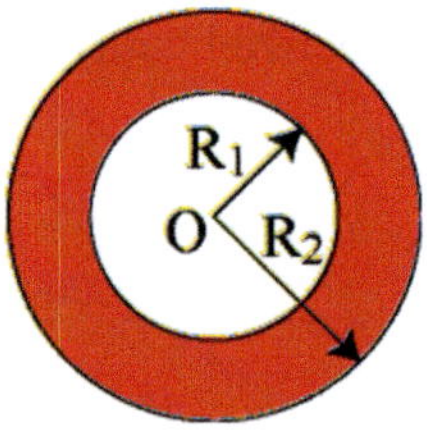

Figure 2.24. *Concentric (solid white area); the reaction being R_2–R_1 = e in the red area*

Upon reaching the surface, which already contains spheres, these voxels will stick to the object being constructed. The common region between the two spheres will be made up of a reactive material that we will be able to polymerize during a subsequent stage. Figure 2.25 is the contact zone between two spheres of the beam R_1 which are stuck together as one with the common "contact" defined by h.

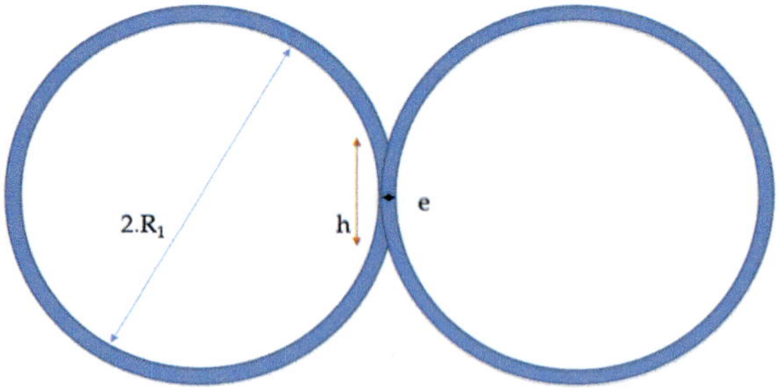

Figure 2.25. *Contact zone between two pre-polymerized spheres*

With h = $2.(R_1.e)^{1/2}$ or in the example dealt with 60 µm (10 µm for e = 1 µm). Over this distance, oxygen, for the same irradiation duration and with the same light power, cannot diffuse sufficiently to prevent both spheres sticking together.

There is therefore a means for this type of mechanism to create random piles of mono-disperse spheres. However, there are several questions which necessitate reflection at this point:

Is there a means of producing controlled droplet sizes (with regard to a given temperature, viscosity, gas flow and other factors)?

Can we be certain that all droplets will come out of the mechanism (without being deposited on the sides)?

2.6.2.2. *Nebulizer*

He [HE 09] produced a thesis around the subject of nebulizers with the view of encapsulation of cells within micro-droplets so that they can survive. The production of spheres was shown in this work with a polydispersity of such spheres from standard systems using micro-emulsions. He proposes the use of microfluidic systems of the type shown in Figure 2.26. This permits the creation of micro-drops of calibrated sizes.

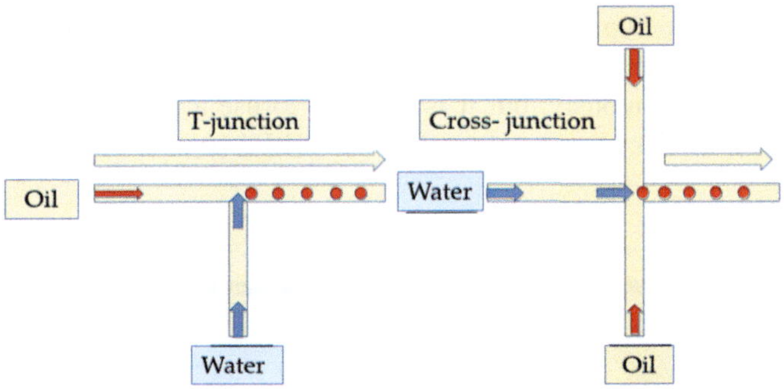

Figure 2.26. *Creation of mono-disperse spheres within microfluidic junctions*

Thorsen *et al.* [THO 01] studied the influence of aqueous phase pressure on the size of spheres for various oil pressures. This illustrated the possibility of creating calibrated microspheres, the size of which may reach dimensions in line with the given proposal (see also Wang *et al.* [WAN 13]). Within such conditions, there is the potential to irradiate monomer droplets containing a photochemical igniter to reach the desired objective. In addition, this system solves the second question set out previously since all of the droplets emerge from the mechanism. However, this forces us to work with two fluids, of which one is generally water. Other systems using two immiscible liquids have also been published (see Morimoto and Takeuchi [MOR 13]).

Recently, Jian *et al.* [JIA 15] showed that it was possible to substitute a gas as a second fluid (the first fluid being water-based) to form droplets of a sub-millimeter size. The principle retained by these authors is shown in Figure 2.27.

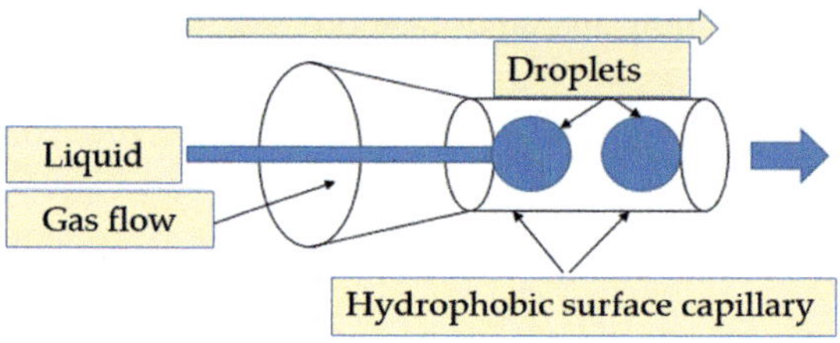

Figure 2.27. *Mechanism used by Jian et al. [JIA 15]*

These authors highlight two systems, one where there may be adhesion of the drop to the surface with the creation of entities which are smaller in size and the other corresponding to a creation of mono-disperse entities. Figure 2.28 illustrates the existence of these two spheres. In these authors' experience, the surface of the tubes were treated with hydrophobic substances (octadecyltrimethoxysilane). This system functions continuously, but it is possible to use a more complex pipe with controlled microvalves by means of "refinement" [LIU 10].

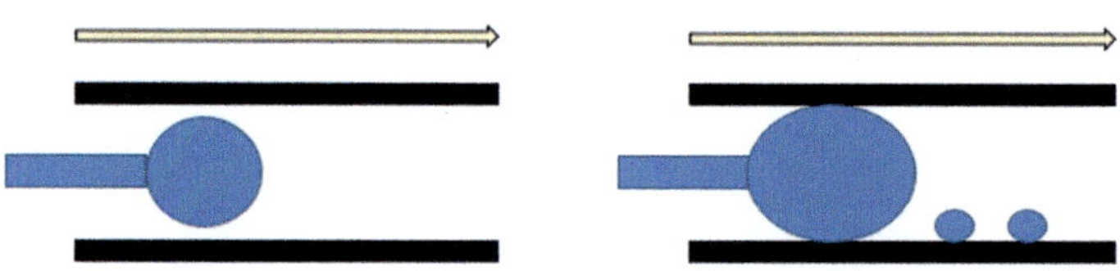

(A) (B)

Figure 2.28. *Highlighting critical areas and droplet stability (A: Increased flow velocity: the drop maintains its integrity; B: Low velocities, the drop adheres to the side with the formation of satellites)*

Abkarian [ABK 11] maintains that there may be no general theory to predict the size of droplets within microfluidic ducts as this would depend on:

– the flow geometry;

– the surface quality of the walls;

– the presence of moistening and surfactant phenomena; and

– the presence of other rheoactive entities (polymers, oligomers and others).

However, this bibliographical approach remains in short supply because of the difficulty of finding information published on this subject, which is of interest for the proposition explained here. We have yet to unlock the set objective. This is namely to deposit identically-sized microspheres on a surface to produce a porous object according to a given 3D technology exploiting µ-fluidic bases. However, according to Abkarian [ABK 11], two specific variables without set values should be taken into consideration in the production of droplets. These are:

– the Capillary number Ca = Viscosity. typical velocity/surface tension; and

– the Weber number We = Density.(Typical velocity)^2. Radius of the duct/ Surface tension.

Within an experimental approach which may be conducted, these components should be increased to produce identical micro-droplets which may be projected outside of the system to create porous objects (see Figure 2.29).

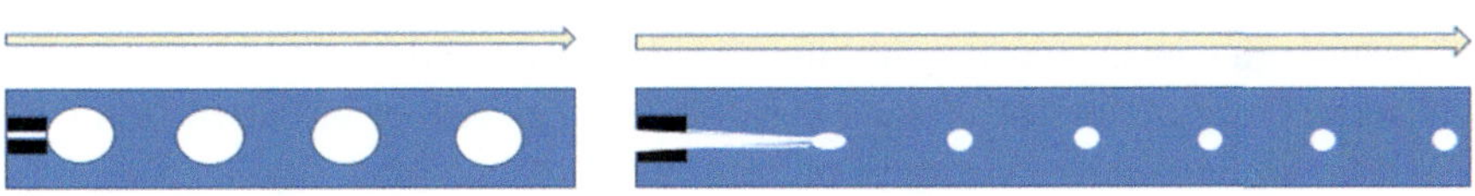

Figure 2.29. *Two possibilities for the production of identically-sized droplets in µ-fluidics (per the ranges of Ca and We)*

Independent of the production of calibrated microspheres, it is possible to photochemically transform monomer droplets using a continuous irradiation device (see Cloteaux *et al.* [CLO 14]). Taking into account the possibility of producing droplets within a gas, droplets may be prepolymerized and ejected in a fine manner through a nozzle placed in the immediate vicinity of the object being constructed. This process appears to have the best advantages to produce a porous solid.

From these mono-disperse spheres, a cluster of the type shown in Figure 2.30 may therefore be created. Calculations as to the sphere distribution were produced by Mari [MAR 07] leading to a density in the order of 0.64 (instead of 0.74 for a face-centered cubic stack) and a certain disorder (amorphous dense stacking).

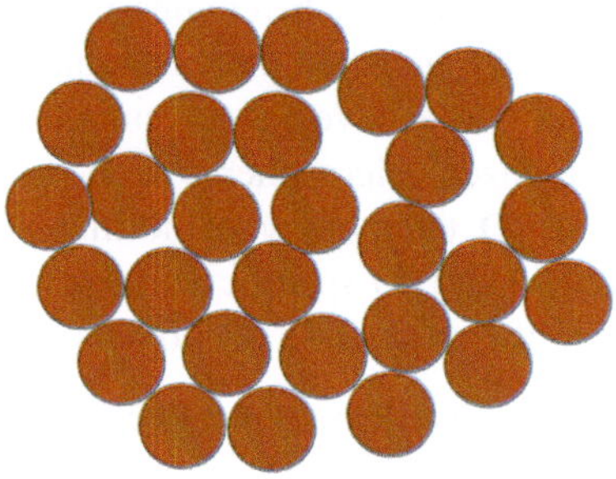

Figure 2.30. *Non-maximum stacking-sphere example*

Movement relating to the ejection mechanism for microspheres relative to the computer-controlled support should enable the production of a given object. Refinements may be envisaged with different types of material to produce appropriate gradients. Lastly, in the final stage, there is the option of diluting/dissolving the material made up of spheres, as Ratner proposed [RAT 09], insofar as, using this polymerization process, there is the possibility of using linear polymers.

2.7. Provisional concept

This section focuses upon a concept which has not been experimentally validated. Table 2.5 collates all of the elements to master to develop this process (classified as + for "easy", = "feasible", - "existence of an impediment", -- "possible large breakthroughs") with comments.

Topic	Accessibility	Commentary
Monomer-initiator couple	++	Known by the term "prior art", the number of products is highly extensive (including biocompatible materials).
Microfluidic system	+	May be produced by micro stereo lithography! These are standard mechanisms which have been published for 10 years.
Production of surface treatments	?	Problem to investigate to avoid droplets formed sticking to the ejector surface.
Polymerization	++	Photo polymerization known within stereo lithography; the calculation of the length of tube and luminous flows to reach Zone II is feasible from knowledge of the movement speed of spheres produced within a given junction.
Surface resolution	–	The spherical particle flow value should be known (laser counting system) to optimize the deposition of particles over the construction surface. The production of non-convex components will necessitate an additional fluid (or a powder) to enable the construction of the object.
Post-processing	+	Post-processing consists of light-induced polymerization either of an object in the course of production (on a "layer after layer" basis), or the object produced in its entirety.
Seeding	?	The possibility of having an additional deposit of living cells, for example situated in a micro-emulsion (tending to be for convex parts)
Production of a negative porous media	+	Standard process

Table 2.5. *Elements to bring under control for development of the process*

2.8. Conclusion

Although are there a wide variety of applicable technologies within the sphere, their current robustness is still not completely established. This is for reasons which are essentially related to manufacturing time if we are aiming to produce three-dimensional objects with an average size of several centimeters. An exception is applicable if we are purely accepting mediocre textures during such production. The issue is considerable as the μ-fluidics market is exploding and technological work resting upon new ideas is probably necessary to show the power of 3D technologies.

That being said, taking a potential example it was possible for us to show that μ-fluidics could be exploited to define one of the 3D printing processes.

2.9. Bibliography

[ABK 11] ABKARIAN M., "La micro-fluidique", available at: http://www.google.fr/url?sa=t&rct=j&q=&esrc=s&source=web&cd=1&cad=rja&uact=8&ved=0CCMQFjAAahUKEwjezoD3wojJAhWCyhoKHcYGBcY&url=http%3A%2F%2Fwww.coulomb.univ-montp2.fr%2Fperso%2Fmanouk.abkarian%2FManouk_Abkarian_Homepage%2FLinks_files%2FMicrofluidique.pdf&usg=AFQjCNHV8W7uk3JyF4T_k68TeHHrmoJSJQ&bvm=bv.107406026,bs.2,d.ZWU, 2011.

[AHM 15] AHMED E.M., "Hydrogel: preparation, characterization and applications: a review", *Journal of Advanced Research*, vol. 6, pp. 105–121, 2015.

[AMI 13] AMINI H., SOLLIER E., MASAELI M. *et al.*, "Engineering fluid flow using sequenced microstructures", *Nature Communications*, vol. 4, doi: 10.1038/ncomms2841, 2013.

[AND 94] ANDRE J.C., CORBEL S., *Stéréo-photo-lithographie laser*, Polytechnica, Paris, 1994.

[AU 11] AU A.K., LAI H., UTELA B.R. *et al.*, "Micro-valves and micro-pumps for BioMEMS", *Micromachines*, vol. 2, p. 179, 2011.

[AU 14] AU A.K., LEE W., FOLCH A., "Mail-order microfluidics: evaluation of stereolithography for the production of microfluidic devices", *Lab on a Chip*, vol. 14, pp. 1294–1301, 2014.

[BAR 04] BAROUD C.N., WILLAIME H., "Multiphase flows in microfluidics", *Comptes Rendus Physique*, vol. 5, pp. 547–555, 2004.

[BEG 14] BEGOLO S., ZHUKOV D.V., SELK D.A. *et al.*, "The pumping lid: investigating multi-material 3D printing for equipment-free, programmable generation of positive and negative pressures for microfluidic applications", *Lab on a Chip*, vol. 21, pp. 4616–4628, 2014.

[BER 11] BERTSCH A., RENAUD P., "Micro-stereo-lithography", in BARTOLO P.J. (ed.), *Stereo-Lithography; Materials, Processes and Applications*, Springer, New York, 2011.

[BET 10] BETTINGER C.J., BORENSTEIN J.T., "Biomaterials-based micro-fluidics for engineered tissues construct", *Soft Matter*, vol. 6, pp. 4999–5015, 2010.

[BON 10] BONYAR A., SANTHA H., RING B. *et al.*, "3D rapid prototyping technology (RPT) as a powerful tool in microfluidic development", *Proceedings Eurosensors XXIV; Procedia Engineering*, vol. 5, pp. 291–294, 2010.

[CAR 06] CARN F., Intégration entre chimie douce et fluides complexes pour la genèse d'architectures poreuses hiérarchisées: synthèses, caractérisations et application, Université Sciences et Technologies, Bordeaux, 2006.

[CAR 12] CARRIER O., De la formation de gouttelettes approche expérimentale à micro-échelle, Thesis, Université de Lorraine, Nancy, France, 2012.

[CHA 13] CHANMANWAR R.M., BALASUBRAMANIAM R., WANKHADE L.N., "Application and manufacturing of microfluidic devices: review", *International Journal of Modern Engineering*, vol. 3, pp. 849–856, 2013.

[CHU 12] CHUNG B.G., LEE K.H., KHADEMHOSSEINI A. *et al.*, "Microfluidic fabrication of microengineered hydrogels and their application in tissue engineering", *Lab on a Chip*, vol. 12, pp. 45–59, 2012.

[CLO 14] CLOTEAUX A., GERARDIN F., THOMAS D. *et al.*, "Fixed bed photo-catalytic reactor for formaldehyde degradation: experimental modeling study", *Chemical Engineering Journal*, vol. 249, pp. 121–129, 2014.

[COX 15] COX S.C., TORNBY J.A., GIBBON G.J. *et al.*, "3D printing of porous hydroxyapatite scaffolds intended for use in bone tissue engineering applications", *Materials Sciences and Applications*, vol. 47C, pp. 237–247, 2015.

[CZY 09] CZYZEWSKI J., BURYNSKI P., GAWEL K. *et al.*, "Rapid prototyping of electrically conductive components using 3D printing technology", *The Journal of Materials Processing Technology*, vol. 209, pp. 5281–5285, 2009.

[DEL 16] DELAPORTE P., ALLONCLE A.P., "Laser-induced forward transfer: a high resolution additive manufacturing technology", *Optics & Laser Technology*, vol. 78, pp. 33–41, 2016.

[DOD 15] DOD, "Process intensification – chemical sector focus; technology assessment", available at: http://energy.gov/sites/prod/files/2015/02/f19/QTR%20Ch8%20-%20Process%20Intensification%20TA%20Feb-13-2015.pdf, 2015.

[DON 15] DONVITO L., GALLUCIO L., LOMBARDO A. *et al.*, "Experimental validation of a simple low cost T-junction droplet generator fabricated trough 3D printing", *Journal of Micromechanics and Microengineering*, vol. 25, pp. 1–11, 2015.

[DRA 15] DRAGONE V., KITSON P.J., ROSNES M.H. *et al.*, "Configurable 3D printed millifluidic and microfluidic "lab-on-chip" reaction-ware devices", available at: http://www.cmac.ac.uk/files/media/Vincenze_Dragone.pdf, 2015.

[EAT 12] EATON S.M., DE MARCO C., MARTINEZ-VASQUEZ R. *et al.*, "Femtosecond laser micro-structuring for polymeric lab-on-chips", *Journal of Biophotonics*, vol. 5, pp. 687–702, 2012.

[ERK 14] ERKAL J.L., SLIMOVIC A., GROSS B.C. *et al.*, "3D printed microfluidic devices with integrated versatile and reusable electrodes", *Lab on a Chip*, vol. 14, pp. 2023–2032, 2014.

[FAL 10] FALK L., DE BELLEFON C., GOURDON C. *et al.*, "Intensification des procédés", *L'Actualité Chimique*, vols 338–339, pp. 101–111, 2010.

[FAL 15] FALK L., "Principes et applications de l'intensification en génie des procédés", available at: http://caser.cnam.fr/evenements/vous-avez-dit-genie-des-procedes/principes-et-applications-de-l-intensification-en.html, 2015.

[FER 15] FERRARI S., LOVERIDGE M., BEATTIE S.D. *et al.*, "Latest advances in the manufacturing of 3D rechargeable lithium microbatteries", *Journal of Power Sources*, vol. 286, pp. 25–46, 2015.

[FOU 16] FOUET M., Micro-fluidique 3D et actionneurs magnétiques. De leur intégration à la préparation d'échantillons biologiques, Thesis, Université de Toulouse, France, 2016.

[FOW 16] FOWLKES J.D., WINKLER R., LEWIS B.B. *et al.*, "Simulation-Guided 3D nanomanufacturing via focused electron beam induced deposition", *ACS Nano*, vol. 10, pp. 6163–6172, 2016.

[GAM 15] GAMM J., "University of Twente 3D print metal structures at 100 nm scale", available at: http://3dprintingindustry.com/2015/06/20/university-of-twente-3d-prints-metal-structures-at-100-nanometer-scale/?utm_source=3D+Printing+Industry+Update&utm_medium=email &utm_campaign=220e370f9e-RSS_EMAIL_CAMPAIGN&utm_term=0_695d5c73dc-22 0e370f9e-64557777, 2015.

[GDR 16] GDR, "GdR nano et micro-fluidique", available at: http://www.gdrmicrofluidique.com/, 2016.

[GEL 15] GELBER M.K., BHARGAVA R., "Monolithic multilayer microfluidics via sacrificial molding of 3D-printed isomalt", *Lab on a Chip*, vol. 15, pp. 1736–1741, doi: 10.1039/c4lc01392a, 2015.

[GON 09] GONZALEZ A., HERVAS M., "Real sample analysis on microfluidic devices", *Talanta*, vol. 74, pp. 342–357, 2009.

[GRI 15] GRIFFIN D.R., WEAVER W.M., SCUMPIA P.O. *et al.*., "Accelerated wound healing by injectable microporous gel scaffolds assembled from annealed building blocks", *Nature Materials*, vol. 14, pp. 737–744, 2015.

[GRO 14] GROSS B.C., ERKAL J.L., LOCKWOOD S.Y. *et al.*, "Evaluation of 3D printing and its potential on biotechnology and the chemical sciences", *Analytical Chemistry*, vol. 86, pp. 3240–3253, 2014.

[GUP 15] GUPTA M.K., MENG F., JOHNSON B.N. *et al.*, "3D printed programmable release capsules", *Nano Letters*, doi: 10.1021/acs.nanolett.5b01688, 2015.

[HAB 16] HABHAB M.B., ISMAIL T., LO J.F., "A laminar flow-based microfluidic tesla pump via lithography enabled 3D printing", *Sensors*, vol. 16, p. 1970, doi: 10.3390/s16111970, 2016.

[HAM 08] HAMMACHER J., FUELLE A., FLAEMIG J. *et al.*, "Stress engineering and mechanical properties of SU-8-layers for mechanical applications", *Microsystem Technologies Micro- and Nano-systems Information Storage and Processing Systems*, vol. 14, pp. 1515–1523, 2008.

[HAS 15] HASHIMOTO Y., MUKAI S.A., SAWADA S.I. *et al.*, "Nanogel tectonics porous gel loading biologics, nano-carriers and cells for advanced scaffolds", *Biomaterials*, vol. 37, pp. 107–115, 2015.

[HE 09] HE P., Conception et réalisation d'un système micro-fluidique pour la réalisation de gouttes calibrées et leur encapsulation, Thesis, Université de Technologie de Compiègne, France, 2009.

[HE 15] HE Y., QIU J., FU J. *et al.*, "Printing 3D microfluidic chips with a 3D sugar print", *Microfluidics and Nanofluidics*, vol. 19, pp. 447–456, 2015.

[HE 16] HE Y., WU Y., FU J.Z. *et al.*, "Developments of 3D printing microfluidics and applications in chemistry and biology: a review", *Electroanalysis*, vol. 28, pp. 1658–1678, doi: 10.1002/elan.201600043, 2016.

[HEN 14] HENGSBACH S., DÍAZ LANTADA A., "Rapid prototyping of multi-scale biomedical micro-devices by combining additive manufacturing technologies", *Biomedical Microdevices*, available at: http://oa.upm.es/35847/1/INVE_MEM_2014_191229.pdf, 2014.

[HIR 16] HIRT L., IHLE S., PAN Z. *et al.*, "Template free 3D microprinting of metal using a force controlled nanopipette for layer by layer electrodeposition", *Advanced Materials*, doi: 10.1002/adma.201504967, 2016.

[HO 15] HO C.M.B., NG S.H., LI K.H.H., "3D printed microfluidics for biological applications", *Lab on a Chip*, vol. 15, pp. 3627–3637, 2015.

[HON 11] HONEGGER T., ELMBERG T., BUTON K. *et al.*, "Visible microlaser two-photon polymerization in a microfluidic cell: a resist study", *Microelectronic Engineering*, vol. 88, pp. 2725–2728, 2011.

[HPC 13] HPC TODAY, "Micro-fluidique: des avancées décisives", available at: http://www.hpctoday.fr/discover/microfluidique-des-avancees-decisives/, 2013.

[HUA 14] HUANG H., DENSMORE D., "Integration of microfluidics into the synthetic biology design flow", *Lab on a Chip*, vol. 14, pp. 3459–3474, 2014.

[HUI 15] HUIS IN'T VELD B., OVERMEYER L., SCHMIDT M. *et al.*, "Micro additive manufacturing using ultra short laser pulses", *CIRP Annals – Manufacturing Technology*, vol. 64, pp. 701–724, 2015.

[HUT 12] HUTH M., PORRATI F., SCHWALB C. *et al.*, "Focused electron beam induced deposition: a perspective", *Beilstein Journal of Nanotechnology*, vol. 3, pp. 597–619, 2012.

[HWA 15] HWANG Y., PAYDAR O.H., CANDLER R.N., "3D printed molds for non-planar PDMS microfluidic channels", *Sensors & Actuators*, vol. 226, pp. 137–142, 2015.

[INS 15] INSTITUT PIERRE-GILLES DE GENNES, "Comprendre la micro-fluidique", available at: http://www.institut-pgg.fr/Comprendre-la-Microfluidique_65.html, 2015.

[JIA 15] JIAN K., LU R.K., DIMITRAKOPOULOS P. *et al.*, "Microfluidic generation of uniform water droplets using gas as the continuous phase", *The Journal of Colloid and Interface Science*, vol. 448, pp. 275–279, 2015.

[JOH 12] JOHNSON R.D., "Custom lab ware: chemical creativity with 3D printing", *Nature Chemistry*, vol. 4, pp. 338–339, 2012.

[JOH 14] JOHNSON-CHAVARRIA E.M., AGRAVIAL U., TANYERI M. *et al.*, "Automated single cell microbioreactor for monitoring intracellular dynamics and cell growth in free solution", *Lab on a Chip*, vol. 14, pp. 1–10, 2014.

[KAM 15] KAMEI K.I., MASHIMO Y., KOYAMA Y. *et al.*, "3D printing of soft lithography mold for rapid production of PDMS based microfluidic device for cell stimulation with concentration gradient", *Biomed. Microdevices*, vol. 17, doi: 10.1007/s10544-015-9928-y, 2015.

[KIT 12] KITSON P.J., ROSNES M.H., SANS V. *et al.*, "Configurable 3D printed multifluidic and microfluidic "lab on chip" reaction ware devices", *Lab on a Chip*, vol. 12, pp. 3267–3271, 2012.

[KRU 14] KRUPADAM R.J., KORDE B.A., ASHOKKUMAR M. *et al.*, "Novel molecularly imprinted polymeric microspheres for pre-concentration and preservation of polycyclic hydrocarbons from environmental sample", *Analytical and Bioanalytical Chemistry*, vol. 406, pp. 5313–5321, 2014.

[LAD 13] LADD C., SO J.H., MUTH J. *et al.*, "3D printing of free standing liquid metal microstructures", *Analytical and Bioanalytical Chemistry*, vol. 25, pp. 5081–5085, 2013.

[LAI 11] LAI H., FOLCH A., "Design and characterization of "single-stroke" peristaltic PDMS micropumps", *Lab on a Chip*, vol. 11, p. 336, 2011.

[LEG 15] LEGRAND P., ANDRE J.C., "Pour une approche 'Safer by design' en nano-médecine", in OMNT, *Micro et nanotechnologies: Avancées, tendances et perspectives*, Grenoble, France, pp. 228–229, 2015.

[LIA 12] LIAO Y., SONG J., LI E. *et al.*, "Rapid prototyping of 3D microfluidic mixers in glass by femtosecond laser direct writing", *Lab on a Chip*, vol. 12, pp. 746–749, 2012.

[LIN 16] LIND J.U., BUSBEE T.A., VALENTINE A.D. *et al.*, "Instrumented cardiac microphysiological devices via multimaterial three-dimensional printing", *Nature Materials*, p. 7, doi: 10.1038/nmat4782, 2016.

[LIU 10] LIU K., CHEN Y.C., TSENG H.R. *et al.*, "Microfluidic device for robust generation of two-component liquid-in-air slugs in individually controlled composition", *Microfluidics and Nanofluidics*, vol. 9, pp. 933–943, 2010.

[LU 15] LU Y., CHANG P.R., ZHANG P. *et al.*, "Porous 3D network rectorite/chitosan gels: preparation and absorption properties", *Applied Clay Science*, vol. 107, pp. 21–27, 2015.

[MAD 96] MADDEN J.D., HUNTER I.W., "Three dimensional microfabrication by localized electrochemical deposition", *Journal of Microelectromechanical Systems*, vol. 5, pp. 24–32, 1996.

[MAL 14] MALINAUSKAS M., REKSTYTÈ S., LUKOSEVICIUS L. *et al.*, "3D microporous scaffolds manufactured via combination of fused filament fabrication and direct laser writing ablation", *Micromachines*, vol. 5, pp. 839–858, 2014.

[MAR 07] MARI R., "Un modèle d'empilement de sphères dures", available at: http://www.ens-lyon.fr/DSM/SDMsite/M2/stages_M2/Mari.pdf, 2007.

[MON 99] MONNERET S., LOUBÈRE V., CORBEL S., "Microstereolithography using a dynamic mask generator and a non-coherent visible light source", *Proceedings of SPIE*, vol. 3680, pp. 553–561, 1999.

[MON 04] MONNERET S., "Micro-stéréo-lithographie de pièces céramiques complexes", *Techniques de l'ingénieur* RE13, p. 11, 2004.

[MOO 11] MOORE J.L., MAC CUISTON A., MITTENDORF J. *et al.*, "Behavior of capillary valves in centrifugal micro-fluidic devices prepared by 3D printing", *Microfluidics and Nanofluidics*, vol. 10, pp. 877–888, 2011.

[MOR 13] MORIMOTO Y., TAKEUCHI S., "3D cell culture based on microfluidic technique to mimic living tissues", *Biomaterials Science*, vol. 1, pp. 257–264, 2013.

[MOS 10] MOSADEGH B., Design and fabrication of microfluidic integrated circuits using normally-closed elastomer valves, PhD Thesis, University of Michigan, Detroit, 2010.

[MOU 14] MOULTON S.E., WALLACE G.G., "3D fabricated polymer based drug delivery systems", *Journal of Controlled Release*, vol. 193, pp. 27–34, 2014.

[MUR 10] MURPHY C.M., HAUGH M.G., O'BRIEN F.J., "The effect of mean pore size on cell attachment, proliferation and migration in collagen-glycosaminoglycan scaffolds for bone tissue engineering", *Biomaterials*, vol. 31, pp. 461–466, 2010.

[OCO 15] O'CONNOR J., PUNCH J., JEFFERS N. *et al.*, "A comparison between the hydrodynamic characteristics of 3D printer polymer and etched silicon microchannels", *Microfluidics and Nanofluidics*, vol. 19, pp. 385–394, 2015.

[ONE 14] O'NEILL P.F., BEN AZOUZ A., VÁZQUEZ M. *et al.*, "Advances in three-dimensional rapid prototyping of microfluidic devices for biological applications", *Biomicrofluidics*, vol. 8, pp. 052112–052112, 2014.

[OH 07] OH S.H., PARK I.K., KIM J.M. *et al.*, "In vitro and in vivo characteristics of PCL scaffolds with pore size gradient fabricated by a centrifugation method", *Biomaterials*, vol. 28, pp. 1664–1671, 2007.

[OHN 08] OHNO K., TACHIKAWA K., MANZ A., "Microfluidics: applications for analytical purposes in chemistry and biochemistry", *Electrophoresis*, vol. 29, pp. 4443–4453, 2008.

[OVS 08] OVSIANIKOV A., VIERTI J., CHICHKOV B. *et al.*, "Ultra-low shrinkage hybrid photosensitive material for two photon polymerization microfabrication", *ACS Nano*, vol. 2, pp. 2257–2262, 2008.

[OVS 11] OVSIANIKOV A., FARSARI M., CHICHKOV B., "Photonic and biomedical applications of the two photon polymerization technique", in BARTOLO P.J. (ed.), *Stereolithography; Materials, Processes and Applications*, Springer, New York, 2011.

[PAR 15] PARDON G., LADHANI L., SANDSTRÖM N. *et al.*, "Aerosol sampling using an electrostatic precipitator integrated with a microfluidic interface", *Sensors & Actuators*, vol. B212, pp. 344–352, 2015.

[RAN 06] RANDOLPH S.J., FOWLKES J.D., RACK P.D., "Focused, nanoscale electron-beam-induced deposition and etching", *Critical Reviews in Solid State and Materials Sciences*, vol. 31, pp. 55–89, 2006.

[RAT 09] RATNER B.D., "Tissue engineering biology-scaffolds-materials science", available at: https://www.youtube.com/watch?v=PNNiK9dgJXo, 2009.

[REN 13] RENAUDOT R., Conception, fabrication de nouvelles puces micro-fluidiques à géométrie programmable et reconfigurable reposant sur les principes d'électro-mouillage et de diélectrophorèse liquide, Thesis, University of Grenoble, France, 2013.

[REY 02] REYES D.R., IOSSIFIDIS D., AUROUX P.A. *et al.*, "Micro total analysis systems. 1. Introduction, theory, and technology", *Analytical Chemistry*, vol. 74, pp. 2623–2636, 2002.

[REZ 13] REZENDE R.A., SCHISHCHEV S.V., KOSYANOV V.A. *et al.*, "An organ biofabrication line: enabling technology for organ printing – Part 1: from Biocad to Biofabricators of spheroids", *Biomedical Engineering*, vol. 47, pp. 116–120, 2013.

[ROT 14] ROTMAN D., "Microscale 3-D printing – inks made from different types of materials, precisely applied, are greatly expanding the kinds of things that can be printed", available at: https://www.technologyreview.com/s/526521/microscale-3-d-printing/, 2014.

[SCH 02] SCHILLING E.A., KAMHOLZ A.E., YAGER P., "Cell lysis and protein extraction in a microfluidic device with detection by a fluorogenic enzyme assay", *Analytical Chemistry*, vol. 74, pp. 1798–1804, 2002.

[SHU 10] SHU X.Y., ZHANG H.H., LIN H.Y. *et al.*, "Experimental study on high viscosity fluid micro-droplet jetting system", *SCIENCE CHINA Technological Sciences*, vol. 53, pp. 182–187, 2010.

[SMA 15] SMALL LAB, "Aerosol jet printing", available at: http://www.small-lab.be/aerosol-jet-printing-0, 2015.

[SPA 14] SPATH S., SEITZ H., "Influence of grain size distribution on workability of granules with 3D printing", *The International Journal of Advanced Manufacturing Technology*, vol. 70, pp. 135–144, 2014.

[SUG 14] SUGIOKA K., XU J., WU D. *et al.*, "Femtosecond laser 3D micromachining: a powerful tool for the fabrication of microfluidic, opto-fluidic, and electro-fluidic devices based on glass", *Lab on a Chip*, vol. 14, pp. 3447–3458, 2014.

[TAB 03] TABELING P., *Introduction à la micro-fluidique*, Belin, Paris, 2003.

[THO 01] THORSEN T., ROBERTS R.W., ARNOLD F.H. *et al.*, "Dynamic pattern in a vesicle-generating microfluidic device", *Physical Review Letters*, vol. 86, pp. 4163–4166, 2001.

[UTK 08] UTKE I., HOFFMANN P., MELNGAILIS J., "Gas-assisted focused electron beam and ion beam processing and fabrication", *Journal of Vacuum Science and Technology*, vol. B26, pp. 1197–1276, 2008.

[VAE 13] VAEZI M., SEITZ H., YANG S., "A review on 3D micro-additive manufacturing technologies", *The International Journal of Advanced Manufacturing Technology*, vol. 67, pp. 1721–1754, 2013.

[VAN 11] VAN KAN J.A., SHAO P.G., WANG Y.H. *et al.*, "Proton beam writing a platform technology for high quality 3D metal molds fabrication for nanofluidic applications", *Microsystem Technologies*, vol. 17, pp. 1519–1527, 2011.

[VIL 04] VILKNER T., JANASEK D., MANZ A., "Micro total analysis systems. Recent developments", *Analytical Chemistry*, vol. 76, pp. 3373–3386, 2004.

[WAL 15] WALCZAK R., ADAMSKI K., "Inkjet 3D printing of microfluidic structures – on the selection of printer towards printing your own microfluidic chips", *Journal of Micromechanics and Microengineering*, vol. 25, p. 085013 (11 pp), 2015.

[WAN 13] WANG J., JIAN Y., NIE J. *et al.*, "Solid photopolymerization and polymer properties of octadecyl-vinyl-ether", *Journal of Photochemistry and Photobiology*, vol. 271A, pp. 105–110, 2013.

[WHI 06] WHITESIDES G.M., "The origins and the future of microfluidics", *Nature*, vol. 442, pp. 368–373, 2006.

[WIK 15] WIKIPÉDIA, "Micro-fluidique", available at: https://fr.wikipedia.org/wiki/Microfluidique, 2015.

[XIA 13] XIA L., LIU X., ZHANG R. *et al.*, "A rapid straightforward and print house compatible mass fabrication method for integrating 3D paper based microfluidic", *Electrophoresis*, vol. 34, pp. 3003–3007, 2013.

[XIN 15] XING J.F., ZHENG M.L., DUAN X.M., "Two photon polymerization microfabrication of hydrogels and advance 3D printing technology for tissue engineering and drug delivery", *Chemical Society Reviews*, vol. 44, pp. 5031–5039, doi: 10.1033/C5CS00278H, 2015.

3

3D Nanomanufacturing, 3D μ-Electronics and μ-Robotics

"What is annoying about modern theories, is not that they are false, but they may be proven true". [ARE 02]

"We are still called to decide what we like best. This may be a theory which is accepted with total confidence, which is built in a manner which makes refutation impossible, through which ethical considerations permeate. Alternatively it could be a theory which is considered with a critical eye, which is capable of improvement, and which leaves us the freedom to organize our lives in the way we find the most helpful". [FEY 99]

"Other blunders punctuate the tumultuous history of 3D printing and its patents. Ranked top is the notification of this technology's creation. On 16 July 1984, French researchers from the "Compagnie Générale d'Électricité", the future Alcatel, lodged with the Institut National de la Propriété Industrielle (INPI – the French National Institute of Industrial Property Rights or Patent Office) their stereolithography patent. However the issue was that the Company did not believe this to be possible, any more than the CNRS did. Moreover it did not pay the renewal fee to ensure that the patent remained valid. Three weeks later the American, Chuck Hull, the founder of the global giant 3D Systems, submitted his patent for 3D printing. Thus it came about that 3D printing became an American invention". [MAK 16]

"It needed to take us a long time to understand that the capability of a given technique was proportional to its inherent inability for verification, and its capacity to surprise us by generating something which is radically new. In fact, although we are not worried when faced with this technique, this is because it is not, of itself, a very revolutionary concept". [FER 11]

"The concept of identity within the idea of self-organization and autonomy is based upon the implied principle of repeated application. The idea constitutes this identity according to one's actions within a particular environment. This form of identity is both active and self-constitutive as such. There is an identity or form of "self" which self-constitutes precisely because it is generated by a given circular process. This nevertheless also depends on a set of external conditions". [MOR 04]

"Not only do we not move towards a position of scarcity, but we move towards a place of sustainable abundance. Not only is our future not obstructed, but we simply stand at the beginning of a new significant innovation cycle... What should we do with these inventions? This is for us to decide. Certainly many good things, and indeed bad things besides. However, more than ever Malthus was wrong, and it is those who inspire us who in fact do deceive us". [SIL 14]

"This baseline epistemological paradigm which he describes as 'empirical epistemology' which is making its way forwards, is more easily formalizing than permanently formalized". [SIM 04]

"Chemical and biological forms are replicated not because they are determined by unchanging laws or by perpetual forms, but owing to a causal influence of prior similar forms. In other words, a given form would reproduce because the particular matter habitually did so...". [SHE 85]

3.1. Introduction

There has been a large quantity of literature on electronic manufacturing system processes going back more than 50 years. Essentially, it concerns 2½-D methods, using subtractive and additive processing of layers of material, relying on recognized lithography principles at the time the given literature was written. Essentially, these are processes with a high added value, exploiting the principles of collective manufacturing and which attempt to respect Moore's law. Wikipedia states [WIK 17], "a common and variable version and without any links to Moore's authentic statements is: 'something' which doubles every eighteen months, this thing being 'power', 'capacity', 'rate', 'the clock rate', and indeed many other variables. However, it is very rarely the density of transistors on a microchip". The reader interested in this field can refer to various works such as those of Bonneau [BON 08], Campbell [CAM 01, CAM 12], Jaeger [JAE 01] and others. The point to consider is that we are looking to achieve objectives the size of which ties in with the nanometric area.

Indeed, the invention of the microscope by Leeuwenhoek during the 17th Century enabled us, owing to the possibility of observing the infinitely small, to exploit this. It did so by going beyond human perception. Since then, we have succeeded in producing near-field optical microscopes that have allowed us to go from a few micrometers to atomic dimensions [JES 16]. Beyond simple observation, it then became possible to move atoms across a given surface [FUE 12, WIL 02, DEN 16]. At the beginning, such work revolved around moving atoms or providing strongly localized energy thanks to ion beams or, more "simply", electron beams to move atoms emanating from precursor gases. Alternatively, researchers would produce a given chemical reaction by processing these crystalline-structured atoms using a process introduced in Chapter 1 known as FEBID (focused electron-beam-induced deposition). This work may be compared, for example, with those of Tseng *et al.* [TSE 05]; Schift [SCH 08]; Bratton *et al.* [BRA 06]; Fowlkes *et al.* [FOW 16]; Randolph *et al.* [RAN 06]; Utke *et al.* [UTK 08] and Létiche *et al.* [LÉT 16]. More sophisticated variants within the nanometric sphere than the 3DP (three-dimensional printing process, defined in Volume 1) have been developed. Technologies relying on AFM (atomic force microscope) systems have been used. This enables us, in practice, to succeed in producing atomic-sized voxels (see, for example, Zhao *et al.* [ZHA 16] and Jesse *et al.* [JES 16]). However, the shape of objects manufactured in this way remains convex. This may constitute a limiting factor (perhaps for now at least) in the significance of current given proposals in the field. Such proposals remain in limited industrial demand.

One example within collective manufacturing concerns the possibility of printing circuits in the near future. Current research concludes that these would have 3D designs in stepped 10-nm structures on silicon discs (wafers). They would be 200 mm in diameter and of 100 nm, placed on such wafers of 300 mm [LAN 15]. Using a pilot production line, the "Inspire" project from CEA-LETI (Laboratoire d'Électronique des Technologies de l'Information – a French research institute for electronics and IT) is integrating lithography by nanoprinting. It is doing so within an automated manufacturing mechanism. It follows that these methods use very high-density energies, indeed having very modest printing rates. These are methods which, however interesting they may be (for example, deposits of a few nanometers), are only able to be included within this work in an incomplete manner. This is due to the spirit in which this work, where we are researching the creation of complex parts, is compiled. In this chapter, the methods both used and contemplated are standard and innovative, but developed incrementally even if they remain particularly delicate to control. This is why this part is devoted to new ideas that are remote from objects produced by specialists in microelectronic printing.

Generally speaking, the main thread running through this work is to consider how to assemble nanomaterials to obtain products that are micromanufactured or nanomanufactured [BRA 16, LIA 11]. Several methods may be subject to researchers' interest. These may include "bonding" methods for nano-sized colloids, welding, deposition, laser-induced conversion (gold particles and nanoparticles for Wilson *et al.* [WIL 08]) or higher energy radiation. All such experiments may be conducted independently of a further one – using technologies for experimenting with nanocomponents in the condensed phase [ASH 86, DHO 06, ILI 02, JUO 08]. These different works fall within the long-term reflection by Pretz [PRE 14], which ranks additive manufacturing of integrated circuits in seventh position within the 10 technologies that are going to "change the world" (or perhaps we should read "might" or "ought to") by 2022.

Amongst the challenges connected with this sphere, the elements set out below have been researched when circumstances allowed:

– resolution: whether we can achieve 3D results for microelectronics technologies (for example, 10 nm with a 1-nm tolerance);

– manufacturing speed: whether this is > 0.5 $mm.s^{-1}$;

– object shape, complexity, practicability of objects relative to processes concerned and depth of "writing" (greater than that which is used in microelectronics);

– manufacturing control;

– process robustness;

– surface finish; and

– collective manufacturing.

3.2. 3D nano-facturing

Daniela Rus succeeded in demonstrating the capacity of robotic voxels to self-assemble into a given shape, by using a type of distributed algorithm, without using a central server, in the manner of cellular automata [SUS 14]. This introduces a form of determinism within the movement of elementary objects (voxels), which form according to the "will" of the designer of a complex component. Current progress within the field of micro-electromechanical systems (MEMS) enables us to contemplate, in the near future, progress in terms of size, but the goal of the nanometric area may not be conceivable. In this part, the idea is to consider whether it is possible, with our current knowledge, to control and manipulate, owing to external inputs (in this case, matter and energy), objects of supramolecular size. This involves exploiting the ability to displace these objects owing to chemical reactions unique to the given objects, and to form complex patterns. Dellaporta and Bedu [DEL 15] illustrated such possibilities by using ion beams with phenanthrene or siloxane as substrates to produce complex parts. There are therefore possible means of action. Thus, in this section, two approaches are envisaged. The first concerns the production of an object using a "bottom-up" method from smart material. The second considers, in light of the elements already presented in Volume 1 and Chapter 1 of this volume, whether it is possible to push additive manufacturing techniques to their limits.

3.2.1. *Smart material: so-called "DNA origami"*

3.2.1.1. *Self-organization*

Advances in physics and chemistry within the sphere of nanotechnologies have already elaborately demonstrated that the properties of the same constituents change according to the scale. At the moment, to the author's knowledge, there is no model capable of transforming from a given scale or organizational structure, for example, molecular, to another level, for example, nanometric [POU 06]. Here, we are tackling a very general theory of self-organization, which transforms from a structure distributed at random (or virtually) to "something" organized. Lehn [LEH 11] says to the extent that we are capable of synthesizing materials with spatial properties, of a physical and chemical nature, manufacturing an object is conceivable. "However, we may also ask whether it might not be possible to develop systems. Within such systems, we could let the object itself select what it needs for manufacture, and therefore progress to a state of self-organization with the selection of the necessary components for construction of the final object"

[LEH 11]. Although in principle we may observe structures, the question posed within the causalist concept of additive manufacturing is to consider the converse issue. That is to say, we precisely define the initial structure of such arrangements which should self-organize to achieve the final structure. Self-organization should thus give the impression that the structure emerges from the combination of interactions between constituent entities (including energy).

In terms of the opposite problem, which applies to the direct manufacturing of objects, it is a matter of being able to create three-dimensional structures, customized if necessary, of varying sizes from a pseudo-homogeneous medium. This is a standard problem "normally" contrary to entropy. It requires us to know what type of energy and which form to have available. This is because the process is anti-dissipative. In fact, there is nothing in the demonstration of the second principle that enables us to assert that there is a causal relationship between entropy variation and evolution towards a chaotic state. However, immediately several entities interact, where the effects are disproportionate to causes, and we might potentially have nonlinear characteristics with specific effects. Chaline says, [CHA 12]: "A system which is heavily non-linear within conventional space may be described by linear laws within scale spaces (scale invariance). We then progress to an even greater level of complexity, with systems which are heavily non-linear within conventional space, and equally non-linear within scale space (covariance scale)". The transition from the single voxel to the given object can therefore not be as easily "programmable" as this.

A definition:

Atlan [ATL 11] defines the position thus: "Self-organization [FOX 09] is a mechanism or a set of mechanisms by which structures are produced at a global level from a system based upon interactions. These take place between the constituents of the structure or structures at a lower level of integration. Interactions between constituents are themselves produced at a local scale, without any reference to a preconceived global structure. This is, in contrast, an emergent property of the system itself, and not a property which is imposed from outside of the system".

This definition leads us to introduce the concept of the emergence of meta-structures, even global structures, from interactions between elementary constituents at the local scale. It can be a question of given self-organizations of structures and/or functions. This aspect will be revisited in-depth in the chapters dedicated to 4D printing and bioprinting, and will not be developed further in this chapter.

Nevertheless, Figure 3.1 explains the principles of “spontaneous” self-organization, where voxels will directly assemble to achieve both a desired shape and a “stimulated” organization. In this case, as shown in the sphere of robotics by Rus [SUS 14], it is anticipated that control will take place outside of the construction of the given object. It is obvious that hybrid situations may also be envisaged.

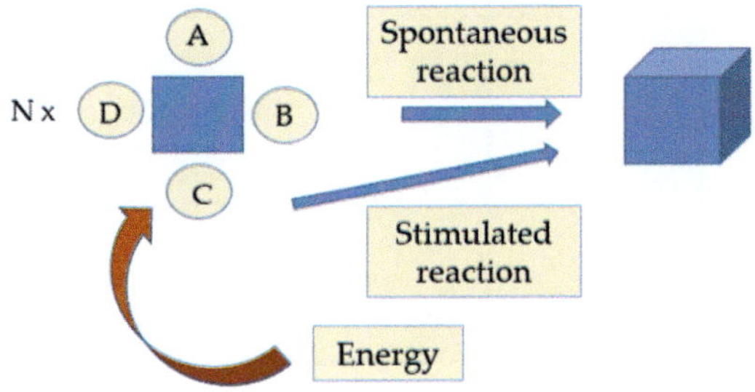

Figure 3.1. *Spontaneous and stimulated self-organizations (A, B, C, D showing the physicochemical potential enabling specific physical and chemical bonds)*

3.2.1.2. *Examples*

Sanchez [SAN 12/13] speaks within his lectures of a genuine “supra-nanoparticulate chemistry”, which is in the process of emerging: from nanoparticles having specific physicochemical properties, it is possible to produce objects which have particular shapes. Within such nano-objects, constituents of the given assembly are analogs of crystalline atom structures. Although nowadays we are witnessing a science in its infancy, the results shown in Figure 3.2 by Kellermeier *et al.* [KEL 12] already prove the significance of this new form of chemistry. It assumes “control of interactions involving attraction (for example, the van der Waals force, hydrophobes, solvophobic substances, hydrogen bonding, capillary forces, dipole forces and magnetic forces). It also similarly assumes control of repellent interactions (electrostatic forces and steric forces, dipole forces, and magnetic forces)” [SAN 12/13]. However, if we process nanoparticles, the size of the objects produced can achieve dimensions of the order of 100 μm. Thus, it is not the nanometric space which is targeted. However, active monitoring around this subject, which comes closer to the case set out below, is relevant. Figure 3.2 gives examples of given structures.

Figure 3.2. *Ordered silica bio-morph assembly (Kellermeier et al. [KEL 12]; reproduced with the kind permission of the publisher)*

Other applications resting on similar foundations were subject to a number of publications (see, for example, Laxminarayana [LAX 12]; Nikitin *et al.* [NIK 10]; Slocik and Naik [SLO 10]; Chiu *et al.* [CHI 11]; Nykypanchuk *et al.* [NYK 08] and Seeman [SEE 04]).

3.2.1.3. *DNA origami*

DNA can take various shapes within different types of assembly as per the example in Figure 3.3, which corresponds to rods which are 170 nm long connected by movable junctions. These can take several conformations [KUZ 14]. As with works on nanoparticle assembly, the DNA origami stems from a technique, which produces arbitrary forms from fragments of this macromolecule. It was developed by Paul Rothemund in [ROT 06]. The chemical properties of the molecules which make up the DNA enable the so-called "Watson-Crick" bonds. These, when created, can model the macromolecule into two-dimensional or three-dimensional structures of nanometric size. The Rothemund technique uses a long chain of virus DNA, with a less complex structure, and a recognized sequence. Next, by using a computer, it is possible to locate precise sites where the DNA macromolecule should be "folded" to achieve the required shape. This then leads to synthesizing small molecules, which will specifically focus on the designated sites, for so-called "binding" of the viral genome and manufacture a 3D structure.

Only these bonds may be synthesized as the DNA is of a natural origin. The so-called synthesis "arms", through the effect of chemical bonds, cling to the DNA by attracting other molecules, which contribute to the shaping of the central DNA strand. It is from this form of folding that the original name of DNA origami derives. In Rothemund's publication [ROT 06], pictures of the structures that he had produced were shown. These took various forms. In the same context, Han's team achieved the creation of sealed three-dimensional shapes ([HAN 11]; see also Sawyer [SAW 11] and Figure 3.4).

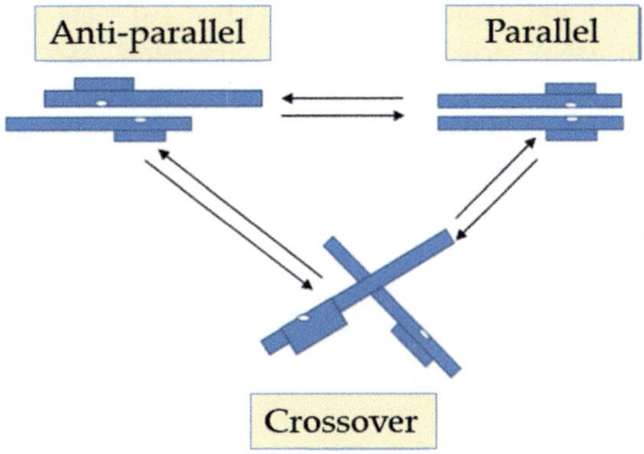

Figure 3.3. *Stimulated alterations of the DNA-rod configuration*

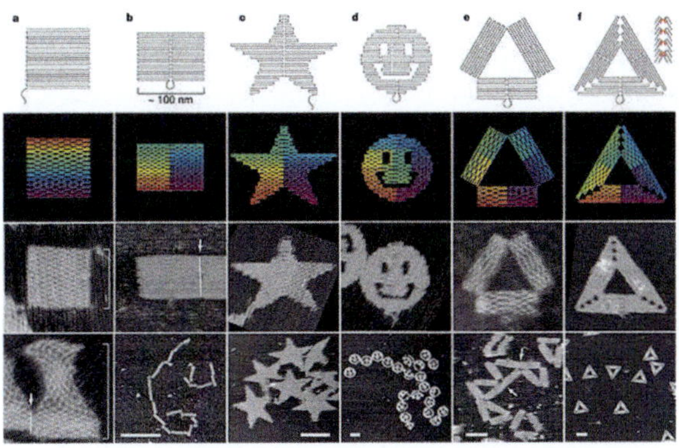

Figure 3.4. *DNA origamis [ROT 06]*

To do this, it is possible to use complex scaffolding made up of concentric DNA rings, connected to each other at strategic locations, by exploiting the DNA's pairing properties. Indeed, each of the bases A, T, C and G is linked in a specific way. The pairings take the form of A with T and C with G, so as to create a double-strand molecule. The rings are linked to each other at so-called "crossover" sites. From this system, manufacturing complete three-dimensional structures becomes possible, with applications ranging from intracellular delivery of drugs [GUI 15], the creation of more effective vaccines, gene therapy, through to microscopic and high-performance information processors. Such a manufacturing process includes the calibration of fluorescent and super-resolution microscopes and other devices [WIK 15].

Sussan [SUS 11] and Veneziano *et al.* [VEN 16] state that it is henceforth possible to create DNA origamis using a computer: "Software having fairly analogous interfaces to those of relatively simple software such as "3D Studio" and "Blender", display 2D and 3D views side by side. Their modus operandi takes account of the specific characteristics of DNA chains. You begin by outlining the sketch of the shape that you wish to bring about. Next, we describe the possible bonds between the different points on the scaffolding. The program will then generate the list of constitutive molecular components of the necessary chemical bridges. This will enable the operation. There is nothing in the real world to synthesize the whole scaffolding! The problem is linked to the relevant chemistry. It is not so easy to describe an origami of this type in virtual terms. As always in biology, unforeseen interactions can prevent molecular chains from producing the desired shape. Thanks to this technique, a team from Stockholm (in Sweden) constructed a sphere, a spiral, a bottle, and a DNA printing structure of the so-called Stanford bunny. This is a test model used within the sphere of the synthesis of the given image", (see Figure 3.5, photo kindly supplied by Bjorn Hogberg [BJO 17]). See also Guitard [GUI 15] and Benson *et al.* [BEN 15].

Figure 3.5. *Representation of 3D nano-structures*

We readily understand that this is a field which will be enriched over the next few years. Moreover, Wu *et al.* [WU 15a] had previously shown that there exists the possibility of producing origamis using proteins. Furthermore, apart from the process's appeal, there are at least two limits to the promotion of these works on the production of objects. One is their complexity that remains modest, and above all, their size that remains nanometric, indeed micron-sized. "However, who knows?" Nevertheless, the generic concept associated with origamis is starting to be exploited to produce micro-surgical components [HOL 16].

3.2.2. *Return from additive manufacturing to standard methods*

Apart from atom displacement technologies from near-field microscopic systems, the emerging techniques are limited in number. Amongst these, some original technologies are set out below. Working with micrometric dimensions, Vaezi *et al.* prepared a summary of usable processes in 2013 [VAE 13], which however result from smaller targeted dimensions in this section.

3.2.2.1. *Electro-deposition*

Je *et al.* [JE 17] suggest various methods of (electro)-deposition of the type shown in Figure 3.6 in their article. It revolves around introducing a fluid containing metal ions through a "Nano pipette", which enables a metallic deposition by means of electroplating. It may also be a question of other biological or organic forms of deposition, by using other methods for material conversion. Although the openings should be a size of the order of a fewmicrometers, the mechanisms must respect a number of constraints, including the absence of any vibrations. When the process concerns droplets containing a metallic ion, for which the conversion into metal may be induced by a source of energy, such as an electric current, there remain problems to resolve. These include the positioning of the drop together with surface tension and wettability aspects, the object's complex form and the velocity of metal deposition.

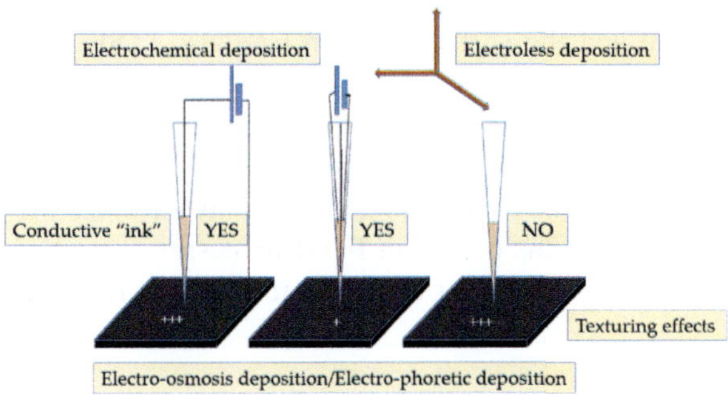

Figure 3.6. *(Electro)-deposition of metal and/or organic compounds on a nanoscopic scale*

Several types of depositions were able to be produced using various resolutions as collated in Table 3.1.

Size (voxel or width of the deposition line)	Species deposited and resolution; in brackets (nm for the first two lines, µm for the latter)
< 100 nm	Carbon nanotubes (90); mercaptohexanoic acid (40); poly-pyrrole (100); copper sulfate (80); polystyrene nanoparticles (70), gold (120–200); copper (100–400); platinum (150)
Between 100 and 1,000 nm	Bacteria (800); protein (200); carbon nanotube composite (800); PMMA (700); graphene oxide (300); DNA (500)
> 1 µm	TMTPA – 4,4 4 –trimethyl-triphenylamine (4); gold nanoparticles (1)

Table 3.1. *Example extracts from Je et al. [JE 17] of resolutions for the processes suggested within their publication*

COMMENT.– Electro-Hydrodynamic Deposition (EHD):

By using a significant electric potential, a conducting fluid can pass through a nozzle of inner radius between 100 nm and 100 µm, molded to form a fluid trickle or a set of micro-droplets that can be deposited on a support. This type of deposition was studied by Onses *et al.* [ONS 15] and is the subject of Figure 3.7 where a µ-droplet forms, the size of which depends simultaneously on the tension between the fluid and the substrate and the surface tension (see also [TSE 14, MAS 16, HAN 15]). Various types of materials have been used with varying degrees of success: liquid metals, metallic nanoparticles, graphene, carbon nanotubes, hydrogels, polymers and others. For the moment, this technology only enables the production of convex parts. However, using multi-nozzles, we may envisage the deposition of a soluble material in a solvent (as occurs in FDM processes – see Volume 1 – and in electronics using sacrificial layers) at the same time as the relevant material. Once the part is produced, we eliminate the soluble compound to then make the final object available. Without using sacrificial layers, non-metric resolutions were obtained using this type of mechanism.

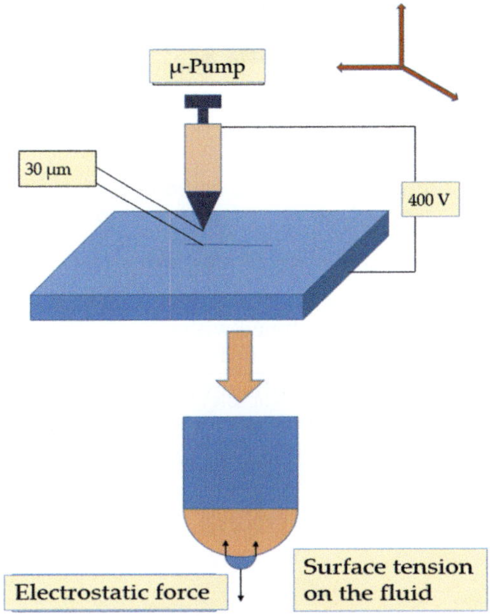

Figure 3.7. *The principle of depositions using electro-hydrodynamics*

3.2.2.2. *Deposition through a nanotube*

Bekarevich *et al.* [BEK 16] demonstrated that it was possible to introduce metal atoms (gold in this case) in the form of nanoparticles within a carbon nanotube and then to use the given organized structure, such as a syringe, to deposit atoms over a particular surface (see Figure 3.8). However, the issue of recharging atoms remains an even more delicate problem. It is necessary to overcome this to ensure continuous deposition. Fan *et al.* [FAN 15] use a micro-tank attached to the "syringe", which constitutes part of the solution for the present problem. They favor the movement of metallic nanoparticles with the assistance of an electric field.

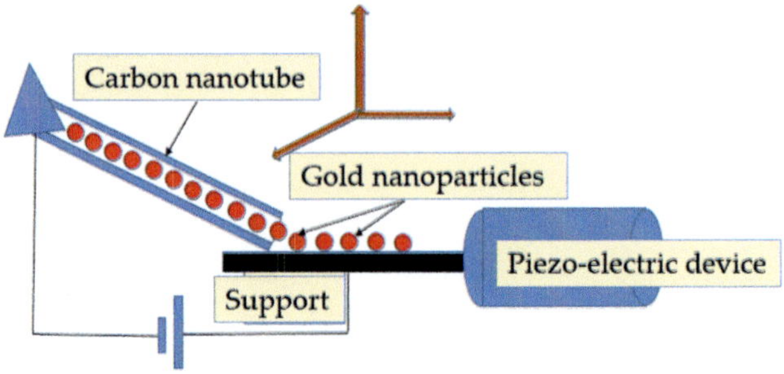

Figure 3.8. *Use of a carbon nanotube such as a syringe to deposit metals*

3.2.2.3. *Focused electron beam*

Figure 3.9, stemming from the works of Fowlkes *et al.* [FOW 16] and Huth *et al.* [HUT 12], recalls the principle of FEBID processes (for "Focused Electron-Beam-Induced Deposition"). This works by precursor gases being led to the vicinity of the electron beam and decomposed into atoms, which can settle in the immediate vicinity of the area where they are produced. Resolutions of a few hundred nanometers can be obtained using this process.

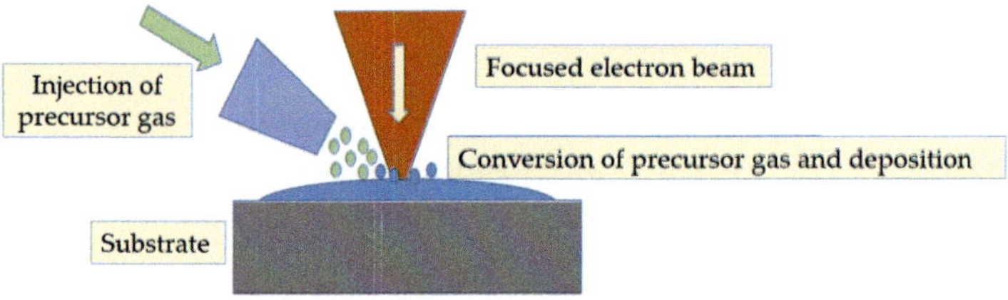

Figure 3.9. *Reminder of the FEBID process (see previous chapter)*

3.2.2.4. *Fused deposition*

The fairly precise deposition of nanofibers is made possible by using an electric field that avoids the problems of Coulomb repulsion. From the mechanism shown in Figure 3.10, it was possible to produce the so-called "nanowalls" using an oscillation movement [LEE 14]. This is a so-called electrospinning method pushed to its limits [LEE 15, TEO 05].

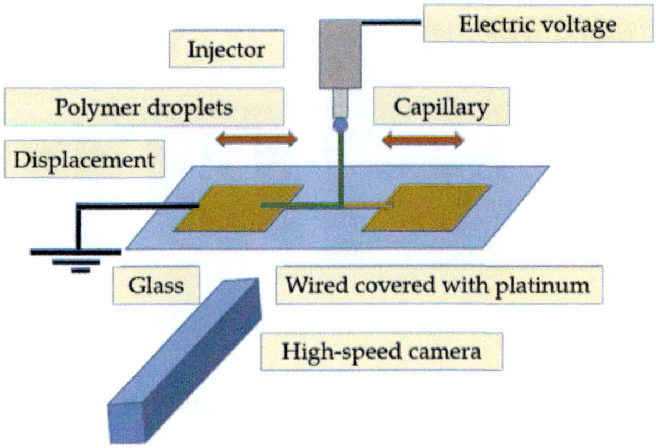

Figure 3.10. *Deposition of nanowires to produce a vertical stacking arrangement (the capillary is 100 µm in size having a 10% solution of polyethylene oxide in water, which leads to diameters which are significantly lower than micrometers after the solvent evaporates)*

3.2.2.5. *Additive–subtractive coupling*

Xiong *et al.* [XIO 12] report, as shown in Figure 3.11, on additive/subtractive coupling processes: a given unit is produced by additive manufacturing (two-photon photo-polymerization) followed by ablation with the aid of a more intense pulsed light source (see also Chu *et al.* [CHU 14] and Zhang *et al.* [ZHA 14]).

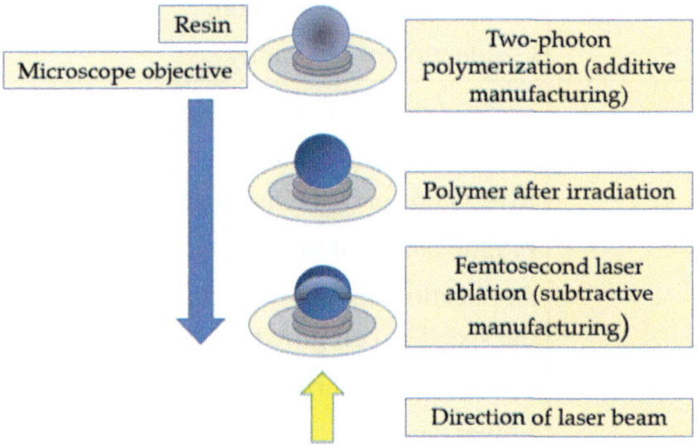

Figure 3.11. *Coupling of additive and subtractive manufacturing*

As a reminder, the pictures in Figure 3.12 are kindly provided by Nanoscribe® (www.nanoscribe.com). This evokes the potential of two-photon absorption mechanisms in terms of resolution.

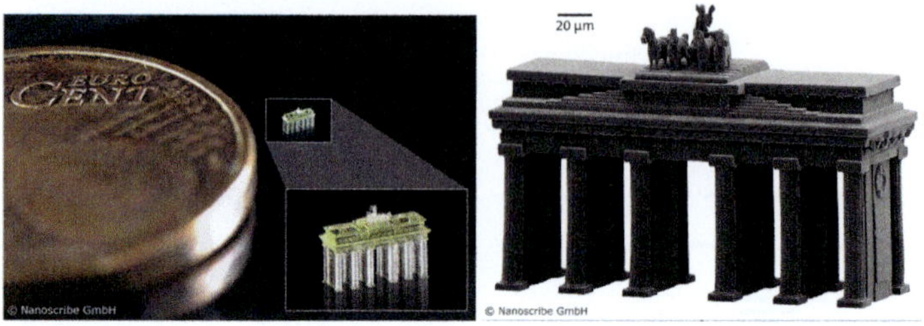

Figure 3.12. *Nano-Brandenburg Gate (photographs kindly provided by nanoscribe)*

3.2.3. *Comment: nanomaterials and additive manufacturing*

In the preceding descriptions, the possibility of creating 3D nanometric-sized structures, indeed sub-micrometric-sized structures, was sought. We achieved this using a bibliographical approach. The majority of scientific articles focused essentially on the use of nanomaterials to produce components of the largest possible sizes. Table 3.2, inspired by Campbell *et al.* [CAM 11], indicates what the significance of these additives might be in terms of the consequences on the quality of the objects manufactured.

Nanoparticle materials	Effects
Metals	Sinter temperature reduction; Increase in density; Lower levels of distortion; Conduction of electricity
Carbon	Very significant improvement in electrical and heat conductivities; Effects on surface tension and fracture; Fragile nature of objects; Application in bio-printing owing to improvement in the speed of cell growth; Poorer quality surface finish; Lower density
Ceramics and semiconductors	Improvement of mechanical properties; Improved rigidity but with an increased fragility (silica); Improvement of sintering properties (alumina)

Table 3.2. *Additive manufacturing and nanomaterials*

Although not an exhaustive list, amongst recent references, the reader interested in the field may consult the following articles:

– Notteau [NOT 16], Bregeras [BRE 16], Lucas [LUC 16] and Victor [VIC 16] on metallic nanoparticles;

– Pina *et al.* [PIN 15] relating to nanocomposites for bone healing procedures;

– Gulati *et al.* [GUL 16] for titanium nanoparticles within the same sphere;

– Hu *et al.* [HU 14] for the production of metal polymer composites;

– Ross *et al.* [ROS 14] on polymers;

– Kwon *et al.* [KWO 16] on electronics;

amongst other works.

Within the current ongoing work in the field, it is possible to envisage nano-inks containing various types of materials, the excellent adhesion properties of which we are well aware. On the one hand, it is possible to easily bond nano-voxels and, on the other hand, by changing nano-ink, there is a possibility of producing gradients of materials inside an object manufactured by additive manufacture [KON 16].

3.2.4. *Conclusion*

While reading these few sentences devoted to additive manufacturing within nanometric areas, a number of observations can be made. Above all, we seek to push these techniques to their limits (simply because they have proved themselves within other contexts). Interesting opportunities appear to be looming with EHP processes, and could be worth considering as they advance.

Figures 3.13 and 3.14, stemming from the works of Engstrom *et al.* [ENG 14], give a few indications as to the performance and costs of production machinery, which is essentially $2^1/_2$D.

In these two figures, the symbols that are used represent: STM: scanning tunneling microscope; AFM: atomic force microscope; OT: optical tweezers; IJP: inkjet printing; EHD: electro-hydrodynamics; EBL: electron-beam lithography; DLW: direct laser writing; PL: photolithography; NIL: nanoimprint lithography; FUVL: far UV lithography; UVDL: UV-depth lithography.

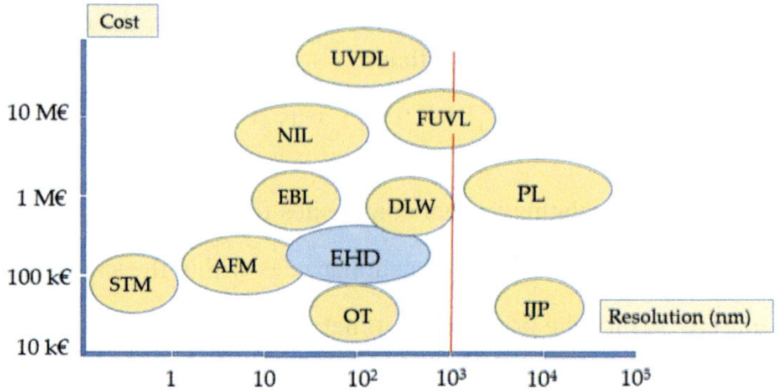

Figure 3.13. *Relationship between cost and resolution*

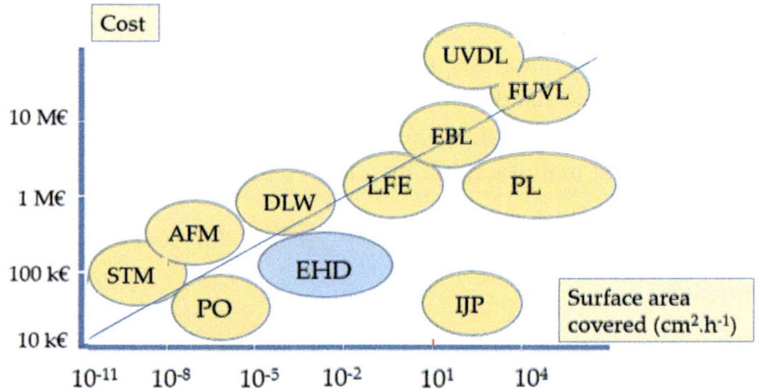

Figure 3.14. *Relationship between cost and coverage velocity*

For reasons relating to manufacturing time, except for specific situations, the choice to further exploit atom manipulation technologies does not appear relevant at present. In Figure 3.14, the diagonal reveals the "the best relationship between quantity covered and cost": technologies that are situated below this line are, in principle, favored. On analyzing both figures, EHD appears to have simultaneously an "acceptable" cost, a good resolution and a satisfactory productivity level. It appears to be a method of favor, all the more so as it had already demonstrated that genuine 3D objects may be created using this technique.

The other emerging sphere that appears to hold both creativity and the future (which is however somewhat removed from the author's knowledge base) is that of "smart" material within the DNA origami. This does have aspects of self-organization. This view sets down that the control of the structure of voxels, in terms of local reactivity to achieve a desired spatial organization, is left to chemists. However, cross-fertilization with specialists in processes and nanotechnologies may enable the involvement of research within the sphere of stimulated self-organization. In what we might probably view as an engineered second stage, it would be advantageous to consider how to achieve the so-called "optimized engineered material" aspects, instead of simply contenting ourselves with opportunities. This is actually quite a good position to be in.

The exploration of this subject, an actual genuine work in progress, poses a certain number of questions that must be dealt with:

– What type of manufacturing model/systems should be retained?

– Can it be tested with the help of existing knowledge?

– If yes, how can this be refined?

– What will its sensitivity to the initial conditions be (multi-stability)?

– How will we control regulatory processes for voxel displacements?

– What will the concentration effects of voxels on the manufacture be?

– How should the given structure be stimulated?

– What knowledge should be exploited around self-organization mechanisms (direct, indirect, cooperative, stimulated and other knowledge)?

– What experiments should be programed to make progress in the field and on which type of nanometric materials?

– Do we have a sufficient generic nature available to widen the application of a potential model?

This open and highly interdisciplinary field of investigation seems entirely to be able to provide methods to produce nanometric objects.

3.3. 3D µ-electronics

Analyzing the few examples set out from the beginning of this volume, what we note is that, for many, the technological origin of processes explained stems above all from know-how within microelectronics. Indeed, additive manufacturing technologies are generally 2½-D [MOS 10, YU 15]. They can, so as to achieve the third dimension, rely on accumulated know-how. In this part, the objective is to envisage whether, apart from the technological presentations that have already been described, it is possible to come up with more original 3D diversification methods.

Within the first stage, a number of annual publications on the subject using two key terms were considered. These were "microelectronics" and "3D printing". There is a net increase in the annual rate, which is still very modest, within this particular sector of additive manufacturing. The rate of publications is around 0.5% of the total publications on additive manufacturing, excluding nanowires, nano-antennae and other similar devices emerging from the field under discussion, for several main reasons discussed as follows:

– the issue of collective manufacturing versus individual elements in additive manufacturing;

– the issue of nanometric resolution versus micrometric resolution. Omitted from such considerations are two-photon absorptions with an obvious advantage to microelectronic technologies and the technologies set out in section 3.2 which have not yet been stabilized;

– the purity of materials and processes to compare with that practiced within "standard" microelectronics;

– and other examples in this field.

This situation is confirmed by the low impact of this specific domain in terms of European financial support relating to additive manufacturing, currently at 1.4% of all such support given (see [EST 16]).

3.3.1. *2D or 3D electronic circuits*

In this case, it is not necessary to take into consideration the selective criteria mentioned earlier in their entirety. The question generally set for scientists in the field is the ability to effect a conducting deposition on surfaces which are not necessarily flat, but which have small dimensions. If such a surface is flat, it is a question of standard processes that come out of the subject covered. However, Optomec [OPT 15] is marketing a 3D printing technology known as Aerosol Jet.

This allows the deposition of printed conductors on the surface of 2D and 3D components (which are centimeter-sized). For its part, Nano Dimension, an Israeli company, has developed inkjet inks made up of nanoparticles that are used within printed circuit manufacturing. This is demonstrated by their application to printed circuits [MÉL 17]. This is illustrated using the example given in Figure 3.15.

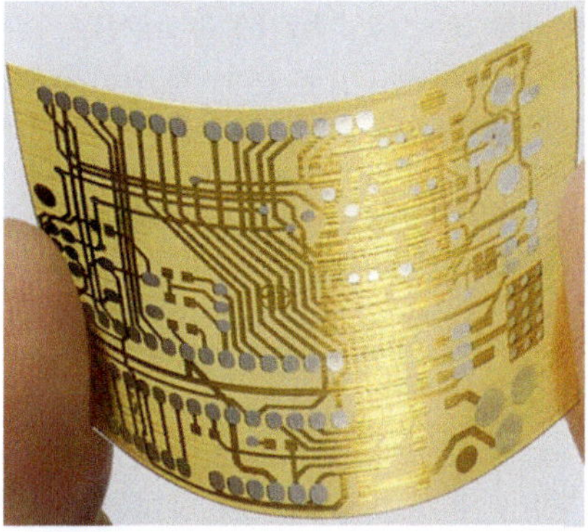

Figure 3.15. *"3D" printed circuit*

Working within the same sphere, Cai *et al.* [CAI 16] have developed a printing process using aerosol spray, enabling the option to work with several materials (indeed with many print heads). The resolutions recorded by these authors are in x,y terms, of the order of 20 μm and of a depth of 0.7 μm.

3.3.1.1. *Applicable spheres*

An article by Lewis and Ahn [LEW 15] was found within a scientific review. This concerned 3D electronics manufacturing by producing diodes using quantum dots. However, the complexity of the component is far from being clear in relation to the work of μ-electronics experts! The examples, albeit in modest numbers, of fairly simple additive manufacturing are found within the literature (see, for example, [LAS 15]). Another way to approach the issue is to consider that it is difficult to work within electronics with additive manufacturing machinery, using only a single material. In the example given by Wu *et al.* [WU 15b], the structure is produced within an insulation material (typically a polymer), and then, a conductor material is introduced. This enables the production of electrical parts or basic components (which is even more remote from the microelectronics sphere). Various types of materials can be used (see, for example, Leigh *et al.* [LEI 12] and Woodward *et al.* [WOO 15]).

Other examples exist. These include those targeting the production of radio-frequency antennae [ADA 15] with appropriate surfaces for frequency selection, within the microwave sphere [BAR 15]. Alternatively, there are flexible antennae produced by metal injection [FER 16], terahertz components [KAU 15, YUD 15] and sub-millimeter antenna [NAY 14, LE 15]. There are diffraction gratings and lenses within the terahertz sphere [SQU 15], waveguides [MOS 15] or resonators [IFT 15]. Sequential methods are used by Li *et al.* [LI 14] for the production of Helmholtz coils and other devices. These objects which have electronic purposes (but not actually within microelectronics!) use conductive "inks" (see, for example, PV Nano Cell [PV 16]), metals or metal alloys with low melting points and non-conducting resins. These are additive manufacturing bicomponents [WAN 14]. Krivec *et al.* [KRI 17] use additive manufacturing technologies for metallic depositions and packaging applied to acoustic transducers (see also Bailey *et al.* [BAI 16]).

If we accept the need to move beyond standard 3D printing technologies in order to attempt to break down a few barriers to form links between the needs of microelectronics, the scope for application may notably widen. This has been achieved by the team led by Niklaus [NIK 14]. They used specific sources to print silicon, one of the most conventional electronic materials. The realization principle is shown in Figure 3.16 [NIK 12].

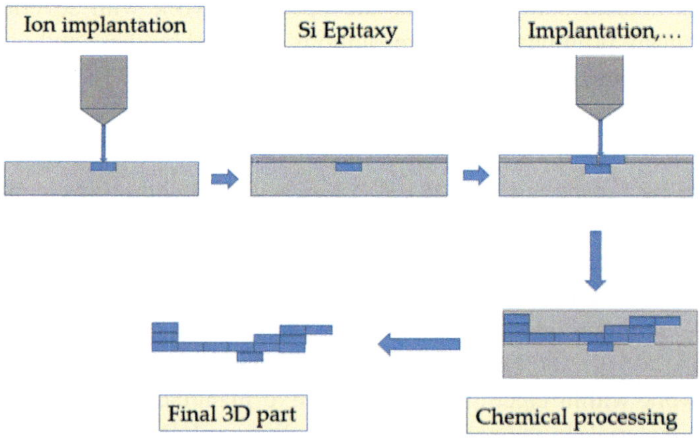

Figure 3.16. *Silicon-based additive manufacturing*

The use of fairly conventional technologies in microelectronics (ion implantation), and technologies which are well-controlled within technological platforms, may enable the desired reconciliation with success. This would be entirely compatible with need (in terms of resolution at the very least with an order of magnitude of 50 nm) [FIS 12].

Other technologies such as LIFT (laser-induced forward transfer) may be exploited within additive manufacturing, even if its use is, above all, of the $2^1/_2$D type [YU 15, DEL 16, PIQ 15]. It is worth noting that numerous nanoparticulate metals and structures can be used for this type of transfer (see, for example, Breckenfeld *et al.* [BRE 15] and Florian *et al.* [FLO 15]).

3.3.1.2. *Materials*

3.3.1.2.1. Conductive materials

By giving control back to chemists, it is possible to find simple plating processes such as that proposed by Ikeda *et al.* in 2001 [IKE 01]. It is possible to make use of, for example, the oxidation–reduction properties of titanium oxide irradiated within a UV light source, to enable a metallic copper deposition (see Figure 3.17). The metallic copper has a surface electrical conductivity of 1.126 10^{25} S. This is sufficient for an electrolytic deposition process enabling the increase in the thickness of the metallic layer (see also Akamatzu *et al.* [AKA 05] and Clarke [CLA 17]). Other metals may be deposited from this TiO_2 coupling and metallic ions: platinum, gold and palladium (see [MAI 11]) or by direct deposition [VIC 16, ROV 16].

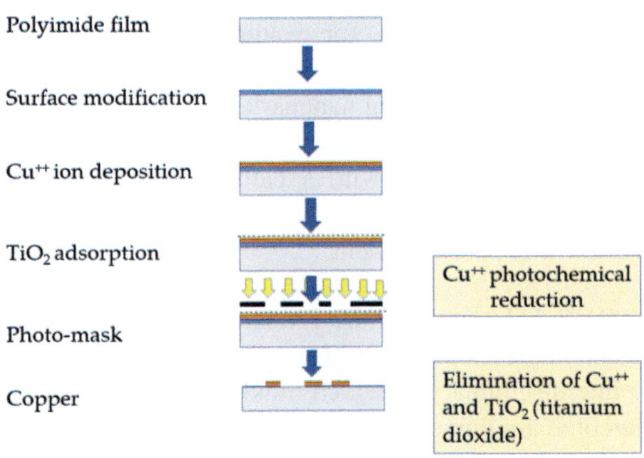

Figure 3.17. *Deposition of copper through the intermediary of photo-catalysis of TiO_2 (resolution 1 μm)*

It is possible for even simpler 3D structures to be created using bi-material systems [JOH 07]. Conductive inks (for example, those of eight-voxel inks) with a silver base metal having highly increased conductive properties (up to 2.10^6 $S.m^{-1}$) relative to carbon-containing composite materials [MOL 15] can be used.

Conductive inks constitute an interesting form of technology, which however are still inadequate to print wire lines with a high level of precision and resolution. Their principle limitations are linked to the mechanical movement of print heads which, operating at a high speed, limits both precision and printing resolution. This process can only, for the moment, transfer low-viscosity inks. That limits the capacity to print thick conductive structures for applications necessitating high currents. The risk of clogging up print heads is significant, the cost of inks is high and the process requires an annealing step. The inkjet process is entirely relevant for certain types of depositions. However, these disadvantages can limit the development of given applications.

Lastly, according to Graphene Lab Inc. [GRA 16], there is a possibility of producing magnetic filaments for various types of electro-technical applications. For Llewellyn-Jones *et al.* [LLE 16], composite materials with specific acoustic and mechanical properties are manufactured owing to the direction of the fibers included within the material. The publication by Vaezi *et al.* [VAE 13] sets out the directions of the various 3D technologies mentioned in Volume 1 with their potential for electrical and electronic applications, for example, sensors and actuators. In the majority of cases, the resolutions are a few dozen micrometers. Recently, INM [INM 16] published its works on using nanoparticles containing silver oxide which, upon receiving light, produce a form of latent image (see the traditional photography processes). The revelation happens with a bath of dissolved silver salts, which enables the release of silver atoms, also allowing for the production of conductive layers, without any form of post-treatment.

Quéré *et al.* [QUÉ 16] state that the present low-impact situation of additive manufacturing technologies is explained by the necessary reflection upon the development of interdisciplinary research linked to creativity. Either way, the difficulties to overcome are numerous before we can begin to achieve integration of 3D technologies within the sphere.

3.3.1.2.2. Insulation–conduction coupling

The implementation of electrical conductive structures has been set out in Volumes 1 and 2. The simplest principle is shown in Figure 3.18. This uses fusible polymers containing electrical conductive particles [AHN 14]. Indeed, within the present works, it is not envisaged that the means to dispose of insulating and/or conducting zones, including this type of process, should be defined. However, this operation is conceivable with polymer blends and insulating materials (for example, ceramics), which, when appropriately shaped, may be used with a bi-material 3D printer. After de-binding and sintering, it may enable the use of 3D conducting and/or insulating zones simultaneously. It is then possible to stabilize the shape of the object being constructed using, in addition, a third sacrificial mixture. This only contains a polymer or a low-cost charge (which may be eliminated) represented by the white voxels in the figure. After production, excess powder is separated from the part which is then subject to heat treatment.

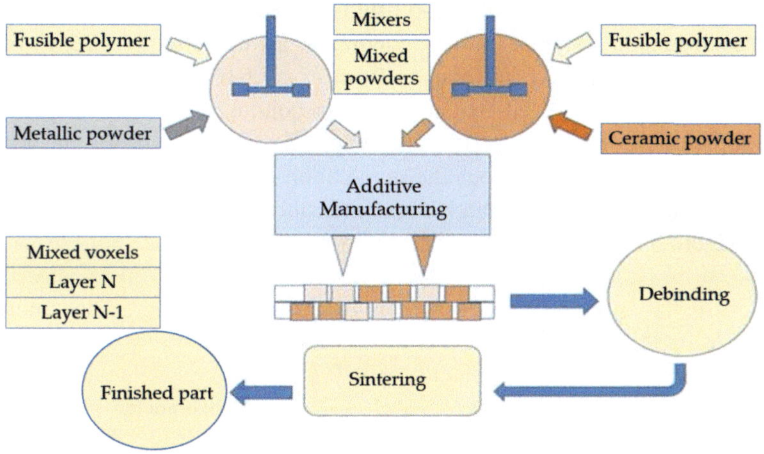

Figure 3.18. *Composite materials and achieving a conductor and insulator blend*

It is also possible to envisage the exploitation of a realization principle of a given object without a fusible polymer present:

Each voxel is produced using a photo-polymerizable resin (stereolithography). To this end, each voxel is produced on the surface of a paste containing the resin, the photochemical initiator and the charge for an induced light conversion. This involves a thermal initiator for an induced polymerization reaction, for example, using an infrared laser, such as with the process using a localized polymer fusion. It is an advantage to use a third paste, which does not contain an initiator. This unconverted paste is separated (recovered) from the composite object after it has been manufactured by 3D printing.

In such a case, we would have a multi-injector system of dry metallic and/or ceramic powders, with an *in situ* laser sintering, such as that undertaken within the CLAD process (direct additive laser manufacturing). "Graded materials with a given function are produced by simultaneous differential injection and controlled by powders, owing to the use of several powder distributors (two or more). It is then possible to create unique alloys, tailored and non-commercialized such as binary alloys or even far more complex high-entropy alloys" (IREPA-LASER – a French industrial research and development company and training center in laser materials – 2017 [IRE 17]).

With polymers such as poly-(bis-alkyl-thio-acetylene), a UV laser irradiation can convert an insulator polymer into a conductor as the slightly dated works of Baumann and Barjon [BAU 96] and de Roth *et al.* [ROT 91] attest. This means might be considered provided that the various conductivities are adapted to the desired application. Hu *et al.* [HU 14] illustrate the possibility of producing insulator/metal composites without the application in electronics being envisaged. As this idea could have the potential for extension to other uses, their process is shown in Figure 3.19.

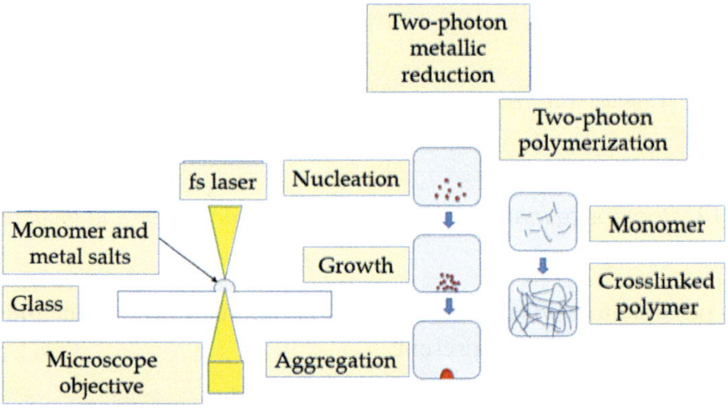

Figure 3.19. *Production of a composite metal polymer by means of a two-photon process*

3.3.2. *Subtractive/additive coupling*

As Figure 3.20 illustrates, there is a possibility of producing complex shapes of which the surface covered by conductive deposits is subject to a suitable laser pulse treatment. This processing produces a localized ablation on the surface of the conductive material. The object may take any shape whatsoever, but it should enable the laser beam to reach its surface.

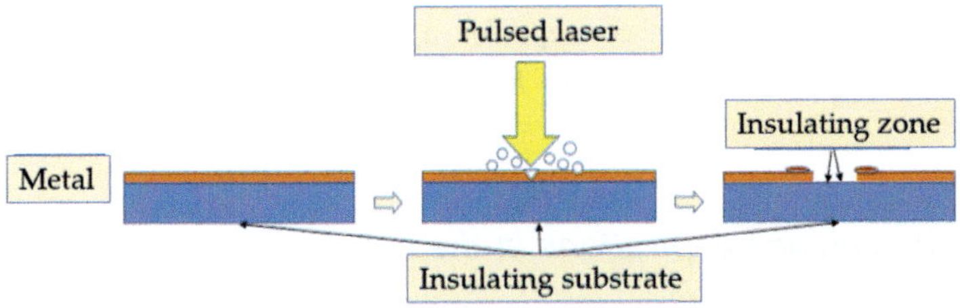

Figure 3.20. *Influence of the laser processing upon the composite surface*

In respect of small-sized parts, such as those used in electronics, it appears useful to contemplate additive and subtractive technologies connected with multi-materials as shown in Figure 3.21, inspired by Ahn *et al.* [AHN 15].

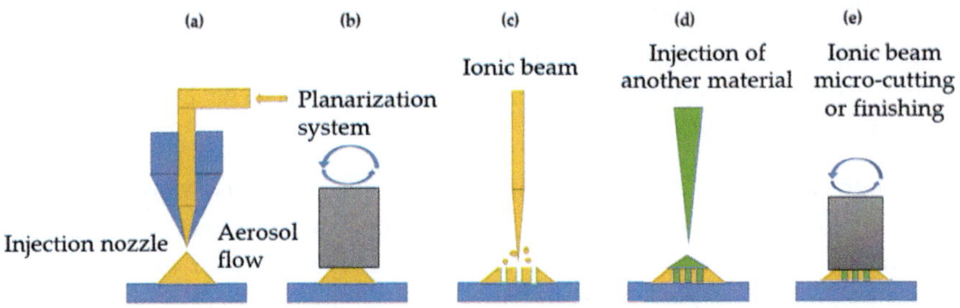

Figure 3.21. *Additive and subtractive technology couplings and materials for electronics [AND 17]*

3.3.3. *μ-Electronics*

The process of 3D printing may provide a noteworthy contribution in the production of complex-shaped supports, as well as printed circuits (the possibility of metallic deposition using special inks, containing, for example, highly concentrated silver) [ESP 14]. With this dynamic, which remains modest, a research effort should emerge to develop 3D collective additive manufacturing technologies. Photonic methods could be advantageously used to reach this objective.

COMMENT.–

In early 2016, Skibba [SKI 16] announced that the University of Stanford in the USA had produced a proof-of-concept for a 3D electronic structure. This used carbon nanotubes and cooling systems to evacuate the calories induced by system operation.

3.3.4. *Conclusion and aspirations in the sphere*

Several observations may be made following this bibliographical analysis (of several dozen publications). These relate, as mentioned, to additive manufacturing connected to the properties of electrical conduction and insulating support, not to mention, of course, μ-electronics:

– first of all, the feasibility is largely proven, even if the manufacturing mechanisms are quite complex;

– the choice of conductive materials is modest;

– the bibliography indicates a fairly modest rate of article production, with however an exhaustive review by Vaezi *et al.* [VAE 13]. The bi-material aspect appears to be a sphere to be promoted;

– although effective 3D productions of various supports were able to be produced using bi-photon processes, to date simultaneous or sequential manufacturing of insulating/conductive couplings does not appear to have been the subject of given publications. This situation necessitating reflection may constitute an original creative axis and a conveyor for additive manufacturing applied to electronics;

– there remains everything or almost everything yet to do in the sphere of microelectronics (and in particular additive manufacturing for semiconductors).

3.4. Actuators and μ-robots

For Remilleret [REM 15], PLEN 2 is a robot that may be printed using additive manufacturing via an open-sourced means for self-assembly. It has an estimated cost of the order of €1,000. "Ultimately, Plen, the Japanese company behind the project, plans for customization to take precedence over original designs and to create a genuine link between human beings and robot humanoids". From this point of view, it is a matter of producing components by 3D printing, without major technological innovation (see also Pimenta [PIM 15]; de Matharel [DE 14]; Atlantic 3D [ATL 15];

Stroud *et al.* [STR 13]; Mehta and Rus [MEH 14]; Yim *et al.* [YIM 07] and Kotay *et al.* [KOT 98]). From a point of view which is close to scientific orthodoxy, the number of publications remains modest, although the possibility of complex parts is to the advantage of 3D technologies. We can cite the works of Megaro *et al.* [MEG 15] from ETH-Zurich in Switzerland using examples, which is exploiting this advantage (see also Cali *et al.* [CAL 12]), for their production of a movable arm. This could be used in the long term for robot manufacturing. In addition, it has parts for artificial hands [3D 15] and other opportunities for exploitation. Indeed, it is far easier to find scientific and technological articles, where robots are developed for additive manufacturing (see, for example, Keating and Oxman [KEA 13]), than publications concerning aspects of μ-actuators and μ-robots!

To try to go beyond what "standard" 3D technologies can provide, the bibliographical approach has sought to ensure the emergence of some ideas or trends, but the so-called "fishing" in this field was meager. Having a significant contribution, 4D printing will make up the second part of Volume 3, constituting a unique way to approach the subject. It can be a question of, as Figure 3.22 shows, simple solutions and traditional actuators produced by additive manufacturing (Peele *et al.* [PEE 15]; see also Walters *et al.* [WAL 10] with a principle of the pneumatic actuator using origamis as Martinez *et al.* describe [MAR 12]). In this figure, the movement to the left or to the right depends on the difference in gas pressure between the two actuator compartments.

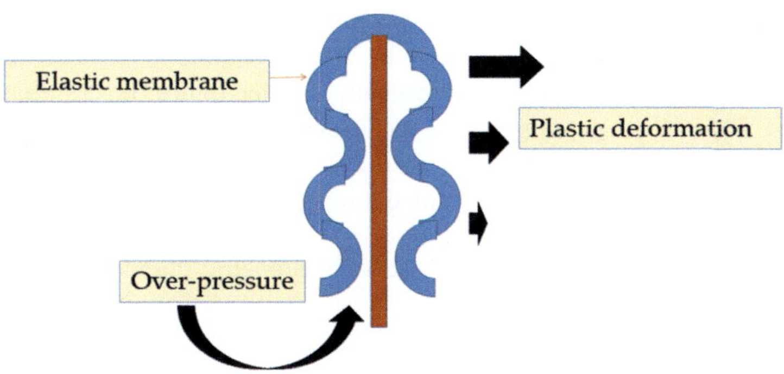

Figure 3.22. *3D pneumatic actuator [AND 17]*

In the same spirit, the works of Hwang *et al.* [HWA 15] lead to usable robotic clamps using an identical operation principle to that used by Peele *et al.* [PEE 15]. The production of micro-actuators is possible through additive manufacturing as the works of Sreetharan *et al.* attest [SRE 12].

COMMENT.– μ-batteries

A technique deriving from holography was used to design a lithium-ion battery with a 4 mm^2 surface and a 10 μm thickness. The battery performance exceeds that of current electrical storage. The integration of such batteries on an electronic microchip is envisaged. Applications might, for example, include portable or implanted medical devices. The process of stereolithography uses interferences, which, within energy zones, lead to a hardening of the given resin enabling deep small-sized engraving. The object is covered with nickel and the polymer is dissolved to manufacture entirely metallic electrodes. "The resulting miniature battery has a power storage capacity of 650 mW. Battery hours are in cubic centimeters with the delivery of a maximum power of 36 watts per centimeter cube" [EXT 15]. Using near-field microscopic technology, it was possible for Létiche *et al.* [LÉT 16] to achieve this same performance by designing a 3D structure based on micro-tubes and with solid electrolytes, but on smaller scales.

3.5. Conclusion

The reduced size aspect may constitute a future means to the extent that other manufacturing techniques cannot provide similar or superior performances. Although for Ghibaudo and He (2016) [GHI 16], electronic printing is likely to reintroduce economic activity in the West, there is still much to invent. This will ensure that additive manufacturing finds a credible position within all of the various technologies within the domain. Although in 2016, according to Combier [COM 17] and Clarivate Analytics [CLA 16], patents in electronics on a global scale represented 29% of all patents within the technology sector, there is no French company in the global Top 100. How then can innovative operations between the academic sector and industries be promoted through French research?

One conceivable possibility is the use of nanometric components to develop materials possessing new properties such as flexible synthetic semiconductors [GAI 16, KWO 16]. Why not then seek to use a breakthrough-based approach to connect 3D printing and organic semiconductors? The problem to deal with will be the evacuation of calories, for which the processing might be conceived by a dispersive structure in respect of the heat involved in the reaction. This particular technology remains yet to be devised.

3.6. Bibliography

[3D 15] 3D PRINT, "3D printed powered exoskeleton", available at: https://www.youtube.com/watch?v=t7T-S3UFR-0, 2015.

[ADA 15] ADAMS J.J., SLIMMER S.C., LEWIS J.A. *et al.*, "3D-printed spherical dipole antenna integrated on small RF node", *Electronics Letters*, vol. 51, pp. 661–662, 2015.

[AHN 14] AHN D., SEO C., PARK T. *et al.*, "Micromachining of stainless steel–polymer composites using nanosecond and femtosecond UV lasers", *The International Journal of Advanced Manufacturing Technology*, vol. 74, pp. 1691–1699, 2014.

[AHN 15] AHN S.H., YOON H.S., JANG K.H. *et al.*, "Nanoscale 3D printing process using aerodynamically focused nanoparticle (AFN) printing, micro-machining and focused ion beam (FIB)", *CIRP Annals – Manufacturing Technology*, vol. 64, pp. 523–536, 2015.

[AKA 03] AKAMATSU K., IKEDA S., NAWAFUNE H., "Site-selective direct silver metallization on surface-modified polyimide layers", *Langmuir*, vol. 19, pp. 10366–10371, 2003.

[AKA 05] AKAMATSU K., KIMURA A., MATSUBARA H. *et al.*, "Site-selective direct photochemical deposition of copper on glass substrates using TiO_2 nanocrystals", *Langmuir*, vol. 21, pp. 8099–8102, 2005.

[AND 17] ANDRÉ J.C., "Impression 3D: niches applicatives porteuses", *Techniques de l'ingénieur*, 2017.

[ARE 02] ARENDT H., *La condition de l'Homme moderne*, Pocket, Paris, 2002.

[ASH 86] ASHKIN A., DZIEDZIC J.M., BJORKHOLM J.E. *et al.*, "Observation of a single-beam gradient force optical trap for dielectric particles", *Optics Letters*, vol. 11, pp. 288–290, 1986.

[ATL 11] ATLAN H., *Le vivant post-génomique ou qu'est-ce que l'auto-organisation*?, Odile Jacob, Paris, 2011.

[ATL 15] ATLANTIC 3D, "Découvrez nos applications", available at: http://www.atlantic-3d.fr/, 2015.

[BAI 16] BAILEY C., STOYANOV S., TILFORD T. *et al.*, "3D-printing and electronic packaging", *Pan Pacific Microelectronics Symposium (Pan Pacific)*, pp. 1–7, 2016.

[BAR 15] BARTON J.H., GARCIA C.R., BERRY E.A. *et al.*, "3D printed all-dielectric frequency selective surface with large bandwith and field of view", *IEEE Transactions on Antennas & Propagation*, vol. 63, pp. 1032–1039, 2015.

[BAU 96] BAUMANN R., BARJON J., "Conducting polymer patterns via laser processing", *Applied Surface Science*, vol. 106, pp. 287–292, 1996.

[BEK 16] BEKAREVICH R., TOYODA M., BABA S. *et al.*, "Refilling of carbon nanotube cartridges for 3D nanomanufacturing", *Nanoscale*, vol. 8, p. 7217, 2016.

[BEN 15] BENSON E., MOHAMMED A., GARDELL J. *et al.*, "DNA rendering of polyhedral meshes at the nanoscale", *Nature*, vol. 523, pp. 441–444, doi: 10.1038/nature 14586 and available at: http://news.cision.com/karolinska-institutet/r/3d--printouts--at-the-nanoscale-using-self-assembling-dna-structures,c9807824, 2015.

[BON 08] BONNEAU O., *Technologie microélectronique – Du silicium aux circuits intégrés*, Ellipses, Paris, 2008.

[BRA 06] BRATTON D., YANG D., DAI J. *et al.*, "Recent progress in high resolution lithography", *Polymers for Advanced Technologies*, vol. 17, pp. 94–103, 2006.

[BRA 16] BRANT A., SUNDARAM M., "A novel electrochemical micro additive manufacturing method of overhanging metal parts without reliance on support structures", *44th Proceedings of the North American Manufacturing Research Institution of SME, Procedia Manufacturing*, vol. 5, pp. 928–943, available at: http://www.sme.org/namrc, 2016.

[BRE 15] BRECKENFELD E., KIM H., AUYEUNG R.C.Y. *et al.*, "Laser-induced forward transfer of silver nanopaste for microwave interconnects", *Applied Surface Science*, vol. 331, pp. 254–261, 2015.

[BRE 16] BREGERAS G., "Poly-Ink, la pépite française qui dope l'électronique imprimée", available at: http://www.lesechos.fr/idees-debats/sciences-prospective/0211500398098-poly-ink-la-pepite-francaise-qui-dope-lelectronique-imprimee-2043389.php?z6PWSM3W4iv59 PYw.99, 2016.

[CAI 16] CAI F., CHANG Y.H., WANG K. *et al.*, "Low-loss 3d multilayer transmission lines and interconnects fabricated by additive manufacturing technologies", *IEEE Transactions on Microwave Theory and Techniques*, vol. 64, pp. 3208–3216, 2016.

[CAL 12] CALI J., CALIAN D.A., KLEINBERGER R. *et al.* "3D-printing of non-assembly, articulated models", available at: http://citeseerx.ist.psu.edu/viewdoc/download?doi=10.1.1.363.9695&rep=rep1&type=pdf, 2012.

[CAM 01] CAMPBELL S.A., *The Science and Engineering of Microelectronic Fabrication (The Oxford Series in Electrical and Computer Engineering)*, Oxford University Press, Oxford, 2001.

[CAM 12] CAMPBELL S.A., *Fabrication Engineering at the Micro- and Nanoscale (The Oxford Series in Electrical and Computer Engineering)*, Oxford University Press, Oxford, 2012.

[CAM 11] CAMPBELL T., WILLIAMS C., IVANOVA O. *et al.*, Could 3D Printing Change the World? Technologies, Potential, and Implications of Additive Manufacturing, strategic Foresight Report, Atlantic Council, Washington, available at: http://www.cbpp.uaa.alaska.edu/afef/Additive%20MFG%20.pdf, 2011.

[CHA 12] CHALINE J., "La relativité d'échelle dans la morphogenèse du vivant: fractal, déterminisme et hasard", *Annales de Chirurgie Plastique Esthétique*, vol. 5, pp. 517–526, 2012.

[CHI 11] CHIU C.Y., LI, Y., RUAN L. *et al.*, "Platinum nanocrystals selectively shaped using facet-specific peptide sequences", *Nature Chemistry*, vol. 3, pp. 393–399, 2011.

[CHU 14] CHU W.S., KIM C.S., LEE Y.T. *et al.* "Hybrid manufacturing in micro/nano-scale: a review", *International Journal of Precision Engineering and Manufacturing-Green Technology*, vol. 1, pp. 75–92, 2014.

[CLA 16] CLARIVATE ANALYTICS, "Top 100 Global Innovators Report", available at: http://top100innovators.stateofinnovation.com/sites/default/files/content/top100/L178_Cvt_Top_100_Innovators_Report_008.pdf, 2016.

[CLA 17] CLARKE C., "Chinese research into liquid metal and 3D printing for functional electronics", available at: https://3dprintingindustry.com/news/chinese-research-liquid-metal-3d-printing-functional-electronics-107340/?goal=0_695d5c73dc-6e9c964dfb-64557777, 2017.

[COM 17] COMBIER E., "La France solide troisième au Top 100 de l'innovation", available at: http://www.lesechos.fr/idees-debats/sciences-prospective/0211680742895-la-france-solide-troisieme-au-top-100-de-linnovation-2056316.php, 2017.

[DE 14] DE MATHAREL L., "Poppy, le robot open source fabriqué avec une imprimante 3D", available at: http://www.usine-digitale.fr/article/poppy-le-robot-open-source-fabrique-avec-une-imprimante-3d.N249523, 2014.

[DEL 16] DELAPORTE P., ALLONCLE A.P., "Laser-induced forward transfer: a high resolution additive manufacturing technology", *Optics & Laser Technology*, vol. 78, pp. 33–41, 2016.

[DEL 15] DELLAPORTA H., BEDU F., "Construction d'objets à l'échelle submicronique", available at: http://liris.cnrs.fr/~cnriut08/actes/articles/112.pdf, 2015.

[DEN 16] DENG J., ZHANG L., DONG J. *et al.*, "AFM-based 3D nanofabrication using ultrasonic vibration assisted nanomachining", *Journal of Manufacturing Processes*, vol. 24, pp. 195–202, 2016.

[DHO 06] DHOLAKIA K., REECE P., "Optical micromanipulation takes hold", *Nano Today*, vol. 1, pp. 18–27, 2006.

[ENG 14] ENGSTROM D.S., PORTER B., PACIOS M. *et al.*, "Additive nano-manufacturing – a review", *Journal of Materials Research*, vol. 29, pp. 1792–1816, 2014.

[ESP 14] ESPALIN D., MUSE D.W., MACDONALD E. *et al.*, "3D printing multi-functionality: structure with electronics", *The International Journal of Advanced Manufacturing Technology*, vol. 72, pp. 963–978, 2014.

[EST 16] ESTEBAN-MUNIZ G., Additive manufacturing and 3D printing technologies in the EC, Workshop, Fabrication additive pour l'électronique, CNRS, Paris, France, 11 January 2016.

[EXT 15] EXTREMTECH, "This holographic micro-battery is just 10 μm thick", available at: http://www.extremetech.com/extreme/205764-this-holographic-microbattery-is-just-10-micrometers-thick, 2015.

[FAN 15] FAN Z., TAO X., FAN X. *et al.*, "Nanotube fountain pen: towards 3D manufacturing of metallic nanostructures", *Carbon*, vol. 86, pp. 280–287, 2015.

[FER 11] FERONE G., VINCENT J.A., *Bienvenue en transhumanie; sur l'Homme de demain*, Grasset, Paris, 2011.

[FER 16] FERRERO F., Fabrication d'antennes 3D miniatures par injection de métal liquide dans une structure réalisée par une imprimante 3D, Workshop, Fabrication additive pour l'électronique, CNRS, Paris, France, 11 January 2016.

[FEY 99] FEYERABEND P., *Une connaissance sans fondements*, Danoïa, Paris, 1999.

[FIS 12] FISCHER A.C., LYUBOV M.B., RIKERS Y.G.M. *et al.*, "3D free-form patterning of silicon by ion implantation, silicon deposition and selective silicon etching", *Advanced Functional Materials*, doi: 10.1002/adfm.201200845, 2012.

[FLO 15] FLORIAN C., CABALLERO-LUCAS F., FERNÁNDEZ-PRADAS J.M. *et al.*, "Conductive silver ink printing through the laser-induced forward transfer technique", *Applied Surface Science*, vol. 336, pp. 304–308, 2015.

[FOW 16] FOWLKES J.D., WINKLER R., LEWIS B.B. *et al.*, "Simulation-guided 3D nanomanufacturing via focused electron beam induced deposition", *ACS Nano*, vol. 10, pp. 6163–6172, 2016.

[FOX 09] FOX-KELLER E., "Organisms, machines, and thunderstorms, a history of self-organization", *Historical Studies in the Natural Sciences*, vol. 39, no. 1, pp. 1–31, 2009.

[FUE 12] FUECHSLE M., MIWA J.A., MAHAPATRA S. *et al.*, "A single-atom", *Nature Nanotechnology*, vol. 7, pp. 242–246, 2012.

[GAI 16] GAIGNAIRE A., "Ingénieurs du futur recherchés", available at: https://www.lesaffaires.com/dossier/profession-ingenieur/ingenieurs-du-futur-recherches/591065, 2016.

[GHI 16] GHIBAUDO T., HE R., "Impression 3D & électronique imprimée associées", available at: http://cerig.pagora.grenoble-inp.fr/memoire/2016/impression3D-electronique-imprimee.htm, 2016.

[GRA 16] GRAPHENE LAB. INC., "3D printed nano-composites", available at: http://www.graphene3dlab.com/s/home.asp, 2016.

[GUI 15] GUITARD A., "Nouvelle technique pour former des nanostructures 3D d'ADN", available at: http://www.diplomatie.gouv.fr/fr/politique-etrangere-de-la-france/diplomatie-scientifique/veille-scientifique-et-technologique/suede/article/nouvelle-technique-pour-former-des-nanostructures-3d-d-adn, 2015.

[GUL 16] GULATI K., KOGAW M., PRIDEAUX M. *et al.*, "Drug-releasing nano-engineered titanium implants: therapeutic efficacy in 3D cell culture model, controlled release and stability", *Materials Science and Engineering C*, vol. 69, pp. 831–840, 2016.

[HAN 11] HAN D., PAL S., NANGREAVE J. *et al.*, "DNA origami with complex curvatures in three-dimensional space", *Science,* vol. 332, pp. 342–346, 2011.

[HAN 15] HAN Y., WEI C., DONG J., "Droplet formation and settlement of phase-change ink in high resolution electro-hydrodynamic (EHD) 3D printing", *Journal of Manufacturing Processes*, vol. 20, pp. 485–491, 2015.

[HOG 17] HOGBERG B., "DNA Origami", Photograph Kindly Transmitted, Karolinska Institute, Stockholm, Sweden, 2017.

[HOL 16] HOLLINGSHEAD T., "Tiny origami-inspired devices opening up new possibilities for minimally-invasive surgery", available at: https://news.byu.edu/news/tiny-origami-inspired-devices-opening-new-possibilities-minimally-invasive-surgery#sthash.2aylPHcl.dpuf, 2016.

[HU 14] HU Q., LIU Y., HE Y. *et al.*, "Fabrication of 3d polymer-metal nano-composites In a single step by two-photon induced polymerisation and metal salt reduction", available at: http://sffsymposium.engr.utexas.edu/sites/default/files/2015/2015-85-Hu.pdf, 2014.

[HUT 12] HUTH M., PORRATI F., SCHWALB C. *et al.*, "Focused electron beam induced deposition: a perspective", *Beilstein Journal of Nanotechnology*, vol. 3, pp. 597–619, 2012.

[HWA 15] HWANG Y., PAYDAR O.H., CANDLER R.N., "Pneumatic microfinger with balloon fins for linear motion using 3D printed molds", *Sensors & Actuators*, vol. 234, pp. 65–71, 2015.

[IFT 15] IFTEKHARUL-HAQUE R., OGAM E., LOUSSERT C. *et al.*, "Fabrication of capacitive acoustic resonators combining 3D printing and 2D inkjet printing techniques", *Sensors*, vol. 15, pp. 26018–26038, 2015.

[IKE 01] IKEDA S., AKAMATSU K., NAWAFUNE H., "Direct photochemical formation of Cu patterns on surface modified polyimide resin", *Journal of Materials Chemistry*, vol. 11, pp. 2919–2921, 2001.

[ILI 02] ILIC B., CZAPLEWSKI D., ZALALUTDINOV M. *et al.*, "Fabrication of flexible polymer tubes for micro and nano-fluidic applications", *Journal of Vacuum Science & Technology B*, vol. 20, pp. 2459–2465, 2002.

[INM 16] INM, "Hannover Messe: Photochemical metallization allows the manufacture of touchscreens in a single step", available at: http://www.leibniz-inm.de/en/2016/04/hannover-messe-photochemical-metallization-allows-the-manufacture-of-touchscreens-in-a-single-step/, 2016.

[IRE 17] IREPA-LASER, "Procédé CLAD", available at: http://www.irepa-laser.com/fr/fabrication-additive/fabrication, 2017.

[JAE 01] JAEGER R.C., *Introduction to Microelectronic Fabrication: Volume 5 of Modular Series on Solid State Devices*, Pearson Higher Ed, New York, 2001.

[JE 17] JE J.H., KIM J.M., JAWORSKI J., "Progression in the fountain pen approach: from 2D writing to 3D free-form micro/nanofabrication", *Small*, vol. 13, 1600137, 2017.

[JES 16] JESSE S., BORISEVICH A.Y., FOWLKES J.D. *et al.*, "Directing matter: toward atomic-scale 3D nano-fabrication", *ACS Nano*, vol. 10, pp. 5600–5618, 2016.

[JOH 07] JOHANDER P., HAASL S., PERSSON K. *et al.*, "Layer manufacturing as a generic tool for microsystem integration", *4M2007 Conference Proceedings*, Borovets, Bulgaria, 2007.

[JUO 08] JUODKAZIS S., MIZEIKIS V., MATSUO S. *et al.*, "Three-dimensional micro- and nano-structuring of materials by tightly focused laser radiation", *Bulletin of the Chemical Society of Japan*, vol. 81, p. 411, 2008.

[KAS 02] KASAP F., *Principles of Electronic Materials and Devices*, McGraw Hill, New York, 2002.

[KAU 15] KAUR A., MYERS J.C., GHAZALI M.I. *et al.*, "Affordable terahertz components using 3D printing", *65th IEEE Electronic Components Technology Conference*, pp. 2071–2076, 2015.

[KEA 13] KEATING S., OXMAN N., "Compounds fabrication: a multi-functional robotic platform for digital design and fabrication", *Robotics & Computer Integrated Manufacturing*, vol. 29, pp. 439–448, 2013.

[KEL 12] KELLERMEIER M., CÖLFEN H., GARCÍA-RUIZ J.M., "Silica biomorphs: complex biomimetic hybrid materials from "Sand and Chalk", *European Journal of Inorganic Chemistry*, vol. 32, pp. 5123–5144, 2012.

[KON 16] KONG Y.L., GUPTA M.K., JOHNSON B.N. *et al.*, "3D printed bionic nano-devices", *Nano-today*, vol. 11, pp. 330–350, 2016.

[KOT 98] KOTAY K., RUS D., VANA M. *et al.*, "The self-reconfiguring robotic molecules: design and control algorithm", available at: http://citeseerx.ist.psu.edu/viewdoc/download?doi=10.1.1.42.6594&rep=rep1&type=pdf, 1998.

[KRI 17] KRIVEC M., ROSHANGHIAS A., ABRAM A. *et al.*, "Exploiting the combination of 3D polymer printing and inkjet Ag-nanoparticle printing for advanced packaging", *Microelectronic Engineering*, vol. 176, pp. 1–5, 2017.

[KUZ 14] KUZUYA A., WATANABE R., HASHIZUME M. *et al.*, "Precise structure control of three state nanomechanical DNA origami devices", *Methods*, vol. 67, pp. 250–255, 2014.

[KWO 16] KWON J., TAKEDA Y., FUKUDA K. *et al.*, "Three-dimensional, inkjet-printed organic transistors and integrated circuits with 100% yield, high uniformity, and long-term stability", *ACS Nano*, vol. 10, pp. 10324–10330, 2016.

[LAN 15] LANDIS S., "Le CEA-LETI prépare l'industrialisation de la lithographie par impression", available at: http://www.usine-digitale.fr/editorial/le-cea-leti-prepare-l-industrialisation-de-la-lithographie-par-impression.N337639, 2015.

[LAS 15] LASZCZAK P., TRANG L., BADER D.L. *et al.*, "Development and validation of a 3D printed interfacial stress sensor for prosthetic applications", *Medical Engineering & Physics*, vol. 37, pp. 132–137, 2015.

[LAX 12] LAXMINARAYANA G.K., Approches colloïdale et bio-inspirée en nano-plasmonique, Thesis, Paul Sabatier University, Toulouse, France, available at: http://thesesups.ups-tlse.fr/1723/1/2012TOU30103.pdf, 2012.

[LE 15] LE T., SONG B., BAHR R.A. *et al.*, "A novel strain sensor based on 3D printing technology and 3D antenna design", *IEEE 65th Electronic Components & Technology Conference (ECTC)*, pp. 981–986, 2015.

[LEE 15] LEE J.W. "3D nano-printing technologies for tissue engineering applications", *Journal of Nanomaterials*, vol. 2015, Article ID 213521, p. 14, available at: http://dx.doi.org/10.1155/2015/213521, 2015.

[LEE 14] LEE M., KIM D.Y., "Toward nanoscale three-dimensional printing: nano-walls built of electro-spun nanofibers", *Langmuir*, vol. 30, pp. 1210–1214, 2014.

[LEH 11] LEHN J. M., "Par-delà la synthèse: l'auto-organisation", *Comptes Rendus Chimie*, vol. 14, pp. 348–361, 2011.

[LEI 11] LEIGH S.J., PURSELL C.P., BOWEN J. *et al.*, "A miniature flow sensor fabricated by micro-stereolithography employing a magnetite/acrylic nanocomposite resin", *Sensors and Actuators*, vol. 168A, pp. 66–71, 2011.

[LEI 12] LEIGH S.J., BRADLEY R.J., PURSSELL C.P. *et al.*, "A simple, low cost conductive material for 3D printing of electronic sensors", *PLoS ONE*, vol. 7, e49365, doi: 10.1371/journal.pone.0049365, 2012.

[LÉT 16] LÉTICHE M., EUSTACHE E., FREIXAS J. *et al.*, "Atomic layer deposition of functional layers for on Chip 3D Li-ion all solid state microbattery", *Advanced Energy Materials*, vol. 2016, 1601402, doi: 10.1002/aenm.201601402, 2016.

[LEW 15] LEWIS J.A., AHN B.Y., "Three dimensional printed electronics", *Nature*, vol. 518, pp. 42–43, 2015.

[LI 14] LI L., ABEDINI-NASSAB R., YELLEN B.B., "Monolithically integrated Helmoltz coils by 3D printing", *Applied Physics Letters*, vol. 104, 253505, 2014.

[LIA 11] LIA L., HONG M., SCHMIDT M. *et al.*, "Laser nano-manufacturing – State of the art and challenges", *CIRP Annals – Manufacturing Technology*, vol. 60, pp. 735–755, 2011.

[LLE 16] LLEWELLYN-JONES T.M., DRINKWATER B.W., RASK R.S., "3D printed components with ultrasonically arranged microscale structures", *Smart Materials and Structures*, vol. 25, 02LT01, doi: http://dx.doi.org/10.1088/0964-1726/25/2/02LT01, 2016.

[LON 16] LONJON C., "3D printing for electronics: what's the next revolution?", available at: https://www.sculpteo.com/blog/2016/12/07/3d-printing-for-electronics-industry-whats-the-next-revolution/?utm_medium=email&utm_campaign=NL%20US%2071216&utm_content=NL%20US%2071216+CID_d914f8c6597202306458e9bfedf465f6&utm_source=Campaign%20Newsletter&utm_term=Learn%20More, 2016.

[LUC 16] LUCAS T., "Trois technos pour imprimer en 3D des métaux sans poudre", available at: http://www.usinenouvelle.com/editorial/trois-technos-pour-imprimer-en-3d-des-metaux-sans-poudre.N465383, 2016.

[MAI 11] MAICU M., HIDALGO M.C., COLON G. *et al.*, "Comparative study of the photo-deposition of Pt, Au and Pd on pre-sulphated TiO_2 for the photocatalytic decomposition of phenol", *Journal of Photochemistry and Photobiology*, vol. 217A, pp. 275–283, 2011.

[MAK 16] MAKERY, "L'histoire accidentée de l'impression 3D", available at: http://www.makery.info/2016/05/03/lhistoire-accidentee-de-limpression-3d/, 2016.

[MAR 12] MARTINEZ, R.V., FISH C.R., CHEN X. *et al.*, "Elastomeric origami: programmable paper-elastomer composites as pneumatic actuators", *Advanced Functional Materials*, vol. 22, pp. 1376–1384, 2012.

[MAS 16] MASHAYEKHI M., WINCHESTER L., EVANS L. *et al.*, "Evaluation of aerosol, superfine inkjet, and photolithography printing techniques for metallization of application specific printed electronic circuits", *IEEE Transactions on Electron Devices*, vol. 63, pp. 1246–1253, 2016.

[MEG 15] MEGARO V., THOMASZEWSKI B., NITTI B. *et al.*, "Interactive design of 3D-printable robotic creatures", *ACM Transactions on Graphics (TOC) – Proceedings of the ACM-SIGGRAPH Asia*, vol. 34, pp. 216:1–216:9, 2015.

[MEH 14] MEHTA A.M., RUS D., "An end-to-end system for designing mechanical structures for print-and-fold robots", *2014 IEEE International Conference on Robotics and Automation (ICRA)*, doi: 10.1109/ICRA.2014.6907044, 2014.

[MÉL 17] MÉLANIE, "Nano Dimension: la fabrication additive au service de l'électronique imprimée", available at: http://www.3dnatives.com/nano-dimension-electronique-28032017/, 2017.

[MOL 15] MOLITCH-HOU M., "CEO Jennifer Lewis on the Future of Electronics 3D Printing & Voxel8's Huge $12M Funding", available at: http://3dprintingindustry.com/2015/07/24/voxel8-ceo-jennifer-lewis-on-how-12m-in-funding-will-fuel-the-future-of-electronics-3d-printing/, 2015.

[MOR 04] MORENO A., "Auto-organisation, autonomie et identité", *Revue internationale de philosophie*, vol. 2, pp. 135–150, 2004.

[MOS 10] MOSADEGH B., Design and fabrication of microfluidic integrated circuits using normally-closed elastomer valves, PhD Thesis, University of Michigan, Detroit, 2010.

[MOS 15] MOSCATO S., BAHR R., LE T. *et al.*, "Additive manufacturing of 3D substrate integrated waveguide components", *Electronics Letters*, vol. 51, pp. 1426–1427, 2015.

[NAY 14] NAYERI P., LIANG M., SABORY-GARCIA R.A. *et al.*, "3D printed dielectric reflectarrays: low-cost high-gain antennas at sub-millimeter waves", *IEEE Transactions on Antennas and Propagation*, vol. 62, pp. 2000–2008, 2014.

[NIK 10] NIKITIN M.P., ZDOBNOVA T.A., LUKASH S.V. *et al.*, "Protein-assisted self-assembly of multifunctional nanoparticles", *Proceedings of the National Academy of Sciences of the United States of America*, vol. 107, pp. 5827–5832, 2010.

[NIK 12] NIKLAUS F., GYLFASON K.B., FISCHER A.C. *et al.*, "Cutting the cost of micro and nano manufacturing by 3D printing of silicon", available at: http://www.micronanomems.com/filestore/pdf/M&M%20KTH.pdf, 2012.

[NIK 14] NIKLAUS F., "The next generation of 3D printers", available at: https://erc.europa.eu/erc-stories/next-generation-3d-printers, 2014.

[NOT 16] NOTTEAU Y., "Tout ce qu'il faut savoir sur les imprimantes 3D métal !", available at: http://www.additiverse.com/tout-ce-quil-faut-savoir-sur-les-imprimantes-3d-metal/, 2016.

[NYK 08] NYKYPANCHUK D., MAYE M.M., VAN DER LELIE D. *et al.*, "DNA-guided crystallization of colloidal nanoparticles", *Nature*, vol. 451, pp. 549–552, 2008.

[ONS 15] ONSES M.S., SUTANTO E., FERREIRA P.M. *et al.*, "Printing mechanisms, capabilities, and applications of high-resolution electro-hydrodynamic jet printing", *Small*, vol. 11, pp. 4237–4266, 2015.

[OPT 15] OPTOMEC, "Production grade 3D printers", available at: http://www.optomec.com/, 2015.

[PEE 15] PEELE B.N., WALLIN T.J., ZHAO H. *et al.*, "3D printing antagonistic system of artificial muscle using projection stereolithography", *Bioinspiration & Biomimetics*, vol. 10, 055003, 2015.

[PIM 15] PIMENTA J., "InMoov, le premier robot humanoïde que vous pourrez imprimer en 3D directement depuis chez vous", available at: http://soocurious.com/fr/inmoov-le-premier-robot-humanoide-que-vous-pourrez-imprimer-en-3d-directement-depuis-chez-vous/, 2015.

[PIN 15] PINA S., OLIVERA J.M., REIS R.I., "Natural-based nanocomposites for bone tissue engineering and regenerative medicine: a review", *Advanced Materials*, vol. 27, pp. 1143–1169, 2015.

[PIQ 15] PIQUÉ A, KIM H.A., RAYMOND C.Y. *et al.*, "Laser induced forward transfer (LIFT) of congruent voxels", *Applied Surface Science*, doi: 10.1016/j.apsusc.2015.09.005, 2015.

[POU 06] POUTEAU S., "Biotechnologies: la vie en morceaux et l'éthique en péril", *Journal International de Bioéthique*, vol. 17, pp. 15–27, 2006.

[PRE 14] PRETZ K., "Ten Technologies That Could Change the World By 2022", available at: http://theinstitute.ieee.org/ieee-roundup/members/achievements/10-technologies-that-could-change-the-world-by-2022, 2014.

[PV 16] PV NANOCELL, "The clean fuel behind digital printing", available at: http://www.pvnanocell.com/, 2016.

[QUÉ 16] QUÉRÉ Y., LAUR V., MAALOUF A. *et al.*, Potentiel des technologies additives pour la conception ou le prototypage d'antennes et de composants, Workshop, Fabrication additive pour l'électronique, CNRS, Paris, France, 11 January 2016.

[RAN 06] RANDOLPH S.J., FOWLKES J.D., RACK P.D., "Focused, nanoscale electron-beam-induced deposition and etching", *Critical Reviews in Solid State and Materials Sciences*, vol. 31, pp. 55–89, 2006.

[REM 15] REMILLERET T., "PLEN2: Le robot open-source qui naît grâce à l'impression 3D", available at: http://www.objetconnecte.net/plen2-robot-open-source-impression-3d/, 2015.

[ROS 14] ROSS C.A., BERGGEN K.K., CHENG J.Y. *et al.*, "Three-dimensional nanofabrication by block copolymer self-assembly", *Advanced Materials*, vol. 26, pp. 4386–4396, 2014.

[ROT 91] ROTH H.K., BAUMANN R., SCHRODNER H. *et al.*, "Laser induced electrical conductivity in poly(bis-alkylthioacetylene) and its applications", *Synthetic Metals*, vol. 41, pp. 141–144, 1991.

[ROT 06] ROTHEMUND P.W.K., "Folding DNA to create nanoscale shapes and patterns", *Nature*, vol. 440, pp. 297–302, 2006.

[ROV 16] ROVELO P., "Amorphous metals now 3D-printed for the first time", available at: https://3dprintingindustry.com/news/amorphous-metals-now-3d-printed-first-time-82555/, 2016.

[SAN 12/13] SANCHEZ C., "La chimie mésoscopique : du nanomatériau au mésocristal", available at: https://www.college-de-france.fr/site/clement-sanchez/course-2012-2013.htm, 2012–2013.

[SAW 11] SAWYER E., "DNA origami", available at: http://www.nature.com/scitable/blog/bio2.0/dna_origami, 2011.

[SCH 08] SCHIFT H., "Nanoimprint lithography: an old story in modern times? A review", *Journal of Vacuum Science & Technology B*, vol. 26, pp. 458–480, 2008.

[SEE 04] SEEMAN N., "Nano-construction en ADN", *Pour la Science*, vol. 321, pp. 46–53, 2004.

[SEO 15] SEOL S.K., KIM D., LEE S. *et al.*, "Electrodeposition-based 3D printing of metallic microarchitectures with controlled internal structures", *Small*, vol. 32, pp. 3896–3902, 2015.

[SHE 85] SHELDRAKE R., *Une nouvelle science de la vie*, Ed. du Rocher, Monaco, available at: http://www.astrologie-quantique.info/Rupert-Sheldrake_Une-Nouvelle-Science-de-la-Vie.pdf, 1985.

[SIL 14] SILBERZAHN P., "200 ans après, le malthusianisme a toujours tort: Vers une société d'abondance durable", available at: https://www.contrepoints.org/2014/09/23/182060-200-ans-apres-le-malthusianisme-a-toujours-tort-vers-une-societe-dabondance-durable, 2014.

[SIM 04] SIMON H.A., *Les sciences de l'artificiel*, Folio-Essais, Paris, 2004.

[SKI 16] SKIBBA R., "Stanford-led skyscraper-style chip design boosts electronic performance by factor of a thousand", available at: http://news.stanford.edu/news/2015/december/n3xt-computing-structure-120915.html, 2016.

[SLO 10] SLOCIK J.M., NAIK, R.R., "Probing peptide–nanomaterial interactions", *Chemical Society Reviews*, vol. 39, pp. 3454–3463, 2010.

[SRE 12] SREETHARAN P.S., WHITNEY J.P., STRAUSS M.D. *et al.*, "Monolithic fabrication of millimeter scale machines", *Journal of Micromechanics and Microengineering*, vol. 22, 0550267, 2012.

[SRE 12] SREETHARAN P.S., WHITNEY J.P., STRAUSS M.D. *et al.*, "Monolithic fabrication of millimeter scale machines", *Journal of Micromechanics and Microengineering*, vol. 22, 0550267, 2012.

[STR 13] STROUD A.B., MORRIS M., CAREY K. *et al.*, "MU-L8: the design architecture and 3D printing on a teen-sized humanoid soccer robot", *Proceedings 8th Workshop on Humanoids Soccer Robots, 13th IEEE-RAS International Conference on Humanoids Robots*, p. 5, available at: http://www.eng.mu.edu/abwilliams/heirlab/papers/HSR13_Williams.pdf, 2013.

[SUS 11] SUSSAN R., "L'origami d'AND assisté par ordinateur: vers une nanotechnologie de garage", available at: http://www.internetactu.net/2011/05/10/lorigami-adn-assiste-par-ordinateur-vers-une-nanotechnologie-de-garage/, 2011.

[SUS 14] SUSSAN R., "Retour sur la matière programmable", available at: http://www.internetactu.net/2014/07/11/retour-sur-la-matiere-programmable/, 2014.

[TEO 05] TEO W.E., KOTAKI M., MO X.M. *et al.*, "Porous tubular structures with controlled fiber orientation using a modified electrospinning method", *Nanotechnology*, vol. 16, pp. 918–924, 2005.

[TSE 05] TSENG A.A., NOTARGIACOMO A., CHEN T., "Nanofabrication by scanning probe microscope lithography: a review", *Journal of Vacuum Science & Technology B, Nanotechnology and Microelectronics: Materials, Processing, Measurement, and Phenomena*, vol. 23, pp. 877–894, 2005.

[TSE 14] TSE L., BARTON K., "A field shaping print-head for high-resolution electro-hydrodynamic jet printing onto non-conductive and uneven surfaces", *Applied Physics Letters*, vol. 104, 143510, 2014.

[UTK 08] UTKE I., HOFFMANN P., MELNGAILIS J., "Gas-assisted focused electron beam and ion beam processing and fabrication", *Journal of Vacuum Science & Technology B*, vol. 26, pp. 1197–1276, 2008.

[VAE 13] VAEZI M., SEITZ H., YANG S., "A review on 3D micro-additive manufacturing technologies", *The International Journal of Advanced Manufacturing Technology*, vol. 67, pp. 1721–1754, 2013.

[VEN 16] VENEZIANO R., RATANALERT S., ZHANG K. *et al.*, "Designer nanoscale DNA assemblies programmed from the top down", *Science,* doi: 10.1126/science.aaf4388, 2016.

[VIC 16] VICTOR, "Un procédé de jet d'encre métallique dévoilé lors du salon Formnext", available at: http://www.3dnatives.com/xjet-impression-3d-07112016/, 2016.

[WAN 14] WANG L., LIU J., "Compatible hybrid 3D printing and nonmetal inks for direct manufacture of end functional devices", *Science China Technological Sciences*, vol. 57, pp. 2089–2091, 2014.

[WAL 10] WALTERS P., ROSSITER J., MCGORAN J., "Smart materials and novel actuators: creative applications in art and design", available at: http://www.uwe.ac.uk/sca/research/cfpr/research/3D/research_projects/smart_materials_novel_actuators.html, 2010.

[WIK 15] WIKIPÉDIA, "Origami ADN", available at: https://fr.wikipedia.org/wiki/Origami_ADN, 2015.

[WIK 17] WIKIPÉDIA, "Loi de Moore", available at: https://fr.wikipedia.org/wiki/Loi_de_Moore, 2017.

[WIL 08] WILSON J.B.K., HEGG M., MIAO M. *et al.*, "Scalable nano-particle assembly by efficient light induced concentration and fusion", *Optics Express*, vol. 16, pp. 17276–17281, 2008.

[WIL 02] WILSON M., KANNANGARA K., SMITH G. *et al.*, *Nanotechnology: Basic Science and Emerging Technologies*, CRC Press, New York, 2002.

[WOO 15] WOODWARD D.I., PURSELL C.P., BILLSON D.R. *et al.*, "Additively-manufactured piezoelectric devices", *Physica Status Solidi A*, vol. 212, pp. 2107–2113, 2015.

[WU 15a] WU F., VAN SCHIE B.G.C., KEYMER J.E. *et al.*, "Symmetry and scale main protein patterns in shaped bacterial sculptures", *Nature Nanotech*, vol. 10, pp. 719–726, 2015.

[WU 15b] WU S.Y., YANG C., HSU W. *et al.*, "3D printed microelectronics for integrated circuitry and passive wireless sensors", *Microsystems and Nanotechnology*, vol. 1, 15013, 2015.

[XIO 12] XIONG W., SHEN ZHOU Y., NAN HE X. *et al.*, "Simultaneous additive and subtractive three-dimensional nanofabrication using integrated two-photon polymerization and multiphoton ablation", *Light: Science & Applications*, vol. 1, e6, doi:10.1038/lsa.2012.6, 2012.

[YIM 07] YIM M., SHEN W.M., SALEMI B. *et al.*, "Modular self-reconfigurable robot systems: challenges and opportunities for the future", *IEEE Robotics Automation Mag*, vol. 14, pp. 43–52, 2007.

[YU 15] YU C.C., CHEN H.L., "Nanoimprint technology for patterning functional materials and its applications", *Microelectronic Engineering*, vol. 132, pp. 98–119, 2015.

[YUD 15] YUDISTIRA H.T., TENGGARA A.P., OH S.S. *et al.*, "High-resolution electrohydrodynamic jet printing for the direct fabrication of 3D multilayer terahertz metamaterial of high refractive index", *Journal of Micromechanics and Microengineering*, vol. 25, 045106, 2015.

[ZHA 10] ZHANG Y.L., CHEN Q.D., XIA H. *et al.*, "Designable 3D nanofabrication by femtosecond laser direct writing", *Nano Today*, vol. 5, pp. 435–448, 2010.

[ZHA 14] ZHANG Z., YUAN L., LEE P.D. *et al.*, "Modeling of time dependent localized flow shear stress and its impact on cellular growth within additive manufactured titanium implants", *Journal of Biomedical Materials Research*, vol. 102B, pp. 1689–1699, 2014.

[ZHA 16] ZHAO J., SWARTZ L.A., LIN W.F. *et al.*, "3D nanoprinting via scanning probe lithography delivered layer-by-layser deposition", *ACS Nano*, doi: 10.1021/acsnano.6b01145, 2016.

PART 3

How Should We Go That One Step Further?

4

A Short Reflection on Spheres to Explore Their Conditions for Achieving Success

"Everything that the most subtle and ingenious of men can do to imitate or aid Nature, which is accustomed to be understood as Magic, to the extent that we have discovered various resources and means that they practice to deal with these extraordinary processes. What we can particularly note amongst us with the invention of Canons and Printing, and upon the discovery of this "new world", the people of which primarily believe that our ships were made by Magic, our archways by enchantment, and that the Spaniards were devils who came to destroy them with the thunder and lightning of their arquebuses and pistols" (Naudé [NAU 25] originally written in ancient French).

"With the advent of artificial intelligence, the machine is not me, I am not the machine, and the machine may cogently argue thus: neither am I it. We stand together. [The computer] integrates a form of action which not only ensures our own behavior but induces new behaviors" (Berger [BER 94]).

"Theory, is when we know everything and yet nothing works. Practice is when everything functions and nobody knows why. Here, we have brought together theory and practice, and – guess what – nobody knows why!" (Einstein [EIN 07]).

"With a civilization saturated by technology, there is no adventure, there only remains the risk of being behind" (Sloterdijk [SLO 06]).

"We live by ideas, morals, sociologies, philosophies and a psychology which belongs to the 21st Century. We are our own great-grandfathers" (Pauwels [PAU 72]).

"A man of reflection cannot simply educate managers, he/she must also convince them that the best ideas are not those that we contribute, but those that we generate: appropriation is essential to move from the phase of anticipation through to action...We must know how to amalgamate reason and passion to succeed in action" (Godet [GOD 01c]).

"The future is there, indeed more present than the present itself...the eyes of the future the look of which haunts us and of which our era will be the subject. It will moreover be the culpable object. They find us out – both our failure and our culpability. Already dead, already a thing, when we have yet to live – our era is the only one in history. People will laugh about our ignorance and will get annoyed about our shortcomings" (Sartre [SAR 76]).

"The absence of any prospective view is also often an inadequacy of the administrative political authorities which prevents us from grasping the signs of the times announcing the advent of new concepts" (Papon [PAP 04]).

"It is marketing and innovation which generate results, the rest, when all is said and done, is only cost" (Drucker [DRU 06]).

"Innovation has thus become a form of general horizon and provided conditions for possibilities. A self-fulfilling prophesy within the field of technoscientific knowledge and associated cultures, it directs and increasingly justifies all institutional transformation policies and system standardization dominating the production of legitimate knowledge" (Ruano-Borbalan [RUA 17]).

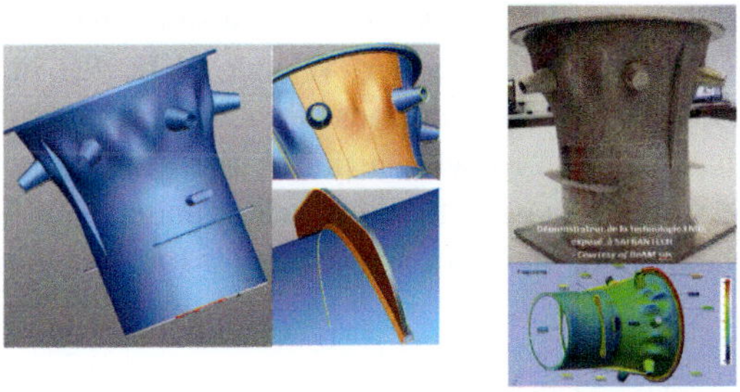

Industrial part (figure kindly supplied by BeAM)

4.1. Introduction

Within the introduction of a given new technology, Jackson [JAC 13] sets out a certain number of classifying criteria, such as:

– prosperity: the continuous production of novelties with, in the background, the concept in which material satisfaction will enable citizen "happiness";

– usefulness: the link between prosperity (or its perception) and the satisfaction supplied by anything which is new (the sense of forging a common identity, predicting the future and other factors);

– realization of full potential: the means through which citizens are capable of producing this. According to the EU (as emphasized by Syntec [SYN 12]), competitiveness is the ability of a given nation for sustainable improvement of its inhabitants' quality of life, and to procure for them a high level of both employment and social cohesion.

However, the change over time of the implementation of new technology, such as the various processes of additive manufacturing with unknown quantities, leads to a form of irreversibility. At present, this precludes a situation where new technology is not taken into consideration. It has necessitated, over approximately 30 years, reaching this state of dynamic equilibrium, which is doubtless a temporary set of circumstances. It may only have been a "flash in the pan" and be collapsing; however, the sum of economic and social advantages has produced an increasingly integrated form of technology, within this particular market. One of the reasons for this situation arising is based upon the fact that we reason around emerging technologies using conservative outlooks, indeed those of the past. This prevents us from envisaging new structures. Moreover, Wiener [WIE 60] in 1960 wrote: "We cannot transfer to new machines assumptions which were axiomatic for the previous generation. I find that I am faced with a public the attitude of which, regarding machinery, developed on the basis of an imperfect understanding of the structure and the operating mode of modern machinery" (at the time, this involved cybernetics).

Forsé [FOR 89] envisaged the dissemination of innovations such as 3D printing from a particular perspective, being that of nonlinear thermodynamics, in which the stability of a system faced with innovation may or may not lead to its emergence. He wrote: "Although the system's kinetic equations are stable compared to 'innovation', the latter will not be able to spread. Conversely, if kinetics occurs fairly quickly, innovations will be restricted and the system will be modified in its entirety by the end of the process. Yet, kinetics operates inversely to the compartmentalization of the system and the paradigm of dissipative structures. This explains the greater or lesser degree of innovation dissemination through the greater or lesser degree of system integration". Thus, the presence of nonlinear interactions within populations affected by additive manufacturing has determined the forms of development and differentiation. These nowadays make up a given social phenomenon. This aspect was widely considered and discussed in Chapter 2 of this volume.

Additive manufacturing technologies fall within these novelty criteria (by strongly renewing production methods and by providing new commitments, sometimes very highly ambitious plans) of self-development and belonging to a form of intellectual elite. Having said that, this is not simply a "digital" elite, concerned with producing "software", as users. However, the majority of 3D machines – the "hardware" – are (still) produced across the Atlantic.

The search for new technologies, such as those of 3D printing, was the result of a demand for a somewhat slower development in the beginning. This relied upon the logic of the original offer, instigated later on by observing the appealing effects for society as a whole. First, these are in terms of the creation of new markets owing to complex lower-cost production, and now, increasingly, in terms of the creation of new functionalities. Indeed, all the technologies, the nature of which is to destabilize the established order, are subject to criticisms and blockages. This is particularly on the part of those who were previously able to benefit from this pre-existing order. However, once it is "accepted" because it has certain obvious advantages, it finds a place with the vast body that constitutes the production system. Figure 4.1, partially inspired by Manyika *et al.* [MAN 13], links the number of scientific and technological papers belonging to a major area of the potential market (in a "Log–Log" plot). There are two main observations which may be made:

– the absence of a direct correlation between this number and the economic impact (it is thus not simply that 3D printing is oversold);

– the positioning in terms of impact is still modest for 3D and 4D technologies (see both Volumes 2 and 3) compared to other major spheres.

Mélanie [MÉL 17] suggests that the current market would only be €4.8 billion/year with a margin of error of +/– 1.2 billion. This is further removed from the potential estimate; with an increase in turnover of the order of 20%, which may grow over time. This is because it is estimated in this data that the figure may reach €20 billion in 2020, there being a modest coverage of personal 3D printers of the order of 6%.

However, in view of the infancy of additive manufacturing technologies, nothing has yet stabilized. However, this figure is a snapshot of the current view with the positioning of "overselling" of 3D technologies, located below the diagonal.

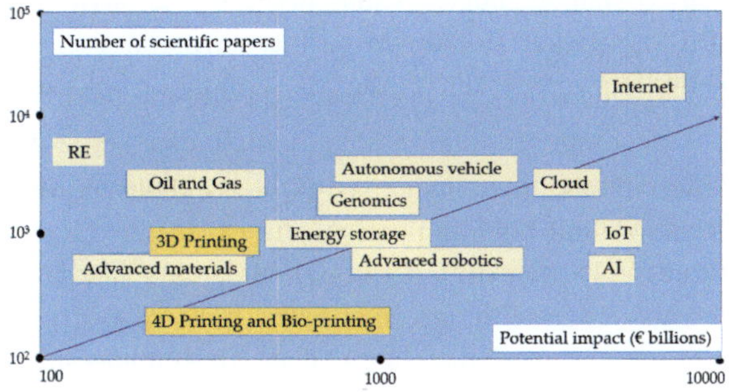

Figure 4.1. *Approximate relationship between the number of publications and the economic impact of major technologies (RE: Renewable Energies; IoT: Internet of Things; AI: Artificial Intelligence)*

The basic concept of additive manufacturing indeed rests upon the conversion of matter fixed in space, without the elimination of matter having localized energy inputs and matter enabling this conversion to occur. In the course of the various chapters of Volumes 1 and 2, a logical progression has been set out. This is summarized in Figure 4.2, which expresses various overlaps and essentially sets out to examine improvement frameworks so that technology achieves its full potential in society (see also Thompson *et al.* [THO 16], Lieneke *et al.* [LIE 15], Weller *et al.* [WEL 15], Arndt *et al.* [ARN 16], DGE [DGE 16] and others).

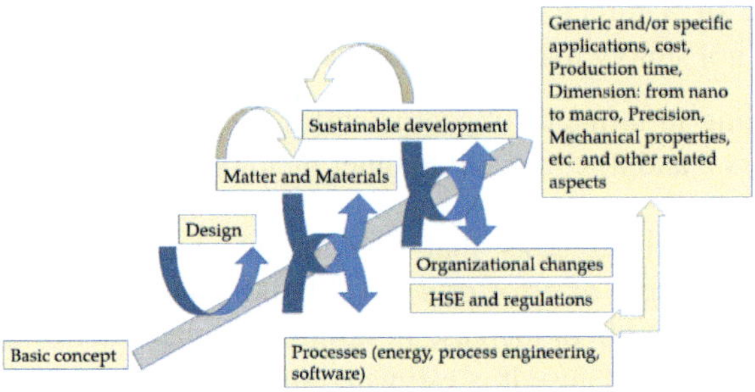

Figure 4.2. *Interrelationships between affected spheres within additive manufacturing (HSE: Hygiene Safety and Environment)*

The dissemination of a given innovation such as additive manufacturing, according to Gaglio [GAG 11a, GAG 11b], proceeds from a "snowball effect" and is represented by an S-shaped curve (this involves a slow start, exponential acceleration, then a significant slowdown with a normalization at a high level of ownership). Stakeholders involved in the dissemination process may take the following "path":

– developing individuals' knowledge as to the existence of a given novelty (knowledge, information and communication phase). In our modern societies, this information is increasingly transmitted through the mass media. This phase goes beyond the simple knowledge of the given novelty. It also concerns the modes for its use, the so-called "how-to-knowledge" as well as the fundamental principles of its internal operation, these being the "knowledge principles";

– the persuasion phase follows. This is highly dependent on the given relational environment in which the novelty is operating. It is generally divided into three sub-stages. These are stimulation on the part of the given professional circle, the introduction of the novelty within the given social system and its legitimization;

– the concluding stage is the arrival of the moment for the individual decides to adopt the given novelty itself. Here, the given decision is individual, but falls within a collective movement, which may reassure and/or strengthen the individual on his chosen path.

The S-shaped curve, referred to by Gaglio, enables the visualization of the dissemination of a given novelty, which loses "novel" status in becoming commonplace. In the course of the process, five types of players follow one another, which are characterized in the following manner [GAG 11a, GAG 11b, TAY 14, ROG 03]:

– pioneers (2.5%) are rebellious and take risks. They go against majority opinions and are not always listened to or indeed taken seriously;

– innovators (13.5%) adopt a given new behavior, but have greater social status than the pioneers. They are thus generally copied by others;

– the early majority (34%) behave as individuals and are described as pragmatists and moderately open to progress. They do not wish to shoulder the burden of the risk associated with pioneer adoption. This is why they are partisans of given wait-and-see policies;

– the late majority (34%) is fairly conservative, skeptical and ill at ease, when faced with new ideas;

– the laggards (16%) only adopt the novelty at the end of its cycle. They have recourse to traditional values.

This evolutionary process integrates dissemination times, just as it does the introduction and adaptation of the new product. The process also does so in the deployment of new uses in applicable niches, communication and the development of networks in which manufacturers and/or the media invest.

The scientific approach, normally carried out in anticipation, is curious to identify, as far as possible, rigorous and legitimate frameworks by which to examine the complexity of the additive manufacturing sphere. Researchers in this "sphere" do not claim to hold the so-called "truth" in their quest for new or more efficient systems. Apart from some proofs of concept, according to Vautier *et al.* [VAU 16] "They observe phenomena, defining variables, and constructing concepts with these variables, and the relationships between the latter. They are able to establish these relationships by taking measurements, and completing experiments, and analyzing data gathered both within nature and in society. They next go back to their professional community, the sole entity able to judge their work scientifically, to set out their work to this community, and subject it to critical review". Although the work provides contributions to the overall 3D printing project, it may be likely to be resumed for applications in the near or distant future, or indeed not be taken into consideration for the usual reasons which apply to innovation. Indeed, Guzzetti

[GUZ 98] said that they are only supporting elements for decision-taking (see Figure 4.3). Within these processes, where researching a possible application avenue, it is important to find all the upstream and downstream partners within joint processes, to achieve competitiveness objectives (which are not in this case solely scientific).

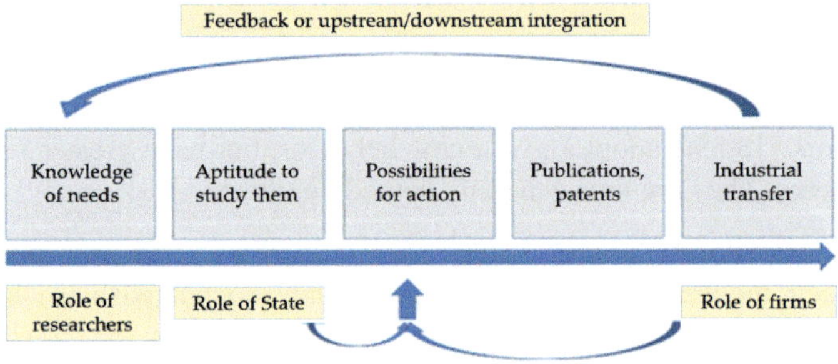

Figure 4.3. *Scientific approach in additive manufacturing*

Three-dimensional printing may call into question the concept of the logistics chain and its consequences may be numerous. These include just-in-time production, a reduction in storage, transport, energy and associated costs. This new technology is shaping up to be, in the near future, extremely promising as has been shown throughout this work. However, taking into account the multiple risks induced by the latter necessitates moving from an analysis of competing technological progress, up to the point of over-communication, which muddies the waters (having aspects of a speculative bubble). There is a need to go from a continuous or "disruptive" approach to the improvement of processes in the broad sense. However, Rosenberg [ROS 15] stated that the only nations which will win, are those which master the various stages of the industrial and societal value chain (see Figure 4.4).

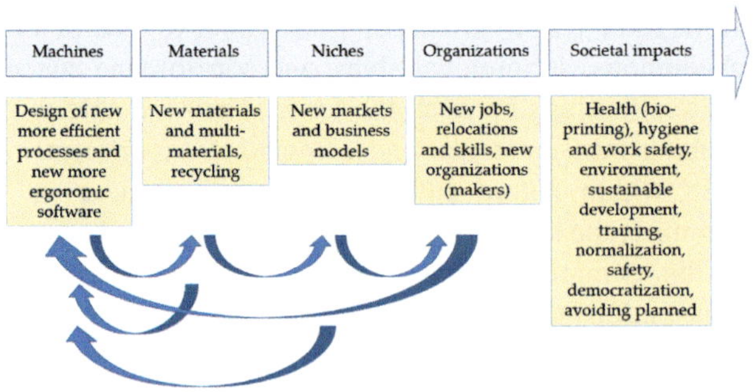

Figure 4.4. *Value chains in additive manufacturing (and interactions between major broad industrial areas)*

In this context, what place is there for research with a practical application if there is no specificity at the local and national scale, relative to other developed countries? This reflection on subsidiarity therefore brings up the issue of using specific components within prediction. These do not only take account of rational aspects, but also risk-taking around scientific subjects, which has not necessarily been envisaged by others. Is this a policy of scientific niches or spheres of excellence which potentially constitute scientific crystallization points and which deserve reflection and a deepening of knowledge?

For consumers, the democratization of 3D printing processes progressively integrates our daily lives. However, a development pathway including education, as Perez [PER 15] indicated, relies upon a means of returning to forms of "physical" reality via 3D printing. The decreased cost of machines participates in the dissemination of technology via Fab-Labs, with open workshops making machine tools for the "self" design of objects (see Volume 1). This trend is therefore yet to be confirmed. However, the new ways of doing things may show at what point the true potential is found in the way that members of these communities seek to "do things differently", permanently questioning industrial operation and maybe change their relationship with the world (see, for example, Fréour [FRÉ 15]). The scope for innovation is above all economic and in this case it is somewhat social.

The existence of increasingly reduced timescales, which are in any case optimized, this important notion of the value chain should favor relationships between the economic sector and research laboratories. In such locations, new knowledge is created. This is a complicated activity in which a discriminating power must be at work to separate the creation of useful knowledge relating to the simple accumulation of scientific data, sometimes considered as somewhat sterile either for or by the visionary engineer, because it is sometimes subject to questions of reliability [MUN 17, BAK 16]. This essential context for company performance assumes within research laboratories, the existence of freedom which is useful for both creativity and the development of updated knowledge (this might help with the performance of future projects). This freedom is relative to the "productivity" increasingly imposed by the requirements of decision makers and by quantitative evaluation research processes. This may lead to a certain follow-my-leader attitude on the part of "doomed" researchers, which increasingly applies within the various programs imposed.

During the various activities, particularly in connection with the economic world, there is no possibility of publishing in part on a "credit" basis. The interaction with a complicit "customer", given his/her interest, but with a firm stand on his commitments, enables an almost continuous questioning of research activities. This opportunity thus enables an almost perpetual questioning of his/her actions, constituting a creativity obligation. Hence, from this flow we have theses, publications and patents. It is a favored route for the examination of nature by science, so that new technology fields are created [AND 16]. Moreover, with that in mind, Lubineau [LUB 16] carried out an investigation around the axes of development desired by the profession (see Figure 4.5).

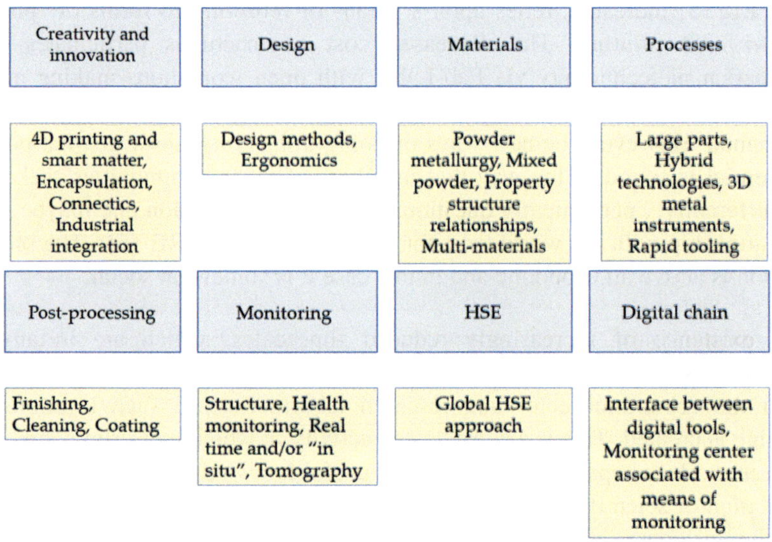

Figure 4.5. *Demands concerning the future of 3D printing*

4.2. Favored spheres of innovation

4.2.1. *How to know where we must anticipate this technology?*

Anticipating the technological future, with a view to being prepared for it (and especially whether or not we are prepared to be subject to it), is one of the keys for long-term sustainability and competitiveness. Taking this requirement into account, within the emerging sphere of additive manufacturing, appears necessary, given the instability of the environment and the speed of technological developments. "Within such a context, the present knowledge, as detailed as it is, proves insufficient to predict the future, such that the systems in which companies develop obey nonlinear dynamics, experiencing bifurcations, and being subject to breakthrough innovations" [MOA 08]. This specific situation also thus encourages a forward-looking perspective [GAU 05]. This is very significant in a world of unprecedented technological and organizational developments, with major social demographic and cultural changes. Although the forward-looking reflection undertaken is imperfect, defined by effective knowledge, by inductive contributions, components linked to chance and chance encounters, the negotiation with the social framework may still be envisaged. However, the definition of an agreement will only be able to be relied upon in terms of principles and procedures enabling current interests to both be heard and make their presence felt, so as to be taken into consideration. This is the reason why the emphasis has been placed in Volumes 1 and 2 upon the aspects of incremental innovation, which may be easily understood and translated into economic success. It is around such bases that we can envisage a "winning" strategy.

With the assessment of possible vulnerability in terms of results (anticipation) or for an agreement around the possible transfer of the aforementioned research works, the position of "strategic" experts is probably rather awkward because the necessary skills are absent or include field data, concerning the observation of effects on economic markets. These data are often not known by these players (and even less so by the author of this work). Their specialisms are nevertheless necessary to select particular designs, the majority of which are chosen on the basis of extrapolation. It is not easy then to define, through a sensitivity study, the weight of various uncertainties around the quality of the opinion within an emerging situation, such as that of additive manufacturing.

According to Moati [MOA 08], it is therefore not surprising that the social demand for given tools enables a strong reduction in future unpredictability. This is in particular on the part of players who are nowadays led to take decisions with wide-ranging implications. These involve significant and irreversible investments, the relevance of which will only be revealed ex-post. This occurs within an environment that will probably be very different to the one which existed at the time the decision was made. This tidal wave has favored the development of economic forecasting [GAU 05]. By exploiting an economic forecast open to technological and social fields worthwhile to society's future, with active monitoring and clear choices, the issue is indeed to participate, through techno-scientific research, in the development of future technological imagination. Indeed, "if we undertake nothing which is completely new, if we do not tirelessly create 'available diversified matter', we find ourselves bereft of substance from which we may choose the best products or services to fit rapidly into a market, which is itself changing at an extremely fast pace" [PET 98].

Yet, within many complex and abundant spheres, such as 3D printing, science proceeds by trial and error. All information which is useful to the media may be exploited to create sensational journalism to such an extent that we more often than not confuse commitments made and reality. In addition, quantitative data on additive manufacturing markets are more than simply uncertainties. It is indeed difficult to rely upon inexact figures to program the future of 3D technologies robustly. What should we do about this? First, within the forward-looking elements, which can/should be obtained, it appears necessary to negotiate calmly, with stakeholders, the strategy that we either desire to or should follow. Indeed more simply, we may choose to observe it having reflected and dared to implement the same. Indeed, nothing is possible within a world which acts upon time, man, strategic choices in terms of research axes, space and other factors, which require too many responses to questions which are often hardly or poorly expressed.

COMMENT.–

We must remember that, through its concept, forecasting should help us "plan for the future", that is to say it "projects the long term within the present, which, as a rule, presupposes all possible hypotheses" [LES 01]. To illustrate the difficulty of a robust approach in this sphere in which to examine the future, Armstrong [ARM 01] identified more than 100 different methodological principles! Even the routes are not certain and, in addition, they are virtually countless! (see also Godet [GOD 01a], Millett and Honton [MIL 91], Plassard [PLA 03], Modis [MOD 92], Guillebeau [GUI 03], and Godet and Durance [GOD 08]). In a summary way, we can distinguish the following groups of methods used either alone or in combination within such foresight activities:

– "crystal ball" (see, for example, Toffler [TOF 71, TOF 83, TOF 91, TOF 06]; de Riedmatten [DE 05]; Gribbin [GRI 98]; Rifkin [RIF 98, RIF 00]; Orwell [ORW 00]; Huxley [HUX 01]; Kaku [KAK 98] and others);

– the Sami technique obliges participants to leave their given roles [GAU 05];

– trend extrapolation [ARM 01];

– consensus methods – "focus groups" [KIT 94, MOR 97, STE 92, GRE 98];

– simulation methods [WAL 08];

– scenarios [GOD 01, VON 89];

– the Decision trees [WAL 08];

– the Delphi methods [BER 67, HEL 67].

These various methods were evaluated by Aaltonen and Barth [AAL 05]. The "crystal ball" methods, resting upon imagination, scare a number of decision makers. This is because using these it is necessary to trust a very limited number of individuals (a high risk) The majority of methodologies used imply that a group can have more legitimacy than a single individual. These are "focus groups" (having public applications), scenarios (for general use) and the Delphi methods (more reassuring). The latter, by relying upon collectives, theoretically reduce the margins for uncertainty. Whatever we say of them they will not enable us to invest in the fields of possibilities other than in an evolutionary approach without using clear-cut breakthroughs, instead of what is produced by a divergent well-informed mind, by exploiting "crystal ball" methodologies (reflecting the works of Alvin Toffler). These methods have been used by Jiang *et al.* [JIA 17]. The following box sets out the main results of their expert assessment.

Production and its location in 2030:

– 50% of industrial activity will be relocated (services in additive manufacturing and specialists);

– there is the possibility of envisaging hybrid additive/subtractive forms of production [URB 17];

– there is the potential for the development of a large number of small- and medium-sized enterprises specialized in the optimal use of machines, with training in ensuring production quality (for example, ISO 9000);

– employment growth will occur [KIA 15];

– within all industries, local production close to customers will be facilitated by additive manufacturing and will increase considerably. At the same time, global production chains will reduce in importance. This will lead to a de-globalization of supply chains;

– the distribution of end products will move in significant volumes towards the sale of digital files for direct manufacturing, in place of the sale of physical products (a similar effect to the MP3 format within music distribution);

– the manufacture of spare parts will be distributed within two systems: less critical parts will be locally produced using additive manufacturing. At the same time, critical parts will be produced in specialized centers, having specific expertise as regards skills/quality control systems. In this process, conventional manufacturing techniques will be used;

– the carbon footprint for manufacturing and transport will be considerably reduced by additive manufacturing.

"Business models" and competitiveness:

– additive manufacturing will change the various sources of competitive advantages in terms of manufacturing capacity and supply chains, forming closer links between customer networks and designers;

– commercial models will not be hugely influenced by additive manufacturing. This is because it is only a production technology simply necessitating new knowledge and skills;

– the conventional measures of "time to market", the "product lifecycle" and "surge capacity", will have reduced as digital products will be in the so-called "continuous-beta phase", and will be subject to frequent design iterations and continuous alterations;

– Germany will be in the top five global players in the development of industrial additive manufacturing technologies and machines, thanks to the producers of existing machines, research institutions and a large number of end users;

– a further feature will be the inclusion within the processes of both the circular economy [DES 17] and also environmental and energy saving aspects [GRI 16].

Consumption and market trends:

– the market share of items made by 3D printing (products and components) compared to articles produced by conventional methods will be significant in all industries;

– a significant number of consumers will use online databases to buy product designs or to access 3D printing designs using "open source" for free;

– the majority of private consumers in industrialized countries will have industrial printers in their homes;

– a significant quantity of objects manufactured will be made of multi-materials and/or contain integrated electronic components. This will enable their use in a wider range of applications;

– human organs will be printed in this way, and will be both viable alternatives to biological human organs and widely used to replace such organs (see Volume 3);

– intellectual property and political strategy;

– the difficulty of protecting conventional intellectual property rights for digital products will bring about a significant use of recent new forms of intellectual property such as the so-called "Creative Commons" or "open source";

– an important regulatory measure will be the regulation of additive manufacturing file-sharing platforms;

– the issues of liability owing to unclear intellectual property rights and the impossibility of both monitoring and taking legal proceedings for intellectual property infringements may place a limit upon the development of additive manufacturing.

Box 4.1. *Main results of the investigation by the Delphi methods from Jiang et al. [JIA 17] and supplementary bibliographical references*

Simultaneously with additive manufacturing development, Li, *et al.* [LI 17] developed algorithms to calculate the influence of the development of additive manufacturing processes, upon the cost of producing components. These authors say that modeling experiments are leading to significant results in favor of 3D technologies (see also Franchetti and Kress [FRA 17]).

4.2.2. *Opportunities*

The two tables below show the comparisons between opportunities and limitations both in technical (Table 4.1) and economic terms (Table 4.2) applied to additive manufacturing, brought together following the various comments set out throughout this work (see also Guo and Leu [GUO 13], Boillat [BOI 15], Weller *et al.* [WEL 15], Deloitte [DEL 14]).

Opportunities	*Limitations*
The complexity of objects is not a limiting factor Possibility of using this aptitude to "redesign" technical objects (for example, in relation to object weight gain) Forecasting software in relation to distortions of the object being produced, depending on the materials used Appealing "recreational" technology Modest maintenance requirements Direct relationship between component design and its manufacture without any intermediate production stages From the nanometer to the decameter Numerous industrial and tertiary applications in promising niches; The absence (in principle) of any waste Recyclability of unused materials (in theory) Low electrical consumption in the processes (to be validated by LCAs – Life Cycle Analyses – across the entire supply chain from the raw material to the finished product) Alteration to the design of a part which is easy to produce (handmade by the designer) Display of the process in real time [TRI 15] Sphere still in the process of emerging with anticipated innovations and new application segments with high expectations in terms of results (see Volumes 2 and 3) Questioning the concept of additive manufacturing.	Manufacturing time relative to the required accuracy Need to have specific materials, adapted to the functionality sought within the given object Necessity in certain processes to introduce supports to enable object construction and the subsequent removal of such supports; Relatively high cost of materials and their molding Surface finish Processes exploiting the size of variable voxels which are highly limited in number Few or indeed no robust processes using multi-materials Lamellar anisotropy Variable quality of objects, depending on the process used Surface finish could be improved Duration of use by the company Capacity utilization rates are sometimes modest Aging of objects; distortions Failure to adapt to serial production, still less to form collective manufacturing Lack of control of hygiene, safety and environmental quality conditions (HSE) Non-immediate use of technology Few individuals trained.

Table 4.1. *Comparison between scientific and technological opportunities offered by additive manufacturing and associated limitations*

Opportunities	*Limitations*
Customized objects Production returns to the national territory Specific operators (i.e. makers or specialized centers) Increasing production feasibility of objects made of high-quality materials for *in situ* production (automobile industry, army, spatial industry and others) Democratization of technology Simplification of innovation of new products Reduction of the number of parts within a given mechanical system Range of production widened Customization of products with feasibility for customer participation in design Reduced cost for complex parts relative to other manufacturing processes ("Complexity costs nothing!") Innovation is possible across spheres where there is a significant market lacking coverage by other technologies (4D printing, Bioprinting and others, but see Volume 3) Organizational alterations.	Marginal costs (specific materials, defective components and other cost factors) Lack of economies of scale Requirement for increasing numbers of dedicated machines (none or few versatile machines for industrial applications) Renewal of machines within continuous innovation at a time when they are not yet profitable Quality control could be improved HSE might be improved upon Copies are easy to produce without appropriate system of checks in place (for example, works of art) Poaching Constantly evolving regulations Constantly evolving standardization Training of operators Structuring of research could be improved upon Organizational alterations – the question remains as to whether the "Uberization" of manufacturing is possible. However, Fab-Labs are developing.

Table 4.2. *Comparison between socio-economic opportunities offered by additive manufacturing and associated limitations*

These limitations constitute, in a way, the beginnings of a specification to achieve strengthened establishment in the spheres of use within additive manufacturing. So as to attempt to draw provisional conclusions, on a highly important aspect concerning the materials–processes coupling, Table 4.3 shows that the various usable materials are in relation to the current additive manufacturing processes ([FIM 15, BER 13], cited by the ITRE Committee – European Parliament's Committee on Industry Research and Energy, 2015). This table illustrates the multiple possibilities, with options which are nowadays very wide in terms of precision, distortion, functionality and naturally costs with relationships between markets, 3D technology and materials. However, there are indeed scientific and technical aspects around these qualities of materials, to which progress still needs to be provided, so that 3D technology achieves its targets, such as are assessed by Manyika *et al.* [MAN 13].

Process	THP	TFP	Metal	Ceramics	Market
Stereolithography		+++	++	++	Prototyping,industrial components, optics andbiomedical
Deposit of molten wire	+++		+	(+ concrete)	Domestic applications, prototyping, biomedical, building and construction
Laser or electronic sintering	++		+++	+++	Industrial components, biomedical and prosthetics
Laser melting	++		+++	+++	Ditto
"Binder jetting"	++ (+waxes)		+++	+++ (+ foundry sand)	Prototyping, art, industrial components and molds
"Material jetting" (see CLIP)			+++		Repair and industrial components
"Sheet lamination", Stratoconception	++ (+paper and wood)		++		Prototyping and industrial components

Table 4.3. *Relationship between materials and processes within additive manufacturing (THP for fusible polymer or thermo-plastics; TFP for thermally fusing polymers)*

4.3. Some conditions to ensure additive manufacturing reaches maturity?

These 3D technologies are developed with six main current preoccupations and involved in more forecast-based aspects and emphasized in the following list:

– showing that additive manufacturing falls within an industrial process enabling a reduction of the "time-to-market";

– finding new technology outlets: everything from children's games to organ printing including nanotechnologies and the building and construction industry;

– innovating to maintain the dynamics and to make technology last: this is an incremental approach to new processes and especially materials and multi-materials. It is a breakthrough-based approach by resuming the concept (4D printing – see Volume 3);

– producing proofs of concept and communicating; the initial demonstrators developing lightweight structures with modest investment but strong levels of creativity;

– going up to the point of being an industrial demonstrator;

– manufacturing end products responding to a functional specification in the lowest number of stages possible.

Currently, we are at an in-between stage, which does not enable us to have an economic rationale for producing a large quantity of components. However, already several thousands of such components are envisaged without creating too much economic risk. When 3D printing does succeed in combining all of the following criteria: "performance, speed, volume, price", an entire industrial sector will, no doubt, suffer huge disruptions. The modernity enabled by 3D printing is characterized by the aspiration to exceed every limit, by the incessant expansion of the zone of possibilities. This is through the world's increasing ownership through technology by creating new sources of value, and as a result (or due to), new skills. This technological stabilization horizon is probably not too far down the road, and as with all the changes, it necessitates awareness raising and anticipation in terms of educational, financial and human investments, indeed cultural alterations [ROC 06].

With regard to information technology and its numerous other applications, it appears that there is a critical moment for change within corresponding organizations, at all the levels, having the capacity to redeploy human resources and to acquire future digital talents. A certain trend towards various forms of origin for "running out of steam" and a loss of dynamics compared to the initial exponential growth in publications and patents (such as sales of new machines) will be apparent. The various skills sought will probably be those of individuals assembling technologies, associated knowledge and the capacity to organize it, develop it and support it [SUC 16]. This is provided that we are stimulating them (or, at least that we are accepting them within the global scientific and technical system). Nevertheless, depending on the technologies taken into consideration, the dynamics vary according to the parameters that we need to explore.

The object in this section is to slightly move away from the purely scientific and technical to evoke the means enabling us to achieve the objective of realizing new ideas, enabling better promotion of 3D technology applications. As we understand, it is a question of considering how it is possible to stimulate incremental research (indeed, ground-breaking inventions), across the entire scientific field. They may also be, as in the United Kingdom, a form of impetus [HAG 16], with an increase of 71% in basic research and 100% in R&D in terms of funding. However, it is instead important to reflect upon a system of research structuring devised at all necessary levels, for actions focused on priority operational targets.

Figure 4.6 taken from the work of Peyré Tartaruga *et al.* [PEY 17] emphasized the traditional means of upstream and basic research (Bohr's model), applied (Edison's model) and intermediate (Pasteur's model). Within this chapter, on the debate about research into 3D printing, it is indeed a matter of researching the alliances between the upstream and downstream sectors, around subjects enabling progress (see Stokes [STO 97]), while widening the approach to other aspects (whether economic, cultural and other factors).

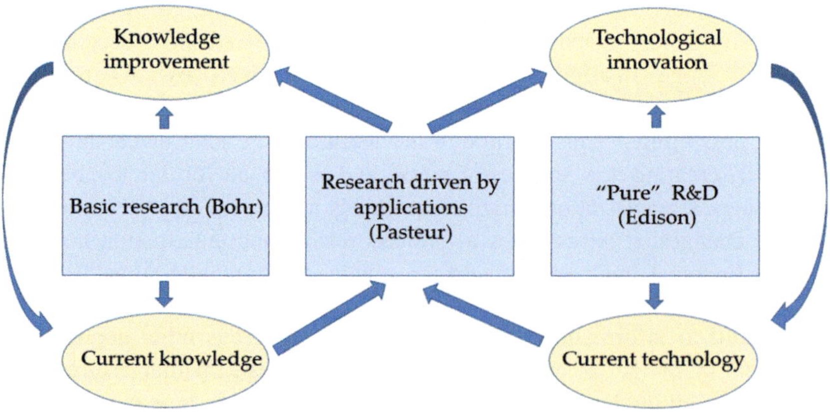

Figure 4.6. *Research models [PEY 17]*

The desired opening is a condition for success, as to construct complex technical objects it is necessary to share knowledge to approach all the aspects of manufacturing [BOT 12]. According to Kaufman [KAU 15] designers know how to "daydream", take time to let their minds wander and "listen to" their intuitions: they take risks and lay themselves open to new experiences and they accept failure as a means to a better recovery. How then should we protect or, better still, support them, so that they are able to make provision in this regard? Independently of this personal aspect, Half [HAL 16] envisaged relationship types which are seen even less frequently within the academic research sphere. He proposed the following:

– put in place "excellent" organization: at present, innovation is teamwork. From an organizational point of view, several trends emerge to keep autonomy favoring the rise of information and the emergence of constructive ideas, establishing working groups to transform an idea into a prototype and thus assess its potential in practice;

– boosting commitment through concerted, open and confident leadership;

– listening and allowing a margin for error while remembering what leads to failure;

– demonstrating agility in decision-making;

– going as far as the proof of concept and indeed the demonstrator;

– based upon various profiles to strengthen the abundance of interactions and, at the same time, accepting the need to take the time so that exchanges are productive;

– working collaboratively.

Interdisciplinarity, such as that recommended in this work, has "earned its stripes" within academic research when this concept was more or less forced to solve problems. With this came the reminder that research works can advantageously have practical benefits. The analysis of science and technology developments is in the process of becoming society's main productive force [HAB 73] being based upon interdisciplinarity [LAF 11].

Interdisciplinarity thus calls into question science which is designed in epistemological terms, such as knowledge by both a given object and a given concept. According to Granger [GRA 86], "science seeks to transpose 'reality' – all reality – in the form of an object open to manipulation likely to explain by means of concepts capable of producing a given representation. Thanks to this it then becomes possible to have 'a precise and penetrating contact' by the fact that it is formed according to abstraction".

"Under this head, science, unlike the interdisciplinarity with which we wish to associate it, deliberately devotes itself to the reduction of the complexity of that which it studies, to have a 'precise and penetrating contact' with its given subject. That is what the aim of science is and it must be concluded that it objects to what provides interdisciplinarity with its *raison d'être*: producing a broad perspective by applying blends of various disciplines. In other words, the means of reduction by which science acts, gives it both its force and its precision. Under this head, reduction in no way signifies fragmenting the object in the extreme, but partially envisaging it to be able to see through it and be perfectly aware of producing a partial but precise view of it. In other words, science does not seek, on the whole, to explain the concept, as interdisciplinarity desires, but to reduce reality in the form of objects which it distinguishes, so as to be able explain them precisely and thus provide sophisticated understanding of reality envisaged in this way" [HAM 13].

4.3.1. *Moreover where does additive manufacturing sit within this interdisciplinarity framework?*

The following tables attempt to propose breakthrough technology and creativity zones which have prevailed within the development of 3D printing. This carries the risk, among other things, of omitting openings linked to the proper control of the chemistry of usable matter and materials within additive manufacturing. To do this, the analysis of the development of various processes progressed to carefully examine the originality and the rupture element on the bases, which are doubtless subjective and subject to the comments set out in Tables 4.4 and 4.5. Two avenues are examined, that which concerns processes (see Table 4.4) and that which concerns applications (Table 4.5 and Figure 4.7). It is not a question, within this approach, of taking account of potential markets or workforce, but of trying to identify "divergent" aspects within the given innovations. The author proposes a creativity index such as is assessed by its sole knowledge (and therefore obviously subject and accepted for possible controversies) ranging from 1 to 5. The value of 1 is a predictable activity (continuity), and the value of 5 relates to a dynamic technological breakthrough.

Topic	*Decades*	*Creativity index*	*Commentary*
Laser stereolithography	1980	4–5	Introduction of the additive manufacturing concept from sound knowledge: CAD, laser, photopolymers having enabled the trialing of other 3D technologies; amplification linked to a chemical chain reaction
Colored objects; laser marking	1990	1–2 3–4	Use of several vats of colored matter; light–matter interaction (the case of titanium oxide); 3DP process
Improvements to the process	1990–2017	1–3	Replacement of galvanometric mirrors by programmable liquid crystal screens, lasers by laser diodes, improvement of the layering process, and indeed other parts of the process
Removal of layers	2010	4–5	Use of the disadvantage of oxygen as an advantage to avoid the provision of monomer layers; the highly ingenious use of Carbon 3D. However, the limits of the system are not yet known (absorbent photochemical initiator in the visible area)
Micromanufacturing and Nanomanufacturing	2000–2017	2–3	Two-photon processes with good spatial resolution, but with no known industrial applications
Frozen monomer	2010	1	Avoiding the provision of layers and supports but using within a difficult routine

Sintering	1990	1–3	Use of a monomer blend – sinterable powder; the object produced is destroyed when hot to eliminate the polymer and rubbed, having the difficulties of layer provision (but without need of supports)
Photo-insolubilization	1990–2015	2	Process already validated. This involves using a soluble polymer made insoluble (or the reverse) by photochemical means (using one-photon and two-photon processes). The problem of light penetration in this type of material is yet to be studied, as is that of the elimination of soluble parts; involving a quantum yield for conversion processes of below one unit
FDM Process (fusion)	1980–2017	2–4	Very low cost "rustic" process which enables a very large number, although not uniquely, of domestic applications; choice of sophisticated materials such as fiberglass containing graphene or biocompatible polymers
Selective laser sintering	1990–2017	3–4	Absence in principle of the implementation of supports; large choice of fuse-wire materials
CLAD and MPA processes	2010	4–5	Provision of matter; the 3D movement of the part under construction enables the use of a base produced by other means and to complete its given mold (beneficial aspects include repairs, production time saved and other factors)
Prefabricated components	2015	2	Assembly of prefabricated components, hence a large production facility (in theory), but the necessity for a component store, which may be a prohibitive aspect

4D Printing	2010	5	Are we able to move from a few demonstrations and a science fiction logic to applications? Numerous barriers to break down (see Volume 3)
"Intelligent" matter	2010	5	A sphere where the imagination entirely has its place, even if there is no urgency for immediate given applications
Self-organization of microrobots	2010	5	Another sphere, very close to the previous one, which also makes people dream

Table 4.4. *Creativity associated with various current or conceivable additive manufacturing processes*

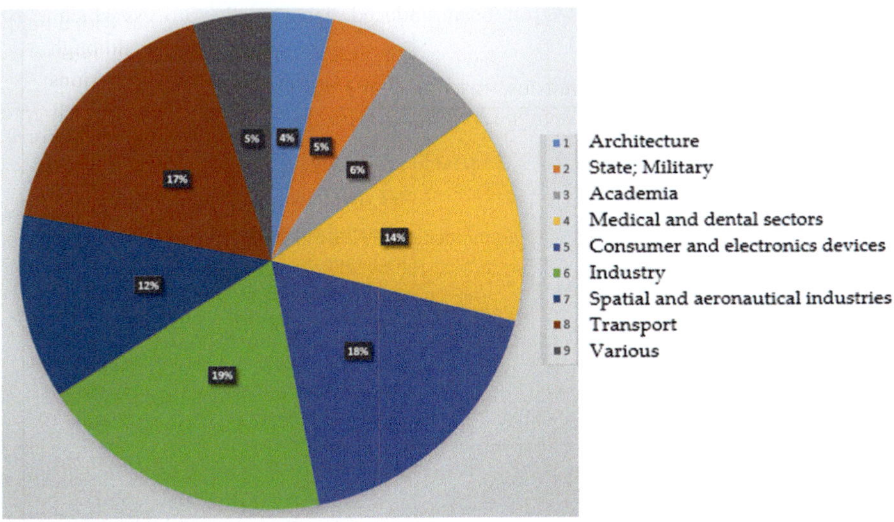

Figure 4.7. *Major spheres of application for 3D printing, on the one hand, according to Bernard and Manil [BER 14] and Earls and Baya [EAR 15], on the other hand. For a color version of the figure, see www.iste.co.uk/andre/printing2.zip*

Topic	*Decades*	*Creativity index*	*Commentary*
Stereolithography and prototype components	1980–2017	1–4	Stabilized process with improvements to low-shrinkage oligomers and potential added mineral charges; excellent three-dimensional precision < 0.1 mm
Nano-stereolithography and micro-stereolithography	2000	1–2	Validated process without many known industrial applications even if remarkable resolutions have been obtained (< 1 μm); impetus for bioprinting (see this volume)
FDM process	1980–2017	1–4	Process ranging from “domestic” applications to bioprinting with very low-cost machines; creativity is found within the development of new materials (graphene; biocompatible materials, glass and others); sometimes with the modest level of precision but with a field of creativity “produced” which is very large (the 3D photocopier)
Blending of powders and metallurgy	1990–2017	1–5	Use of fusible powders possible within a very broad spectrum (with possible need for investigation around the significance of using polydisperse materials; satisfactory accuracy to produce microfluidic components and the intensification of material transformation processes; industrial applications around stabilized processes which are growing rapidly
Bioprinting	2010	2–5	Processes not yet entirely stabilized and investigations into biocompatible materials and across stem-cell printing; an appealing topic for its enormous applications in terms of repair indeed for man’s “improvement” aspects (associated with ethical problems); the transition from the inert matter to living material is a genuine impediment (see Volume 3)

Microfluidics	2010	1–3	See metallurgy; the possibility of innovating by using a number of "intelligent matter" concepts
Intelligent matter; 4D printing	2010	5	Are we able to move from a few demonstrations and a science fiction rationale to applications? There are numerous barriers to break down, even if some experiences, such as "photochemical muscles" or DNA origami, define the means of scientific investigation (see Volume 3)

Table 4.5. *Creativity associated with various applications of current or conceivable additive manufacturing processes*

4.3.2. *Observations*

On interpreting these tables, we observe that Europe is essentially in the position of good-quality research. However, it relies upon innovations which instead take an incremental form, particularly within the development of forms of matter and materials, stemming from their know-how preceding the emergence of 3D technologies. Creativity is mainly located on the other side of the Atlantic (e.g. Caltech, MIT and other organizations) and, to a lesser extent, in the countries of the East (such as Japan and China). The beginnings of a hierarchical transfer are already being felt with scientific publications increasingly from the East. In France, do we not perhaps have the art of excelling in the capacity to miss our chances because creativity and the exploration of the unknown are not favored on the Old Continent? [EUR 12]. Moreover, must we/can we find the right compromise for the cursor within a research group having divergent standpoints, when they are present and indeed conservative? Philippe [PHI 12] wrote on this topic, "Those who within the given organization have ideas to act differently and best divide into two categories. There are those who dare not to and those who dare to. Those who dare not to understand the issues and the significance of new ideas, but they are paralyzed by risk-taking and the fear of causing displeasure. Having never attempted anything, they have never known failure and are therefore free from criticism … they are, you might say, quitters. Those who dare to, the innovators, move forward by disrupting conventional ideas, structures and sometimes procedures. They are associated with fear and incomprehension and are heavily criticized".

Dissimilar from the notion of progress, the notion of change has been imposed, for approximately 30 years, as a positive and enormous value within developed countries. This is within the context of immediacy, creation and networks in which the knowledge economy is based. The idea of permanent change seems to have become a new paradigm, which has failed to give rise to concerns at the same time as numerous hopes. However, at the same time, according to Alexandre [ALE 11]: "As regards industrial matters, the illusion that the stagecoach managers will invent the railways leads to the concentration of technological aids around the large industrial groups, who are strong lobbyists, in the absence of understanding future trends". How can we be creative within this backdrop defined by framework programs induced by the EU and/or the French National Research Strategy? In addition, Vincent Berger, the General Rapporteur of the advisory committee of the Assises de l'Enseignement Supérieur et de la Recherche (this translates as the "Conference for Higher Education and Research"), has commented on this. He says, "We have gone too far in project funding", with projects which return within a given zone of scientific conformity, and do not enable sufficient numbers of researchers to willingly take risks [SCI 14].

What is more, with funding leaving little space for individual initiative, how can we stand out from the crowd in this context? The author was able to observe the development of modeling of what is known (but which needs to be better controlled). He shares Ganascia's reflection [GAN 00]. The latter writes, "Beyond this dematerialization of support, scientific practices undergo profound changes. Researchers are giving up the traditional experimentation laboratory to sit in front of a computer screen. Moreover with these developments, the epistemological status of the experiment changes radically". Thus by seeking to resist a given pressure, that of producing the greatest number of publications, we risk costing researchers the need for contact and encounter with the given matter. This is damaging to additive manufacturing. Correspondingly, we are therefore costing them the possibility of validating an idea in detail, limiting the development upstream of industrial activities stemming from these ideas.

Finally, to add a layer and to justify this approach – at least partly – a company can only develop in relation to its history. This is long for Europe and shorter for the USA. This is because its know-how and inter-personal skills are the components of its socio-economic heritage. Having products in line with a company, its positioning and its culture is a necessity for a coherent strategy. This is because businesses which succeed build their history, not by sheer luck, and do so within an advocated continuity. When you develop a company strategy, you have to think of the products that the company sells, and the markets in which they are sold. This overall coherency constitutes success. Creation must take the form of cement between the past and the present, history and the future. This is, moreover, a definition of modernity which knows how to move with the times. It then becomes the classical territory of incremental innovation, which is seldom, if ever, "disruptive". According to Marchesnay [MAR 13], "Herbert Marcuse recalled that the great strength of organizations is their inertia. The latter is all the more meaningful when it is based upon a broad and far-sighted view, and it comes up against vested interests".

However, the world changes and will do so even more as in the East ways of thinking evolve. Knowledge and skills exist, which are as much ancestral as they are founded in high technology. In Asia, as in the United States of America, implementation and replication are current "business models", but ones which are largely out-of-date, as the integration of novelties in real time within industrial processes is very fast. We can take the example of prototype machine manufacturers and additive manufacturing, aside of a modest number of French, American and Eastern manufacturers. The obvious development and establishing new technologies such as 3D printing within local product offerings with a high added value, with an internationally acknowledged creativity, is likely to continue to appeal to the Western world, including France. What should firms do to stay competitive? "The company strategist, university researchers, within their given duties and with their means and cultures, having been in turn a moderate forward-thinker, a manufacturer, a salesman, a marketing specialist and a financier will have to integrate all of these concepts so as to become [out of necessity] a 'creator of new disruptive trends'. However that assumes, within our country, numerous developments both across cultures and in behaviors" [CHA 14].

Creativity describes the ability of a given individual or a group to imagine or construct and implement a new concept, a new object or to discover an original solution to a problem. It in particular crosses individual creativity with serendipity: the ability to use the various elements discovered, while we were actually looking for something else. To this end, it becomes, such as it is, a method for problem resolution. According to the classical view of creativity based by Guilford [GUI 56] upon the dichotomous principle of divergence/convergence, the creative approach starts by the acknowledgment of a given problem (see Kurzberg and Amabile [KUR 00/01]). The "added value aspect" of Design Thinking adapted to new forms of innovation rests with Marti (2017) around three innovative elements:

– the in-depth knowledge of the needs of end users. This involves integrating a dual movement of overlapping interests, ranging from science's contribution and that of populations seeking move forward "which can be assimilated". This takes account of practices, indeed traditions, their possible given evolutions, and also economic aspects, access to energy and to specialist subject areas, as well as other aspects;

– project management relying upon iteration with the use of prototypes as research tools and technological progress;

– "the concept of Design Thinking rests upon the constitution of workshops, with pluridisciplinary and non-hierarchical teams. Initially the objective is to bring about the maximum number of ideas then, within a second phase to proceed to a filtering process and to define a given direction".

Technological innovations such as those linked to additive manufacturing technically cover new products and/or processes, as well as significant completed technological improvements (see Tables 4.4 and 4.5). These are introduced into the market or used within a given production process. Innovations within this general framework involve several types of activity: scientific research, technological research, organization, expertise, finance and business and many others. Generally, many industrial sectors consider that academic research (and less so with research which proves to be disciplinary) is only very rarely a source of innovative ideas, but that it is a means to resolve problems (see [INS 15]). Although the given field hierarchizes knowledge, it does not envisage, in principle, the governance of links between knowledge and tackling complex problems, which are inseparable from innovation. Hence, from this flow problems of a lack of understanding and the so-called "turf war" which tarnish relationships between specialists from different disciplines [DEC 16]. The existence of know-how within scientific research on the Western scale, a genuine "facility", is therefore useful for the promotion of new ideas but does not constitute, for these sectors, the heart of an operation where innovation is at work.

Within a national territory, currently somewhat devastated from an industrial viewpoint, there is no doubt an obligation to examine a transforming "rupture" power on the part of the world of research and university training by participating in the support of economic redeployments. This is provided that this academic world, including major research bodies and as a result the state, commit within this willingness to effectively support the creation of economic value. These redeployments are authorized by the quality and quantity of the academic training and research organizations who can invest in the development of innovative actions (in relation to the social body) in a reconquest of economic performance, currently in decline in France. Within the years 1985–1990, it was a question of, as was the case in many spheres, the 3D printing process not failing. It provided a means for activities enabling social progress simultaneously, employment within a difficult context and responsibility for industrial development, which is indeed nowadays, at least in part, sustainable. However, Monod [MON 84] wrote this prescient sentence during this period, “Europe remains at a watershed, losing the final vestiges of its former domineering power without having, for all that, the dynamism of a creative power”. There is thus a deep-seated crisis.

Innovative processes articulate the periods of creative divergence with multiple proposals and the periods of convergence leading to feasible scientific projects, which are generally of an interdisciplinary nature. By wishing to “manage” the creative process, the innovator within his/her group should be capable of setting up this link and introducing in a tailor-made fashion forms of productive coordination with meaning and mutual values. It must not solely be limited to organizations which have already proved their worth, but can also advantageously take into consideration the need to work upon means for the actual pooling of interests and purposes yet to be created. This will have to leave the ideas to mature by aiming for an operative efficiency, which implies support for the risk taking and the change in the management of the public research to achieve this objective. This falls within an “adaptive” dynamics with continuous reformulation. It involves actual planning because the future state, which we wish to reach, demands a set of interdependent decisions, that is to say a system of decisions and therefore actual management. This is basically applied to scientific policy and governance – an excellent exploration of complexity. Clearly, this situation necessitates strong support by supervisors to move forward.

Independently of this comment, when the idea is being fleshed out in terms of proof of concept, there is still a long way to go. Silverzahn [SIL 16] states that it is when the hierarchy realizes its existence and, worse still, becomes aware of its importance that it can decide to support it, even urge it on, so that it succeeds on a large scale. However, for the price of its support, senior management will require, and this is perfectly reasonable, that the project is “properly managed”. He goes on, “For good governance, senior management in fact intends that above all order reigns. It must be noted that, born into secrecy, the project has often seen its promoters resort to do-it-yourself (DIY) and performance. Factors would include starting small, using available resources, working within co-creation and other associated parameters. All of this is not very sexy, not very effective, not very ‘blue ocean’ and besides it is too small. How can we have a real impact upon turnover with this form of ‘do-it-yourself?’ No – this is not the answer. You have to think big! However it is especially necessary to shift from do-it-yourself to ‘actual’ management. It is necessary to have better control over costs”. Moreover, according to the DGRI (Directorate-General for Research and Innovation) of the European

Union (2017), France is "appreciated" for its degrees of complexity within means of coordination and financing.

"Exit Do-It-Yourself forms of business and put in place genuine management. Within the logic of rationalization, management will consist of wangling everything specific to the innovative project. It is necessary to connect to the company's IT system and stop using open platforms. We must rely upon organizational functions and reduce project autonomy. It is necessary to align the innovative offer with the existing offer and be more coherent in this regard. We need to know where we are going. We must do what everyone else does. We must simplify everything. The given project, or rather its breakthrough dimension, is dying through the infamous Chinese torture of death by a thousand cuts ("Lingchi"). This consists in successively cutting and removing the body parts and limbs of the person sentenced to death before lastly severing the person's the head. In the end dozens of small alterations, all for good reasons and perfectly defensible, reshape the project to fit it into the organizational mold. However, these alterations transform the breakthrough project into a continuous innovation project, that is to say by extending the actual range (the same thing but better) or more specifically of the current business model. Instead adapting the organization to the innovation, we adapt the innovation to the existing organization by trying to force it into the current framework. This, out of necessity, has the purpose of distorting it" [SIL 16].

The difficulty is that the protection of the project necessitates considerable energy and financial support. These are increasingly difficult to find [LES 16]. The designer who innovates risks ending up both worn-out and cut right out by the system in one form or another. As a result, with a modest research budget, the "Joint Research Center" (JRC) in Seville in Spain has just published a report in which France is in an obvious situation of weakness within Europe, as Figure 4.8 shows, relative to its direct business partners. This is so even if its position is above the average which is 32 using the JRC units. Although France is in fifth position, when placed upon the exponential curve, it is shifted "into a better position" by the results. The first four are broadly situated above the exponential curve. There are therefore reasons to ask a few questions relative to optimization. These would only be for the purpose of the promotion of innovation of additive manufacturing.

We can observe that the Canadian Academies Council (CAC) published a report at the end of 2016, which places France in tenth position within a list of Western countries in terms of overall innovation (see also OECD [OEC 16]; Le Monde [LE 16]; Giret [GIR 17]).

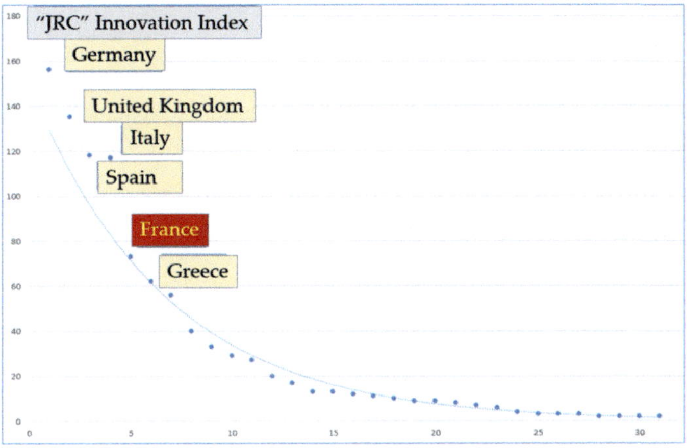

Figure 4.8. *Innovation index taking European countries in descending order*

Florence Jany-Catric cited by Vanier [VAN 16] says, "To be both an efficient and legitimate mechanism, performance should ideally be negotiated within institutional mechanisms, the objective of which may be to supply procedural legitimacy. The latter would come from the fact that these institutional mechanisms would enable the organization of interactions between stakeholders, guaranteeing that the performance mechanism in place is the fruit of a collectively validated acknowledgment or collectively shared obvious facts". In other words, we must investigate whether the managerial tooling of the business world may aid academic research, without seeking to control or manage its operations, even help in assessing its policies. However, Comtesse [COM 16] says that what we should support above all is "disruptive" innovation and leave more predictable "incremental" innovation, in considering that businesses can finance it directly, given that it is less risky. "The state has to finance risk and disruptive innovation; that is if we wish to remain an Innovative Nation of tomorrow".

The aim of this book is not to deepen a debate which is partially outside this subject, even if it may have notable effects on the national position within international competition in additive manufacturing, a debate which we abandoned very early on.

4.3.3. *Some possible solutions?*

With subjects which are still new, such as 3D manufacturing, modes of action are mainly interdisciplinary with the need for convergence to achieve an application-based objective (see Figure 4.9).

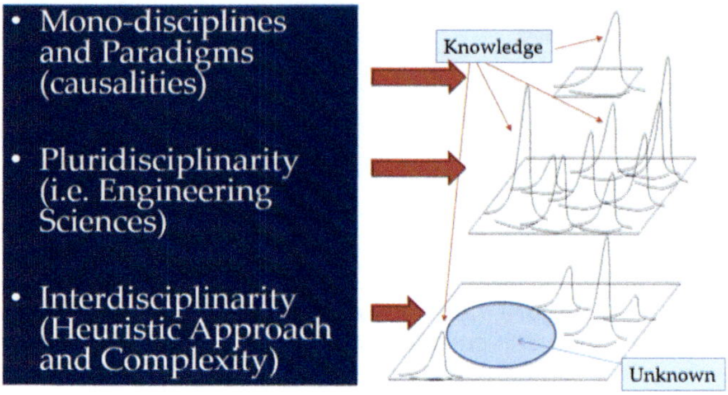

Figure 4.9. *Exploration of complexity*

Within the existing systems or unknown zones, it is necessary to enter into a heuristic risk-driven approach to attempt to find "profitable" approaches. This is the alternative to what often takes place, expecting that others will do it for you. From this figure, the proposals put forward below try to provide the means to favor this form of examination.

– Bottom-up and incremental research:

We must know and control the most recent aspects of scientific achievements, making adapted equipment available to develop incremental scientific and technological innovations.

– Moving towards "an increase in creativity within a heuristic approach":

First, any interdisciplinary process starts from a given aim or view. It is therefore necessary to have designers and/or divergent thinkers within a group of operators enabling the emergence of open and new modes of action. Upon a subject of this type, mastery of the understanding of the actual and possible situation involves updating knowledge around the subjects from given visionary individuals. A means, at least partially, to save time (and devote itself to the fusion of knowledge) is to question disciplinary specialists whose skills could quickly bring this knowledge. The question posed may take the form: "In view of your competences within the sphere of 'XXX', how do you perceive the contribution of your given discipline to…?". Within these conditions, the working group may already have extensive single-discipline contributions to an initial reflection on their interwoven nature.

The diagram for such a process is shown in Figure 4.10.

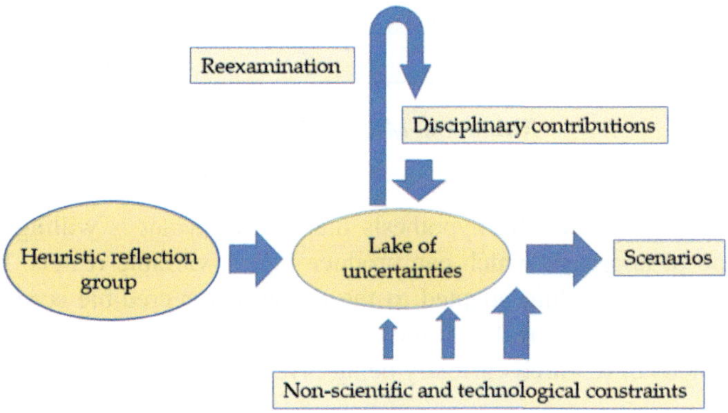

Figure 4.10. *Favoring interdisciplinarity for given action*

The creation of an interdisciplinary group, practicing "divergent" thought and having competences on the given topic, is likely to define, with external contributions what is possible and therefore known, as well as partial or entire areas, in which knowledge is lacking. According to the aims defined by a given "pilot" (an informed driving force), it should be possible to define scientific and technological scenarios adapted to the achievement of the set aim. Alternatively, you should be able to state that you are unable to achieve it giving the necessary explanation as to why it is thus impossible. On the basis of these ideas, it falls to the given applicant (for the hierarchy) to make his/her choices, using his/her decision tree.

– Needs: the latter are shown in summary form in Table 4.6.

Definition by the person requesting the given goals (possibly with a time estimate for reaching these objectives);
Choice of working groups using a given leader;
Framework for each aim: knowledge required; call for disciplinary specialists;
Return and integration of disciplinary knowledge;
Heuristic approach (maieutic) in defining scenarios. Taking account of external constraints;
Researching general topics among the different aims;
Decisions and research of other means for the deployment of scientific and technological works within other niches; other specialized research; openings and cooperation and other activities.

Table 4.6. *Means of action*

This framework devises the hypothesis that the applicant is willing to operate using this type of approach which can produce highly variable results. It can also have possibilities for funding adapted to the proposal for credible scenarios to go beyond, using a given organizational mode that is less "top-down" or "bottom-up", within the methods of research currently being investigated.

4.3.4. *Proposed solutions?*

Alain Juppé and Michel Rocard (ex-French Prime Ministers), in the report *Investir Pour L'Avenir* (which translates as "Investing for the Future", 2009), start their foreword by asking a question, a genuine rhetorical *leitmotif.* This has troubled the French for a long time, "Why are we so good at actual research and yet so weak when it comes to its increase in value?" It was observed, "Three years later, the Gallois report, entitled *Pacte pour la Compétitivité de L'Industrie Française* ("Pact for French Industrial Competitiveness") once again considered this issue. It noted that the weakness of links and exchanges between research and businesses constituted one of the main hindrances to French competitiveness" (MENESR – Ministry for National Education, Higher Education and Research, 2014).

However, in a quasi-simultaneous way, several significant national operations have just been launched. This is despite the national immense complexity of research support mechanisms [BER 16, MEN 14]. These are (lastly) readable operations, but which remain somewhat modest. Some ten billion euros, or half a GDP point, are nowadays devoted to supporting innovation by various public players (mainly the State, the regions and Europe). This is a considerable total relative to the 100 million euros concerned in the following projects:

– The association 4 IRT (French Technological Research Institutes) in an integrated program worth 62 million euros over two years (FIT-Additive Manufacturing, 2016; Primante 3D, 2016) covers process–material themes. Figure 4.11 shows the various couplings subject to importance to the operation innovators (compared this to the proposals of Lubineau [LUB 16] and Figure 4.5).

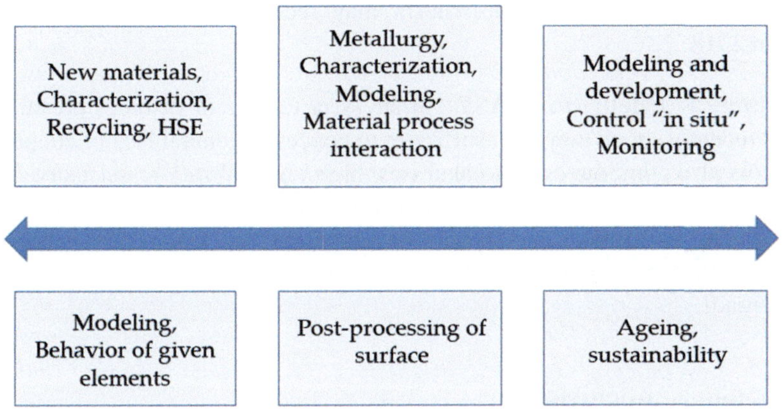

Figure 4.11. *Optimization and industrialization of the 3D printing value chain*

– The CER-eFACTORY project, which is more financially limited (€2.2 million), has the ambition of rationalizing costs, deadlines for the design cycle for new products and increasing the diversity and the quality of conceivable products (top-of-the-range, customization, size and high precision). The target is the production of metallic or metal components [BOU 16].

– A further case study is the SOFIA project "*Solutions pour la Fabrication Industrielle Additive Métallique*" (which translates as "Metallic Industrial Additive Manufacturing Solutions"). The project runs over six years and up to €50 million. It aims for the convergence of competences and experiences of superior players, which cover the entire value chain, from the various powders through to finished components [ZON 16, VIC 16a, VIC 16b].

The State is starting not to be on the sidelines. The French Minister of State for Industry notes, "Additive manufacturing will change many things within the industry, probably everything, although we should be cautious at this stage. France, despite innovative businesses, has a low equipment rate compared to other countries. It is seventh in the world in terms of modern machinery, and fourth in Europe". Christophe Sirugue's ambition is to position France among the top five global additive manufacturing producers [PRO 17]. The idea is to take advantage of PIA ("*Investissements d'Avenir*" – which translates as "Future Investments Programme"), but with the pooling of interest a number of technological platforms. This same source states, the mechanism may receive financial support of €0.6 million in 2018.

However, as Astolfi *et al.* [AST 16] stress, a feeling of disorder prevails. The alliances recently set up may be insufficient to conceal a general lack of cooperation. The authors also point out other weaknesses: high regionalization assistance funding and projects lead to expensive duplication, fragmented sub-contracting faced with a high need for investments, embryonic development and fragmented research. In brief, although hopes emerge, the die has not yet been cast, but we are perhaps on the right road!

4.4. A positive conclusion

We are under no obligation to believe all of the prophecies made by Saam [SAA 15] who wrote: "3D printing does not enable us to beat bar-turning machines or injection molding in the field of high-speed low-cost manufacturing. However printers will improve, indeed more quickly than we could possibly think. The range of printed products will (also) increase and occupy unexpected sectors. We already print electronic circuits, including processors, screens and touch sensors. Other companies are using 3D printing for viruses able to attack carcinogenic cells and bacteria. Completing 3D printing of DNA sequences which are directly injected into the patient is no longer a work of science-fiction". However, whether this takes place in France which is (slightly) resuming effective dynamics or abroad, daily progress is appearing and will continue to improve additive manufacturing processes. This book which represents a snapshot mold of a given situation therefore risks aging very quickly. Moreover, that is the author's wish!

Let's meet up again to discuss the updated position in five years' time (although this is, in probability terms, less and less conceivable for the author) to verify whether the current commitments have been honored or are very different from the projections, arising from both current requirements and research.

What makes particular knowledge become in keeping with the beliefs and rules of a society at a given moment? What makes it increase in value or serves to support the development of political or economic strategies? What are the procedures and other factors to bear in mind? The approaches proposed stress that "techniques, institutional or intellectual material used in the transfer of knowledge constitute one of the major instruments for this form of legitimacy. They also invite us to think about how their nature, constraints and environment are acting, even defining, the models of thought and the knowledge itself" [RUA 17].

4.5. Bibliography

[AAL 05] AALTONEN M., BARTH T., "How do we make sense of the future?", *Journal of Futures Studies*, vol. 9, pp. 45–60, 2005.

[ALE 11] ALEXANDRE L., *La mort de la mort; comment la techno-médecine va bouleverser l'humanité?*, Lattès J.C., Paris, 2011.

[AND 16] ANDRÉ J.C., "Désordres dans la troisième dimension; un cheminement entre formation et interdisciplinarité", *VARIA*, vol. 5, pp. 62–72, 2016.

[ARM 01] ARMSTRONG J., *Principles of Forecasting: A Handbook for Researchers and Practitioners*, Kluwer Academic Publishers, Norwell, 2001.

[ARN 16] ARNDT A., ANDERL R., SCHÜTZER K., "Additive manufacturing – automation in customized production", *21st International Alta Tecnologia Seminar*, 06.10.2016, Piracicaba, Sao Paulo, Brazil, 2016.

[AST 16] ASTOLFI C.P., CONSTANTIN E., MOULET A., "Fabrication additive: mobiliser les forces françaises", *La fabrique de l'industrie*, available at: http://www.la-fabrique.fr/fr/publication/fabrication-additive-mobiliser-les-forces-francaises/, 2016.

[BAK 16] BAKER M., "1,500 scientists lift the lid on reproducibility", *Nature*, vol. 533, 7604 p., available at: http://www.nature.com/news/1-500-scientists-lift-the-lid-on-reproducibility-1.19970, 2016.

[BER 67] BERGER G., *Étapes de la prospective*, Presses Universitaires de France, Paris, 1967.

[BER 94] BERGER R., "Le virtuel jubilatoire: palpitation toujours recommencée", in MARÉCHAL I. A. (ed.), *Sciences et imaginaire*, A. Michel, Paris, 1994.

[BER 13] BERGER R., "Industrie 4.0: L'usine connectée", available at: http://www.gimelec.fr/Actualites/Actualites-Gimelec/Le-Gimelec-publie-Industrie-4.0-les-leviers-de-la-transformation, 2013.

[BER 14] BERNARD A., MANIL P., "Etat et enjeux de la fabrication additive", *Conférence à la Journée IRFU-CEA, Fabrication additive*, Paris, France, December 2014.

[BER 16] BERGER S., DORMAN R., STARBUCK H., "Reforms in the French Industrial Ecosystem", available at: http://www.sf2m.asso.fr/CommissionsThematiques/DocCom Thematiques/RNM_MIT_Final_Summary.pdf, 2016.

[BOI 15] BOILLAT E., "Techniques de fabrication additive", available at: http://www.swissmem.ch/fr/presentations-fabrication-additive.html, 2015.

[BOT 12] BOTTOLIER-DEPOIS F., "FabLabs, Hakerspaces: nouvelles formes d'innovation et militantisme libertaire", Observatoire du Management Alternatif, HEC, Paris, 2012.

[BOU 16] BOURSORAMA, "CERINNOV: lancement d'un projet R&D collaboratif dans l'impression 3D", available at: http://www.boursorama.com/actualites/cerinnov-lancement-d-un-projet-r-d-collaboratif-dans-l-impression-3d-9228ea7d6935751a43d084c34bf5ff00, 2016.

[CAC 16] CAC, "Mise à jour préliminaire des données: rendement et réputation internationale de la recherche au Canada", available at: http://sciencepourlepublic.ca/uploads/fr/assessmentspublicationsnewsreleases/stird2016/st_interimdataupdate_fr_web.pdf, 2016.

[CHA 14] CHANUT-GUIEU C., GUIEU G., "Créativité, PME et entrepreneuriat: des zones d'ombre et de lumière", *Revue internationale P.M.E.: économie et gestion de la petite et moyenne entreprise*, vol. 27, pp. 35–63, 2014.

[COM 16] COMTESSE X., "Une innovation disruptive", available at: http://www.hebdo.ch/les-blogs/un-tsunami-num%C3%A9rique-r%C3%A9volutionne-la-sant%C3%A9/une-innovation-disruptive, 2016.

[CYE 63] CYERT R.M., MARCH J.G., *A Behavioral Theory of the Firm*, Prentice-Hall, Englewood Cliffs, 1963.

[DE 05] DE RIEDMATTEN E., *XXIe siècle: les innovations qui vont changer notre vie*, L'archipel, Paris, 2005.

[DEC 16] DECHAMP G., DELAUNAY C., "Favoriser et gérer l'interdisciplinarité dans l'entreprise: la notion de territoires partagés", *Management & Avenir*, vol. 88, pp. 37–58, 2016.

[DEL 14] DELOITTE, "Du nouveau dans le secteur manufacturier: les répercussions de l'impression tridimensionnelle", available at: http://www2.deloitte.com/content/dam/Deloitte/ca/Documents/insights-and-issues/ca-fr-perspectives-et-enjeux-daffaires-du-nouveau-dans-le-secteur-manufacturier.pdf, 2014.

[DES 17] DESPEISSE M., BAUMERS M., BROWN P. *et al.*, "Unlocking value for a circular economy through 3D printing: a research agenda", *Technological Forecasting and Social Change*, vol. 115, pp. 75–84, 2017.

[DGE 16] DGE, "Technologies clés – Préparer l'industrie du future 2020", available at: http://www.entreprises.gouv.fr/files/files/directions_services/politique-et-enjeux/innovation/technologies-cles-2020/technologies-cles-2020.pdf, 2016.

[DGR 17] DGRI – EU, "Research and innovation analysis in the European Semester Country Reports 2017", available at: https://rio.jrc.ec.europa.eu/en/file/10598/download?token=XYtVdzs0, 2017.

[DRU 06] DRUCKER P., *Innovation and Entrepreneurship*, Collins, New York, 2006.

[DUF 11] DUFOURMENTELLE A., *Eloge du risque*, Payot, Paris, 2011.

[EAR 15] EARLS A., BAYA V., "The road ahead for 3D printers", available at: http://www.pwc.com/us/en/technology-forecast/2014/3d-printing/features/future-3d-printing.html, 2015.

[EIN 07] EINSTEIN A., *L'espoir économique: vers la révolution du quaternaire*, Bourin, Paris, 2007, (cited by DEBONNEUIL M.).

[EUR 12] EUROGROUP CONSULTING, "L'art du management de l'innovation dans le service public", Eurogroup, Paris, France, 2012.

[FIM 15] FIM, "La France dispose enfin d'un document de référence sur l'usine du futur", announcement by Philippe Contet of FIM, available at: http://www.usinenouvelle.com/article/la-france-dispose-enfin-d-un-document-de-reference-sur-l-usine-du-futur-annonce-philippe-contet-de-la-fim.N363680, 2015.

[FIT 16] FIT-ADDITIVE MANUFACTURING, "4 IRT s'associent pour traiter l'ensemble du spectre R&T de la fabrication additive afin de préparer différentes filières à une évolution industrielle", available at: http://www.irt-saintexupery.com/wp-content/uploads/2016/12/Additive-Manufacturing_Fiche_IRTStEx_FR.pdf, 2016.

[FOR 89] FORSÉ M., *L'ordre improbable*, PUF, Paris, 1989.

[FRA 17] FRANCHETTI M., KRESS C., "An economic analysis comparing the cost feasibility of replacing injection molding processes with emerging additive manufacturing techniques", *International Journal of Advanced Manufacturing Technology*, vol. 88, pp. 2573–2579, 2017. doi: 10.1007/s00170-016-8968-7.

[FRÉ 15] FRÉOUR P., "Une prothèse de main robotisée en impression 3D", available at: http://sante.lefigaro.fr/actualite/2015/06/20/23868-prothese-main-robotisee-impression-3d, 2015.

[GAG 11a] GAGLIO G., *Sociologie de l'innovation*, PUF, Paris, 2011.

[GAG 11b] GAGLIO G., LAURIOL J., DU TERTRE C., *L'économie de la fonctionnalité: une voie nouvelle vers un développement durable?*, Octarès, Paris, 2011.

[GAN 00] GANASCIA J.G., "Science x.0", in HAIGNERÉ C. (ed.), *Plaidoyer pour réconcilier la science avec la culture*, Le Pommier, Paris, 2000.

[GAU 05] GAUDIN T., *La prospective*, PUF, Paris, 2005.

[GIR 17] GIRET V., "Le décryptage éco. La France à la traîne de l'innovation", available at: http://www.francetvinfo.fr/replay-radio/le-decryptage-eco/le-decryptage-eco-la-france-a-la-traine-de-l-innovation_2015094.html, 2017.

[GOD 01a] GODET M.,"La boîte à outils de prospective stratégique", *LIPS working papers*, vol. 5, Paris, France, 2001.

[GOD 01b] GODET M., "De la rigueur pour une indiscipline intellectuelle", in LESOURNE J., STOFAÈS S. (eds), *Prospective stratégique d'entreprise*, Dunod, Paris, pp. 101–131, 2001.

[GOD 01c] GODET M., "Décision, prospective, auto-organisation – A propos des, Mélanges en l'honneur de Jacques Lesourne", *Futuribles*, vol. 263, pp. 29–35, 2001. quoted by Cohendet P. (2001).

[GRA 86] GRANGER G.G., "Pour une épistémologie du travail scientifique" in HAMBURGER J. (ed.), *La philosophie des sciences aujourd'hui*, Gauthier-Villars, Paris, 1986.

[GRE 98] GREENBAUM T.L., "10 tips for running successful focus groups", available at: http://www.groupsplus.com/pages/mmn091498.html, 1998.

[GRI 98] GRIBBIN J., *The Future Now: Predicting the 21st Century*, Weidentfeld & Nicolson, London, 1998.

[GRI 16] GRIFFITHS C.A., HOWARTH J., DE ALMEIDA-ROWBOTHAM G. *et al.*, "A design of experiments approach for the optimization of energy and waste during the production of parts manufactured by 3D printing", *Journal of Cleaner Production*, vol. 139, pp. 74–85, 2016.

[GUI 56] GUILFORD J.P., "The structure of intellect", *Psychological Bulletin*, vol. 53, pp. 267–293, 1956.

[GUI 03] GUILLEBAUD J.C., *Le gout de l'avenir*, Le Seuil, Paris, 2003.

[GUO 13] GUO N., LEU M.C., "Additive manufacturing: technology, applications and research needs", *Frontiers of Mechanical Engineering*, vol. 8, pp. 215–243, 2013.

[GUZ 98] GUZZETTI L., "Euroscientia forum", *Commission Européenne*, Brussels, Belgium, vol. 1, pp. 85–86, 1998.

[HAB 73] HABERMAS J., *La technique et la science comme "idéologie"*, Gallimard, Paris, 1973.

[HAG 16] HAGUE R., REEVES P., JONES S., "Innovate UK – mapping UK research and innovation in additive manufacturing", available at: https://www.gov.uk/government/uploads/system/uploads/attachment_data/file/6991/CO307_Mapping_UK_Capability_AM_WL_Print_.pdf, 2016.

[HAL 16] HALF R., "Techniques de management pour doper la créativité des équipes", available at: https://www.roberthalf.fr/blog/techniques-de-management, 2016.

[HAM 13] HAMEL J., "L'interdisciplinarité, manière de faire ou de dire la science?", available at: http://www.espacestemps.net/articles/linterdisciplinarite-maniere-de-faire-ou-de-dire-la-science-2/, 2013.

[HEL 67] HELMER O., *Analysis of the Future: the Delphi Method*, Rand Corporation, Santa Monica, 1967.

[HUX 01] HUXLEY A., *Le meilleur des mondes*, Livre de poche, Paris, 2001 [1932].

[INS 15] INSIS, Rapport de prospective, CNRS, Paris, 2015.

[JAC 13] JACKSON T., *Prosperité sans croissance*, De Boeck, Brussels, 2013.

[JIA 17] JIANG R., KLEER R., PILLER F.T., "Predicting the future of additive manufacturing: a Delphi study on economic and societal implications of 3D printing for 2030", *Technological Forecasting and Social Change*, vol. 117, pp. 84–97, 2017.

[JRC 13] JRC, "Universities and collaborative innovation in EC-funded research projects: an analysis based on innovation radar data", available at: http://publications.jrc.ec.europa.eu/repository/bitstream/JRC104870/jrc104870_innovation-radar_2_final.pdf, 2013.

[KAK 98] KAKU M., *Visions: How Science Will Revolutionize the Twenty-first Century?*, Oxford University Press, Oxford, 1998.

[KAU 15] KAUFMAN S.B., "The emotions that make us more creative", available at: https://hbr.org/2015/08/the-emotions-that-make-us-more-creative, 2015.

[KIA 15] KIANIAN B., TAVASSOLI S., LARSSON T.C., "The role of additive manufacturing technology in job creation: an exploratory case study of suppliers of additive manufacturing in Sweden", *Procedia* CIRP, Elsevier, Procedia CIRP, pp. 93–98, available at: http://dx.doi.org/10.1016/j.procir.2014.07.109, 2015.

[KIT 94] KITZINGER J., "The methodology of focus groups: the importance of interaction between research participants", *Sociology of Health*, vol. 16, pp. 103–121, 1994.

[KOE 11] KOESTLER A., *Le cri d'Archimède: la découverte de l'art et l'art de la découverte*, Les Belles Lettres, Paris, 2011.

[KUR 00/01] KURZBERG T.R., AMABILE T.M., "From Guilford to creative synergy: opening the black box of team-level creativity", *Creativity Research Journal*, vol. 13, pp. 285–294, 2000-2001.

[LAF 11] LAFLAMME S., "Recherche interdisciplinaire et réflexion sur l'interdisciplinarité", *Nouvelles Perspectives en Sciences Sociales*, vol. 7, pp. 49–64, 2011.

[LAG 13] LAGADEC P., *Piloter en univers inconnu*, Préventique, Bordeaux, France, 2013.

[LE 16] LE MONDE, "Hollande et la recherche: aveux à l'Assemblée", available at: http://huet.blog.lemonde.fr/2016/12/11/hollande-et-la-recherche-aveux-a-lassemblee/, 2016.

[LES 01] LESOURNE J., STOFFAÈS C. (eds), *Prospectives stratégiques d'entreprise*, Dunod, Paris, 2001.

[LES 16] LESAGE S.E., HURTEAU P., "L'innovation en recherche universitaire: Austérité et arrimage aux intérêts du secteur privé", available at: http://iris-recherche.s3.amazonaws.com/uploads/publication/file/web_NouveauTexteEtude_RechercheUniversitaire_vli_02.pdf, 2016.

[LI 17] LI Q., KUCUKKOC I., ZHANG D.Z., "Production planning in additive manufacturing and 3D printing", *Computers & Operations Research*, vol. 83, pp. 157–172, 2017.

[LIE 15] LIENEKE T., ADAM G.A.O., LEUDERS S. *et al.*, "Systematical determination of tolerances for additive manufacturing by measuring linear dimensions", *26th Annual International Solid Freeform Fabrication Symposium*, Austin, Texas, USA, available at: http://sffsymposium.engr.utexas.edu/sites/default/files/2015/2015-30-Lieneke.pdf, pp. 371–384, 2015.

[LOU 17] LOUISOT J.M., "Optimiser la prise de risques", *Risk-Assur*, vol. 474, pp. 28–37 available at: https://www.riskassur-hebdo.com, 2017.

[LUB 16] LUBINEAU P., "La Fabrication Additive en France: cartographie des acteurs", available at: http://aif.diwi.org/wp-content/uploads/2016/07/20160719-Etude-Carnot-FIMP-Fab-Add-Cartographie-des-compétences-Diff-AIF.pdf, 2016.

[MAN 13] MANYIKA J., CHUI M., DOBBS R. *et al.*, *Disruptives Technologies: Advances That Will Transform Life, Business, and the Global Economy*, McKinsey Global Institute, San Francisco, CA, 180 p., 2013.

[MAR 13] MARCHESNAY M., "Le système 1 "Parva sed apta": l'innovation dans les PME", in BOUTILLIER S., DJELLAL F., UZUNIDIS D. (eds), *L'innovation: analyser, anticiper, agir*, P. Lang, Brussels, Belgium, pp. 29–62, 2013.

[MAR 17] MARTI J.F., "Design Thinking: un outil qui favorise l'innovation et remet en cause les certitudes dans l'entreprise", available at: http://www.hbrfrance.fr/chroniques-experts/2017/01/13811-design-thinking-un-outil-qui-favorise-linnovation-et-remet-en-cause-les-certitudes-dans-lentreprise/, 2017.

[MÉL 17] MÉLANIE, "L'impression 3D atteindrait $20,5 milliards en 2020 selon Deloitte", available at: http://www.3dnatives.com/deloitte-impression-3d-02012017/, 2017.

[MEN 14] MENESR, "Développement et impact du crédit d'impôt recherche, 1983-2011", available at: http://www.enseignementsup-recherche.gouv.fr/cid80816/developpement-et-impact-du-credit-d-impot-recherche-1983-2011.html, 2014.

[MIL 91] MILLETT S., HONTON E., *A Manager's Guide to Technology Forecasting and Strategy Analysis Method*, Battelle Press, New York, 1991.

[MOA 08] MOATI P., "La prospective sectorielle: les apports de l'approche évolutionniste", *Management & Avenir*, vol. 3, pp. 205–233, available at: www.cairn.info/revue-management-et-avenir-2008-3-page-205.htm, 2008.

[MOD 92] MODIS S., *Predictions: Society Telltale Signature Reveals the Past and Forecast the Future*, Simon & Schuster, New York, 1992.

[MON 84] MONOD T., *L'émeraude de Garamantès*, Babel, Paris, 1984/1999.

[MOR 97] MORGAN D.L., *Focus Groups as Qualitative Research*, Sage, London, 1997.

[MUN 17] MUNAFO R.R., NOSEK B.A., BISHOP D.V.M. *et al.*, “A manifesto for reproducible science”, *Nature Human Behaviour*, vol. 1, p. 0021, available at: http://www.nature.com/articles/s41562-016-0021, 2017.

[NAU 45] NAUDÉ G., De Nipho iudicium in Agostino Nifo, Opuscola moralia et politica Parisiis, sumptibus Roleti le Duc, 1645, (cited by [SCH 16]).

[NOV 13a] NOVEL A.S., “L’imprimante 3D, ce gadget qui change le monde”, available at: http://alternatives.blog.lemonde.fr/2013/03/06/limprimante-3d-ce-gadget-qui-change-le-monde/comment-page-1/, 2013.

[NOV 13b] NOVEL A.S., RIOT S., *Vive la co-révolution; pour une société collaborative*, Manifestô-Gallimard, Paris, 2013.

[OEC 16] OECD, “OECD Science, Technology and Innovation Outlook 2016”, available at: http://www.agefi.fr/sites/agefi.fr/files/fichiers/2016/12/stio_key_messages_backup.pdf, 2016.

[ORW 00] ORWELL G., *Nineteen Eighty-four*, Penguin, London, 2000.

[PAP 04] PAPON P., *Le temps des ruptures*, Fayard, Paris, 2004.

[PAU 72] PAUWELS L., BERGIER J., *Le matin des magiciens*, 1st ed., Folio, Paris, 1972.

[PAU 07] PAULI G., *Croissance sans limites: objectif zéro pollution – Croissance économique et régénération de la nature*, Quintessence, Paris, 2007.

[PER 15] PEREZ A., “3D printing’s role in shapping the future of education”, available at: http://3dprintingindustry.com/2015/07/14/3d-printings-role-in-shaping-the-future-of-education/?utm_source=3D+Printing+Industry+Update&utm_medium=email&utm_campaign=75f7ab45d6-RSS_EMAIL_CAMPAIGN&utm_term=0_695d5c73dc-75f7ab45d6-64557777, 2015.

[PET 98] PETERS T., *L’innovation, un cercle vertueux*, Village Mondial, Paris, 1998.

[PEY 17] PEYRÉ TARTARUGA I.G., CAZAROTTO R.T., BACKX MARTINS C.H. *et al.*, “Innovation and public understanding of science: possibility of new indicators for the analysis of public attitudes to science, technology and innovation”, available at: https://mpra.ub.uni-muenchen.de/76288/1/MPRA_paper_76262.pdf, 2017.

[PHI 12] PHILIPPE J., “L’innovation managériale, comment innover dans l’univers bancaire?”, in EUROGROUP CONSULTING (ed.), L’art du management de l’innovation dans le service public, Eurogroup, Paris, pp. 79–102, 2012.

[PLA 03] PLASSARD F., Rétrospective de la prospective: la prospective dans le domaine des transports; recherché critique, Rapport du PREDIT au Ministère de l’équipement, Paris, France, 2003.

[PRI 16] PRIMANTE 3D, "Fabrication additive métallique: un programme de formation pour les PME!", available at: http://www.primante3d.com/pack-pme-22122016/, 2016.

[PRO 17] PROTAIS M., "L'Etat dévoile sa stratégie pour favoriser l'impression 3D", available at: http://www.usinenouvelle.com/editorial/l-etat-devoile-sa-strategie-pour-favoriser-l-impression-3d.N517559, 2017.

[REI 10] REICH M.R., "Le cas de la pollution du Michigan par le PBB", in AKRICH M., BARTH Y., RÉMY C. (eds), *Sciences, politique et environnement*, Mines ParisTech, Paris, 2010.

[RIF 98] RIFKIN J., *The Biotech Century: Harnessing the Gene and Remaking the World*, Gattancz, London, 1998.

[RIF 00] RIFKIN J., *L'âge de l'accès*, La Découverte, Paris, 2000.

[ROC 06] ROCHER S., MATHÉ J.C., "Contrôler la rupture technologique pour rester leader: une stratégie gagnante? Une illustration au travers du marché des consoles de jeux vidéo", *Management & Avenir*, vol. 4, pp. 9–24, 2006.

[ROG 03] ROGER E., *Diffusion of Innovations*, Oxford University Press, New York, 2003.

[ROQ 07] ROQUETTE M.L., *La créativité*, Que sais-je, PUF, Paris, 2007.

[ROS 15] ROSENBERG J., "L'impression 3D, une révolution industrielle mondiale; contexte, enjeux et opportunités", available at: https://www.arts-et-metiers.asso.fr/manifestation_cr/cr_3054.pdf, 2015.

[RUA 17] RUANO-BORBALAN J.C., "Techno-sciences en société : les voies multiples de la légitimation des savoirs", *Innovations*, vol. 1/2017, pp. 5–15, 2017.

[SAA 15] SAAM C., "La fabrique du futur", available at: http://www.patentattorneys.ch/2015/10/17/la-fabrique-du-futur/, 2015.

[SAI 00] SAINT EXUPÉRY A., *Citadelle*, Folio, Paris, 2000.

[SAR 76] SARTRE J.P., *L'être et le néant*, Gallimard, Paris, 1976, [1943].

[SCH 16] SCHINO L., "La critique libertine de la religion: mécanismes de formation des croyances et psychologie des masses", *ThéoRèmes*, vol. 9, available at: http://theoremes.revues.org/880, 2016.

[SCH 17] SCHMITT L., "L'industrie en pleine révolution", available at: http://www.estrepublicain.fr/actualite/2017/01/19/l-industrie-en-pleine-revolution, 2017.

[SCI 14] SCIENCE EN MARCHE, "Financement des laboratoires", available at: http://sciencesenmarche.org/fr/financement-des-laboratoires/, 2014.

[SIL 16] SILVERZAHN P., "Comment l'innovation est étouffée par ceux qui la défendent", available at: https://www.contrepoints.org/16/05/03/1100-comment-innovation-est-tuee-par-ceux-qui-la-defendent#FQQiKgW5FRimzZ9f.99, 2016.

[SLO 06] SLOTERDIJK P., *Le palais de cristal: à l'intérieur du capitalisme planétaire*, Maren Sell, Paris, 2006.

[STE 92] STEWART D.W., SHAM DASANI P.N., *Focus Groups: Theory and Practices*, Sage, London, 1992.

[STO 97] STOKES D.E., *Pasteur's Quadrant: Basic Science and Technological Innovation*, Brookings Institution Press, Washington, DC, 1997.

[SUC 16] SUCHAUD N., "Prospective: Quel impact l'innovation numérique peut-elle vraiment avoir sur l'économie?", available at: http://www.maddyness.com/prospective/2016/01/04/g9plus-idate-innovation-economie/, 2016.

[SYN 12] SYNTEC, *Compétitivité AAA; proposition pour un nouveau pacte France-entreprises*, Eyrolles – Ed. d'Organisation, Paris, 2012.

[TAY 14] TAYLOR S., "Trend evolution: 3D printing trends", available at: http://3dprintingindustry.com/2014/02/24/trend-evolution-3d-printing-trends-part-1/, 2014.

[THO 16] THOMPSON M.K., MORONI G., VANEKER T. *et al.*, "Design for additive manufacturing: trends, opportunities, considerations, and constraints", *CIRP Annals – Manufacturing Technology*, vol. 65, pp. 737–760, 2016.

[TOF 71] TOFFLER A., *Le choc du futur*, Denoël, Paris, 1971.

[TOF 83] TOFFLER A., *Les cartes du futur*, Denoël, Paris, 1983.

[TOF 91] TOFFLER A., *Les nouveaux pouvoirs: savoir richesse et violence à la veille du XXIe siècle*, Fayard, Paris, 1991.

[TOF 06] TOFFLER A., TOFFLER H., *Revolutionary Wealth*, Knopf, New York, 2006.

[TRI 15] TRIBOT S., "La Multifab, l'imprimante 3D capable de manipuler jusqu'à 10 matériaux", available at: http://www.techniques-ingenieur.fr/actualite/articles/la-multifab-limprimante-3d-capable-de-manipuler-jusqua-10-materiaux-28741/, 2015.

[URB 17] URBANIC S.J., HEDRICK R.W., BURFORD C.G., "A process planning framework and virtual representation for bead-based additive manufacturing processes", *International Journal of Advanced Manufacturing Technology*, vol. 90, pp. 361–376. doi: 10.1007/s00170-016-9392-8, 2017.

[VAN 14a] VANEK J., GARCIA-GALICIA J.A., BENES B. *et al.*, "PackMerger: a 3D print volume optimizer", *Computer Graphics Forum*, vol. 33, pp. 322–332, 2014.

[VAN 14b] VAN CUTSEM M., "Les futurs qui n'ont pas eu lieu sont-ils vraiment passés?", in GUYOT J.L., BRUNET S. (eds), *Construire les futurs; contributions épistémologiques et méthodologiques à la démarche prospective*, Presses Universitaires de Namur, Belgium, pp. 163–180, 2014.

[VAN 16] VANIER J., "Performance ou d'évaluation du développement économique: comment sortir de la confusion?", available at: http://www.lesechos.fr/idees-debats/cercle/cercle-164037-performance-ou-devaluation-du-developpement-economique-comment-sortir-de-la-confusion-2051889.php#ELyERMWF9RDl0e0F.99, 2016.

[VAU 16] VAUTIER J.F., PÉRINET R., MAZRI C., "C'est scientifique, c'est indiscutable?", *Tribunes de la Sécurité industrielle*, vol. 7, p. 4, available at: https://www.foncsi.org/fr/publications/collections/tribunes-securite-industrielle/c-est-scientifique-c-est-indiscutable/tribune-2016-07, 2016.

[VIC 16a] VICTOR, "ECO 3D: L'actu économique de l'impression 3D en novembre", available at: http://www.3dnatives.com/economie-impression-3d-30112016/, 2016.

[VIC 16b] VICTOR, "SOFIA, le plan français dédié à la fabrication additive métal", available at: http://www.3dnatives.com/sofia-fabrication-additive-29112016/, 2016.

[VON 89] VON REIBNITZ U. , *La technique des scénarios*, AFNOR, Paris, 1989.

[WAL 08] WALLONICK D.S., "An overview of forecasting methodology", available at: http://www.stratpac.com/research-papers/forecasting.html, 2008.

[WEL 15] WELLER C., KLEER R., PILLER F.T., "Economic implications of 3D printing: markets tructure models inlight of additive manufacturing revisited", *International Journal of Production Economics*, vol. 164, pp. 43–56, 2015.

[WIE 60] WIENER N., "Some moral and technical consequences of automation", *Science*, vol. 131, pp. 1355–1358, 1960.

[ZON 16] ZONE BOURSE, "Esi: Industriels et académiques s'associent pour lancer SOFIA", available at: http://www.zonebourse.com/ESI-GROUP-5195/actualite/Esi-Industriels-et-academiques-s-rsquo-associent-pour-lancer-SOFIA-23450997/, 2016.

5

Questions of Hope and "Unhope"

"Whether it is a matter of manufacturing or management, we certainly always get involved in a world that we did not create. However, manufacturing hides this relationship with Nature, while management is constrained to take it into consideration. It is from this perspective that the difference between them must be questioned". [LAR 17]

Application to bio-printing?: "Deciphering the enigmas of life and producing a theory that reflects reality is not their primary goal. It is unveiling possibilities rather than the real". [BEN 15]

"The plebeians of research could only count on their own fear of losing their place in this hot, renowned, and compartmentalized trashcan". [CÉL 72]

"Alliances do not interfere with the purity of the circulation that is scientific information, but rather, they accelerate it and precipitate its pulsations". [LAT 07]

"The insane rush where angels fear to venture". [POP 11]

"If it is no longer a matter of conquering the power of fiction, it is always about putting it to the test, about submitting the reasons that we invent to a third party likely to put them at risk". [LEM 02]

"The map is not the territory and, as is known, the word 'dog' does not bark. However, like maps, the word [...] representing reality guides each of us in the construction of his/her own perspective of this reality". [AND 13b]

"A plausible future is announced, of course, where there could be scientists, but where they, more or less skilled employees, could no longer be distinguished from all the others who, each in his/her own position, sell their workforce". [STE 97]

"The entire system [...] is struck with myopia. These are henceforth short-sighted operations, at the mercy of the immediate situation that replaces long-term historical visions or political strategies". [ROS 13]

"Arthur Koestler was not far off when he described the behavior of certain creators in science by comparing them to that of sleepwalkers, stumbling upon major discoveries. However, it is still necessary for the moment to be favorable and the mind prepared, and possibly for chance to come to your aid". [DEG 12]

"No authority decides what a good idea is". [FEY 98]

"His mind slid away into the labyrinthine world of doublethink. To know and not to know, to be conscious of complete truthfulness while telling carefully constructed lies, to hold simultaneously two opinions which cancelled out, knowing them to be contradictory and believing in both of them, to use logic against logic, to repudiate morality while laying claim to it". [ORW 72]

5.1. Introduction

Going through a few dozen pages to present a rather exciting, more than 30-year voyage allows some questions to be asked about the management of research on emerging subjects. For Kaufman [KAU 15], "the creative know how to daydream, to take the time to let their mind wander, and, if possible, to work during the hours that best suit them… They also love to be alone, to better 'listen to' their intuition. They know what they love, fully dedicate themselves to it, and take risks and expose themselves to new experiences, sensations, and states of mind. They know how to observe the world and people, to put themselves in others' shoes, to ask questions, to be curious like a young child who wants to know everything, to accept failure, to start better again. At the same time, they know how to sprinkle friendly and even professional relationships with sarcasm and irony,... Sarcasm could mobilize divergent thought, abstract functions of thought." On the basis of this is the idea that the more sarcastic are the most intelligent... How does all of this "ebullition" produce something useful in the "well-organized research system"?

Independently of this aspect linked to the person, Half [HAL 16] foresaw still infrequent relational methods in the academic domain. He proposed:

– Implementing good organization: Innovation is a team effort today. From an organizational standpoint, several trends emerge: leaving autonomy to encourage an increase in information and the emergence of good ideas, regularly organizing mini-meetings where ideas are presented and submitted to the reactions of each, or even forming working groups to transform an idea into a prototype and thus concretely evaluating its potential;

– Stimulating engagement: The manager's role is to give his/her colleagues a desire to go beyond their "job description" and their career plan. He/she must show that he/she really counts on their diagnoses and their suggestions;

– Listening and granting the right to make mistakes: It is of little importance if the field analysis or the suggestion is not pertinent; what matters is trying. Maybe next time will be better;

– Knowing how to react quickly: Proving speed and deciding to realize "a pilot/demonstrator rather than creating a business model" (like in the French National Research Fund Agency (ANR French National Research Agency) or in the H2020 (Present Framework program from the EU) project!).

In the world of innovation, execution speed is vital. Indeed, it is easier to note the pertinence of an idea when it is implemented and tested by the final users,

– Recruiting different profiles: The more varied the profiles within the team are, the richer the reactions will be and the more likely they are to bring about innovation. This diversity can be expressed in different ways: degrees, sensitivities, ages, professional experiences and so on;

– Working in collaboration: Multiplying the initiatives for a cooperative operation. Lending value to collective intelligence by creating collective ideas where all interested parties can contribute with their expertise.

It can be interesting to pass these propositions through the filter of work on 3D printing and the understanding of the practices and cultures of the academic world. To do this, with the help of engineering student Marine Horckmans, an "anthropological" survey was conducted in 2016 to study the working activities/action modes of a "laboratory" tribe, which will constitute the first part of this final chapter (we must finish by concluding Volume 2).

With an emerging subject like additive manufacturing, there is a need to give free reign to imagination because, at least in what was presented in Volume 2, there is room to set it free to propose action plans; not all of which plans will bring the future, of course. However, how can we help create this future? It is thus a second, more hierarchical tribe that will be studied after the first.

These two explorations will serve as an attempt to bring out some strong points so that we can continue to play with the matter to produce objects that are mobile, living, nanometric and others in every imaginable dimension. Thus, an attempt to give a synthetic standpoint on the vision of non-programming for responsible research [AND 13a] that supports risk-taking is highly demanded.

5.2. The "lab-tribe" (LT) approach

According to Boudon [BOU 09], science, at its origin, is born of a vague "program": "describing the real as it is [...]. Every scientific discipline describes, classifies, and explains." The large missions of scientific research vis-à-vis Industry and Society are defined by Fitreman [FIT 04] by the following modes of action:

– production of scientific knowledge with an aim to develop the culture and emancipation of citizens;

– transfer of knowledge to industries: economic growth, employment, sustainable development;

– communication with society and closer relations between research and citizens. In the European Charter for Researchers [COM 05], several important complementary elements should be remembered:

- research freedom: "Researchers should focus their research work on the well-being of humanity and the expansion of the borders of scientific knowledge, all while enjoying the freedom of thought and expression, as well as the freedom to determine the methods that allow problem-solving according to the recognized ethical practices and principles";

- responsibility: "Researchers ensure that in the case of delegating any aspect of their work, the delegate has the necessary skill";

- professional attitude: "Researchers [...] should ask for all the necessary authorizations before starting their research task or accessing the provided resources";

- good practices: "Researchers should, at all times, adopt safe working methods, in conformance with the national legislation, and notably take the necessary precautions to guarantee health and safety";

- engagement vis-à-vis Society: "Researchers should ensure that their research activities are inclined to society's knowledge as a whole such that they can be understood by non-specialists, thereby improving society's understanding of Science." This engagement must also explore the future [KOS 06].

This ideal, even slightly angelic, situation is limited for different reasons. It is already in opposition to Mahé's views [MAH 02], revisiting those of Bourdieu [BOU 96] that define scientific competition today as a power stake: "purely scientific maximization of profit, i.e. the recognition likely to be obtained from competing pairs with a view to obtain a monopoly on scientific authority, inseparably defined as technical capacity and social power." Thus, for example, Rémi Barré [BAR 04], calling on the general problems of research, proposed once again updating the social contract between the State and scientific research environments. He wrote: "However, it is necessary to observe that the debate is difficult to create and rather gives the feeling of agitation in vain than that of advances towards solutions elaborated from diagnostics and shared visions." Yet, in the absence of true, profound debates by all stakeholders, on the existence of the associated disciplines and corporatisms, this difficulty of evolution has particularly appeared during controversies about the ends of research: growth of knowledge and help in solving problems related to society. For a long time, scientific disciplines have shown their pertinence; if these are kept in the daily lives of researchers, other approaches on boundary objects support more interdisciplinary working, imposing new, open and ephemeral links between stakeholders. This world to be explored by science, common to these two approaches and visited during this survey, could use a more in-depth anthropology because it is complex, and because it must return to the foundations between the goals (which vary according to the level of reading) and the means to attain them, in a domain that strongly interacts with the economic world working in engineering sciences on "science objects".

The difficulty of making the structures of scientific research evolve is not new. Change must be explained as a function of criteria with diverse origins, whose robustness is sometimes arguable, which leads to modest potential added values, as they sometimes are based on foundations that are far from rationality. Indeed, the value of the importance of research activity cannot be shown, for they are also the goals of science itself. In fact, problematic difficulties demonstrating this come from this state of things. Mahé [MAH 02], for example, believed that, "the conservatism of Science is not so much a refusal of change as this need for consensual norms without which science could not be done."

"The utility of the latter [science] comes from evidence, while that of the former comes from experience: all new knowledge proving itself likely to give rise to original applications or progress of existing applications" [PIG 04]. Essentially, the linear stereotype "Science Finds, Industry Applies, Man Conforms" (slogan of the 1933 World's Fair in Chicago, cited by Bourg and Schlegel [BOU 01a]) remains particularly lively because there is a desire to believe in it, given its truly simple and easily understood appearance (particularly for those who have never developed profound interactions with the socio-economy from science). On this subject, should Thierry Gaudin's view [GAU 78] be accepted when he wrote: "In some way, research is in a fetal situation: an umbilical cord regularly feeds it; warm and cozy, it perceives the sound of the muffled, faraway world. From deep within it, it dreams: the confrontations that it imagines are based on principles: it is a matter of absolute combat, reflections of the uniqueness of its feeding. Unsatisfied, it turns over and exhausts itself in intestinal battles with dreamlike stakes; out of a reflex for self-preservation, it becomes incomprehensible and incommensurable, hides away in an extreme status where its suppression would be the same as denying a concept, metaphysical crime"?

5.2.1. *Context elements*

In France, as in many developed countries, the recognition of the importance of science and technology in economic policies, in the innovation process and in daily life, leads to growing attention being paid to the interactions between science and society. The breadth of the challenges our societies face actually accelerate scientific interrogations that go far beyond the single community of researchers, for these challenges directly affect all citizens: this is particularly true for health or the environment – climate change and food serve as remarkable examples.

Simultaneously, scientific advances and technological developments themselves have very concrete, significant impacts on daily life: they therefore create hope, sometimes senseless hope, appropriations, detours, questionings, even oppositions and conflicts, vis-à-vis research, science and technology.

LT activity, considered very satisfactory by scientific evaluators, is developing in a world that has greatly evolved since the emergence of its scientific discipline, without great challenges, as it seems difficult to find major inflections of concepts and methods; instruments and modeling have changed with the objects of study. This situation poses the question of the temporality of research within a paradigm. It could be necessary for a persistent number of serious anomalies to emerge to challenge the present paradigmatic exploration and particularly for a new conceptual work base to emerge to follow it. Under these conditions, the questions that are posed to the laboratory do not concern the theoretical foundations on which it is based, but interrogations of its sociological evolutions to remain a national reference in its domain.

It could be interesting to understand the existence of a certain pettiness of hypotheses, of investments in creative operations with high risks and individual preferences, with the nature of interactions and mutualization of knowledge and so on. By attempting to reflect on the role of exchange devices and force relationships being exercised between personnel both internally and externally, the surveying process hoped to escape pure scientific discourse. In paradigmatic stability, the idea frees an objective function ordering a set of apparently disjointed disciplinary research works from the outside: the applicability of research that deeply shapes the view of researchers toward their own scientific activity.

Independent of this aspect, there is in research a singular element, that of rarity and of originality, which can bring about exchange through the dynamics and attractiveness created. It could have been interesting, then, to examine why and how exchange relationships were developing internally or externally (in the shape of mutualization of knowledge, cross fertilization, client–provider relation, barter or money). On this subject, researchers in their scientific quest must have financial and human support; this problematic thus poses the question of research management (or of the autonomy of the researcher in his/her role as a "virtuous" craftsman) and possible cooperation to help certain some maintain a cap for the "liner" that is a large research unit. In fact, how can the discontinued financing that researchers know in the support of agencies (with success rates of about 10–20%) be smoothed? Then, what forces act to ensure that separate researchers "stay together" and constitute what the CNRS (National Center for Scientific Research) calls a research unit? The survey conducted tried to update recreational social mediations from which these individual-to-individual realizations find themselves transformed and modeled to avoid a possible anarchy, and to get involved in a global compatibility creating meaning. What energy is provided to attain this dynamic state of equilibrium allowing the second principle of thermodynamics to be fought? In the facts, doesn't a conception clearly appear of the research world focused on itself, its promotion and a standardized fetishism of the evaluation that is adapted to it? Don't the relations between actors appear, then, in the background, nearly in antagonism? Is this a detour relative to the mission (insofar as it is defined)?

5.2.2. *Some results*

The author refers to the work of Horckmans [HOR 16] to access the details of his survey (method, data and interview management, cross-referenced questions, comments, etc.). Only results that must help concoct the future of additive manufacturing are presented. (All while knowing that a certain number of responses involved are socially agreed upon).

5.2.2.1. *General vision*

For the question concerning their vision of their research, the results of the interviews presented in Figure 5.1 report on the great yearning for the creation of knowledge, all in a calm framework (the discipline of the research unit's work has been hidden).

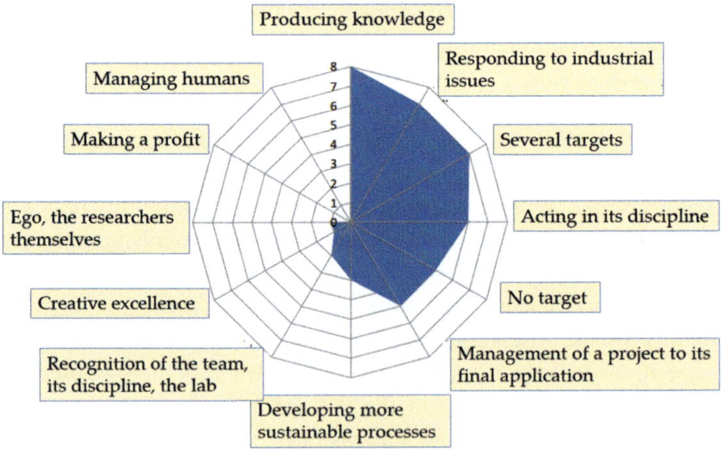

Figure 5.1. *The laboratory's targets*

Scientists have perceived manufacturers' demands as a "threat": that of no longer conducting "disinterested", detached, "pure" science, that of preferring to conduct science that "works" rather than trying to understand it fundamentally. Some researchers could have been "victims" of exploitation by manufacturers by proposing loans at ridiculously low prices. Some teams feel a bit recessed in relation to what they perceive that the rest of the laboratory does, but their daily activity does not allow them to perform reconciliation operations; with insufficient means, they do not choose the direction of their research. Moreover, it is this dependence on financing that sometimes controls the researcher's evolution in relation to the target: it depends on the proportion of industrial contracts (mis-appropriation of a hypothetical goal that is common to the laboratory). This strong dependence on calls for projects could prevent them from having a long-term vision.

In order to evaluate the independence, even the "partitioning" of the teams making up the laboratory, it was a question of knowing the relationships maintained by the axes, the teams and, more precisely, the permanent members among them. Several trends, sometimes paradoxical, emerge from interviews:

– no laboratory member knows what happens in all the other buildings/other teams;

– very few interact outside of work (joint projects);

– some teams are closed off to themselves;

– a stated will to open up more to other components of the unit.

It is a young structure that finds its origin in the fusion of old research groups; some think that internal exchanges will multiply in the years to come.

5.2.2.2. *Supports to creativity*

The following question was asked:

"What would you think of the possibility of allocating a space of freedom to risky projects, a space common to several researchers (e.g. 20% of working time) so that they can enjoy themselves and/or get invested in crazier, riskier projects (in the sense that the result would not necessarily end in something or for it to be unexplored territory) and more personal projects, with time allotted to several members of the laboratory in a synchronous manner so that there could be interdisciplinarity, crossing views on common projects?"

This led to the results from the LT presented in Figure 5.2.

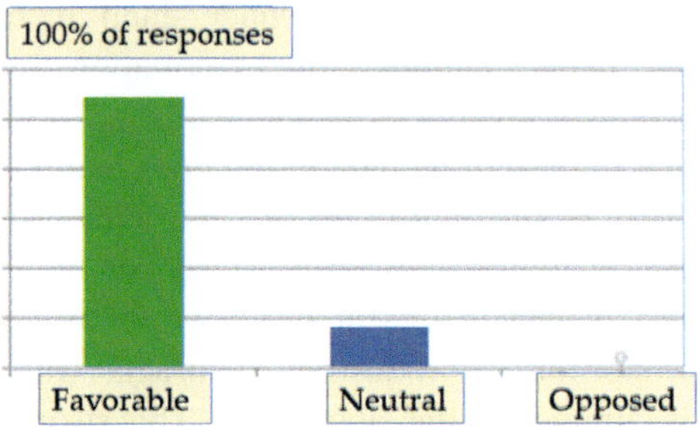

Figure 5.2. *Agreement to support creativity*

The people interviewed showed themselves to be rather interested in the idea of working on an innovative, cross-functional project. It is indeed a means of attempting to decipher the "mysteries" of new integrative subjects ensuring research teams, should they succeed, real international legitimacy rather than a means of properly resolving problems and seizing opportunities. It is necessary to go a bit further by no longer basing oneself on classical inertial foundations (because they have presented their evidence) and by proposing escaping the following conservative spiral: rather than producing knowledge adapted to situations as they take shape, are these adapted to existing knowledge or knowledge as it is foreseen? However, a third of the tribe announced its refusal to contribute to this; 10% proposed contributing to short-term (ST) projects, only getting involved in a part of the project or even not at all. Nearly half ("followers") announced that they would be willing to get involved in possible operations, and 15% even proposed participating in their organization and decision-making if the project suits them. These numbers are encouraging (knowing that, with declarative procedures, it is always necessary to take the difference between saying and doing into consideration and that it will be possible to attract some researchers from the "detached" category (see Figure 5.3)).

The question of the budget has been proposed, and the bar graph shown in Figure 5.4 takes the different propositions back up.

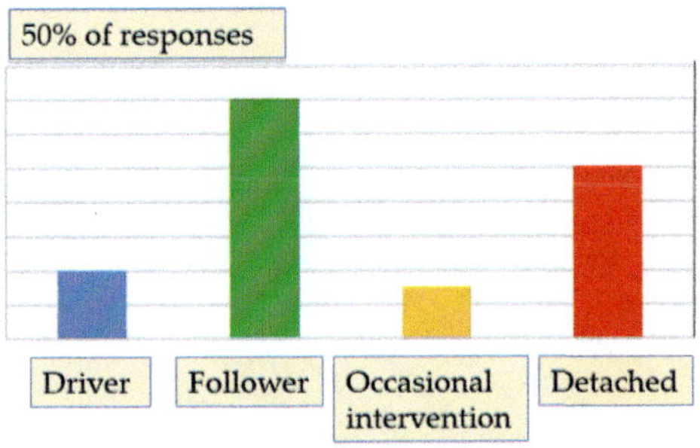

Figure 5.3. *Investment (estimation) in an innovative project breaking with the existing one*

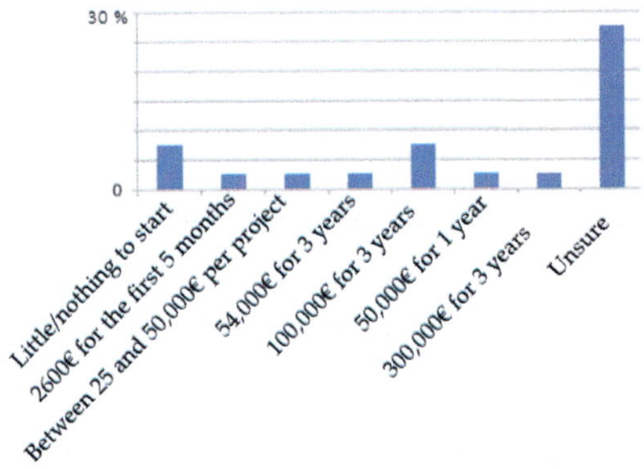

Figure 5.4. *Estimation of the cost of the creative/innovative project (interview extract)*

The average budget is about 70,000 € (but with a large disparity in responses). Those interviewed explained that the cost of the project depended on the domain in which they found themselves. Some researchers also estimate that a doctoral student will have to be included for a risky project, representing a minimum investment of 100,000 €. However, 10% of respondents remarked that to create a forecasting group of scientific debate initiative (maybe the first stage of the process), the necessary material represents a very weak investment. It is simply necessary to accept spending time.

Fears: Independent of this aspect, some express fears about these kinds of cross-functioning actions (see Figure 5.5).

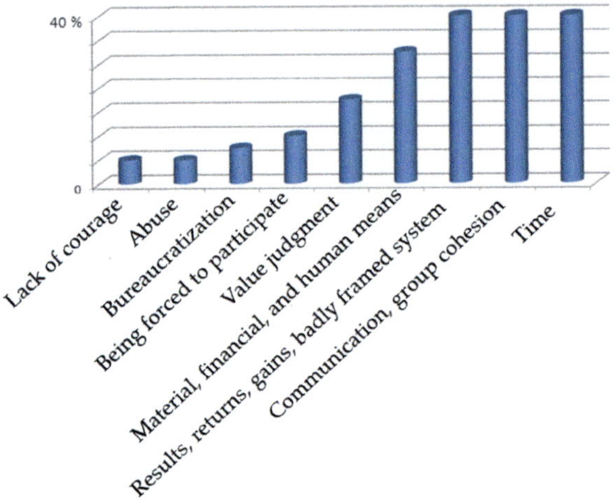

Figure 5.5. *List of fears in relation to the "innovation/ creativity" project versus number of people concerned*

5.2.2.3. *Time to be dedicated to potential breakthroughs*

The lack of time is the most frequently mentioned reason to justify the inability (potential or real, partial or total) of a member of the unit to become invested in a project on a new boundary object. During the interviews, the development of this subject was the object of reflection. Figure 5.6 gives a declarative indication of the (apparently incompressible) average use of time by the interviewed people. On average, this use of time could be divided into three large categories: teaching, research and administrative tasks.

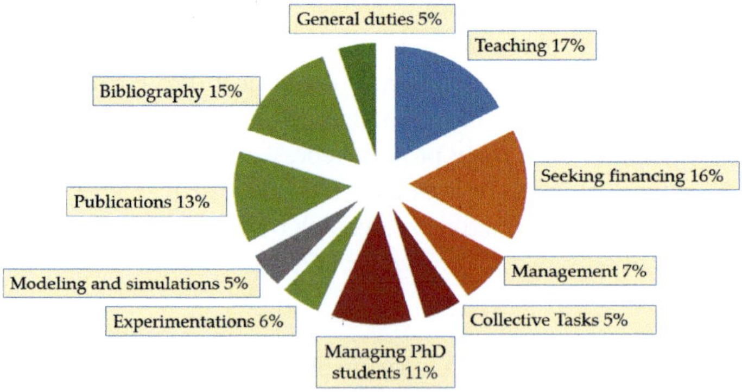

Figure 5.6. *Division of activities as a percentage of total working time (researchers, teacher-researchers and doctoral students combined)*

On average, activity is divided into approximately 1/3 research, 1/3 teaching, 1/3 various administrative tasks (including arranging projects). The teacher-researchers interviewed are the most constrained by the criterion of time optimization, because the activities linked to education represent time they cannot use efficiently in terms of scientific production (in the same way that they make their research valuable by publishing it, for example). This is why they must use their research time in an optimal manner. A researcher who is not also active as a teacher (or who does little teaching, i.e. CNRS researchers) should potentially be more able to sacrifice a part of his/her research time to open up to others (hypothesis confirmed during the interviews).

A certain evolution has been felt over the last 20 years. Even if the relation (changes of obtaining financing/time spent formulating the project) is weak, arranging projects (particularly ANR, H2020) has become researchers' most time-consuming activity. Some maintain that ANR projects will soon be abandoned due to their low likelihood of success, a thesis supported by the fact that some researchers have already abandoned the idea of obtaining financing for this reason. The interviews showed that they were not very inclined to participate in an innovative interdisciplinary project if it had to be developed during extra hours and/or the weekend.

5.2.2.3.1. Abuse, Predation – group cohesion

Some express reservations due to the risk of abuse: fear of having their ideas stolen, dealing with profiteers or an appropriation of others' ideas. How can team spirit be ensured in so that the members will play the game? It is also possible that, throughout the concretization of the project, the initial direction taken may privilege one discipline over another (specialization/strategy; creation of a hierarchy (see Chapter 4)). For some, while in an exploratory phase, it would be necessary to think about ensuring, at the different stages of the project, that the initial specifications are always followed. For example, establishing a strict control every 18 months to verify that the subject is still respectful of the initial specifications (does it include the different disciplines present in the unit?) to avoid deviating to a personal subject. The different tensions that exist in every structure could also compromise the trusting environment of a possible "creativity" open group. Difficulties communicating arise from difficulty on the parts of certain people popularizing their work, in making specialists in other disciplines understand through the vernacular language barrier of certain disciplines. The will to be open essentially aims at a status quo.

5.2.2.3.2. Open science

Recently, scientific environments have seen notable upheavals, particularly concerning the modalities of producing and spreading knowledge. Movements for "open science" and "open data" in favor of free access to publications and research materials, the multiplication of platforms for researcher networking, the growing number of initiatives for participatory sciences or citizen sciences based on the mass use of collaborative platforms and the development of new spaces for scientific communication with more and more diverse audiences actively participate in the re-definition of the scientific uses of the contemporary Web [MIL 15]. These innovations are accompanied by discourses highlighting a certain number of evolutions – even transformations – of scientific activity crystallized today around buzzwords like "science 2.0". This opening does not represent a significant element in the responses and comments of those interviewed.

5.2.2.3.3. To go a bit further

Doesn't this fear reflect a lack of conviviality in the laboratory? It is unthinkable, in a structure the size of a laboratory with a few hundred people in it, to deny the existence of internal conflicts (some were clearly mentioned during the interviews). No one gets along with everyone. Furthermore, is it possible for property right abuses (signing publications one did not contribute to, stealing an idea) to have already taken place, if the fear of this happening again is present? Out of a desire to protect oneself, interactions grow cold (see growing evolution of individualism). Rather than sticking together, won't teams have a tendency to withdraw into themselves (divided waiting rooms, etc.)? This is far from the Fab-Lab spirit described in Volume 1.

However, "we must overcome our fears". What seem to come up time and time again in the undertone of many responses are the courage to throw oneself into a risky project, the courage to step outside one's comfort zone and the courage to be confronted by different points of view. It seems that there is, at the same time, a strong will to establish a "leader" of the project, sometimes brought about by the project itself (desire for the subject to be passionate, eye-catching, attractive): it is thus necessary to have a figure to motivate the researchers, to give them drive. Then, there is the expression of a will for sanctuary for the project: support/protection of its direction, having a foot on the ground, a sufficiently large safety zone, without which no one dares to branch out, to invest himself/herself. The campaign to motivate the "troops", if this type of operation is implemented, will, however, ensure that existing projects maintain their value and that the innovation project is not placed on a pedestal (see above). It is thus a paradoxical injunction at all levels.

OTHER COMMENTS.–

– Abusive bureaucratization: The interviews underscore the fact that a good innovative project should take place with a balance between freedom space and framing. However, the majority of researchers have not developed this idea further. When the possible initiative of a "creativity open group" was mentioned, some people imagined a badly framed debate, an overly "easy-going" atmosphere, a project not being taken seriously, that is informal, lacking distance, debates of "balanced" ideas without their ripening beforehand, whereas in reality, the organization of such a project is founded on a ripening work, one of negotiation, alliances, in "brainstorming", "war-room" and deepening meetings;

– forced implication: By all accounts, to realize a cross-functional project, it is necessary to leave the choice of whether or not to participate to the researchers;

– judgment: Risk of passing judgment on the researcher who has chosen to join (or not) the breakthrough project: risks of frustration, devaluing his/her disciplinary activities;

– return: If someone participates, even in a limited way, in the cross-functional project, it is necessary to ensure him/her some return within his/her team, keeping him/her informed about the evolution of the project. Not setting anyone aside;

– some researchers think that it is useless to want to do things for looks, that it is not necessary to do just whatever, in order to continue doing his/her work correctly and honestly: recognition would naturally come thereafter.

5.2.3. *"Scientific excellence"*

It has been a question of examining if opinions were involved in ethics of scientific of practice formulated several decades ago by Robert Merton [MER 42, MER 68], which is associated with an open discipline such as that practiced in engineering sciences within a unit. Merton defined the scientific ethos using a series of norms, both methodological and ethical, guiding the behavior of scientists. In particular, he distinguished between four ethical norms somewhat constituting "the scientific spirit" [SAI 13]:

– universalism (in opposition to individualism), wanting the same rules to apply for everyone and thus for scientific value to be evaluated on the basis of universal criteria;

– communalism (in opposition to secrets), considering knowledge to be collective goods that must be spread through society;

– selflessness as an attitude aiming to advance knowledge for the sake of knowledge and demanding verification processes;

– organized skepticism, demanding the systematic evaluation of scientific productions and of scientists themselves.

These "ideal" norms have been widely criticized for not being considered effective norms guiding scientists, but rather as "rhetorical resources" acting as an ideology. These ideals are not completely to be found, particularly the ideal of communalism, if only through the strong links that exist between researchers and socioeconomic environments. We see the principle of knowledge as a common good, in particular, reformulated by publications and, to a lesser degree, the demand for verification and evaluation, conducted by "peers" or outside instances of evaluation.

5.2.3.1. *Quantitative evaluation criteria*

Spontaneously, it is often H-Factors [HIR 05] that are mentioned first. However, they face severe criticism, without challenging the important idea of measuring excellence. Because of the evaluation systems, which are based on measurable criteria, some scientists do not dare to take risks: they do what they know how to do, limiting themselves to projects that they are sure to succeed at and for which they have funding.

5.2.3.2. *Qualitative criteria*

In what way does the researcher contribute "new" knowledge? What novelty is contributed? The researcher must have a historic vision of his/her subject, in order to know what his/her work contributes, to emphasize the interest of his/her works, taking the time to reflect on what is being done. "In some domains, innovation is so rare that contributing this verges on excellence" (extract from an interview [HOR 16]). In others, obtaining a result with a significantly lower margin of uncertainty than previous works can also be a sign of excellence.

"Someone who is world renowned by his/her peers or his/her scientific community." The best are those who "are not seen": humble, accessible by everyone (even students), available to others, a person capable of understanding complex phenomena and making them intelligible for the community. Others think that excellence is something intuitive, it is "felt" if the person is good or not in what he/she does, during a conference or by carefully reading innovation works that he/she conducts (extracts from interviews).

5.2.4. *Financing and the orientation of research*

Some scientists from the Laboratory believe that, in light of the number of industrial bids, conducting their research boils down to choosing what they feel like doing from among the propositions. This allows them, among others, not only to open their minds by discovering new issues that are partially distant to them, by facing challenges, but also anchoring themselves in reality (research that is close to the final application). The feasibility of a project (financial means, but also human means with the creation of one or several hypothesis offers) will be a determining factor, much more than the aesthetic aspect, risk or anchoring in the team's topic.

A researcher rarely refuses a project where he/she can employ doctoral students. This represents publications, results, pure research and financial support. In short, "promising doctoral students" is a good way to "impose" a contract upon a researcher. The underlying idea expressed is that if a doctoral student works on the project, the researcher believes that he/she can dedicate more time to his/her own research (without serving manufacturers directly – see Figure 5.6.). Numerous researchers would like to do more fundamental science without needing to seek credit. If money does not influence the work of several members of the unit, it is nevertheless a driving force in their dynamics (with the risk of confusing ends and means).

Nevertheless, some would purely and simply refuse to undertake ANR projects, finding the situation of undertaking a project from industrial contracts more "comfortable": many bids, fewer administrative constraints, neither the State nor the laboratory take much of the money from these contracts, no files to fill out to justify the use of money. Moreover, it is not necessarily fundamental research that drives the application, as there are numerous cases proving that the opposite is possible.

Finally, some researchers make no effort at all to seek financing, either because they count on their colleagues to land contracts and distribute a part of the funds collected (division of tasks) or because their domain only requires very few means (particularly work on a computer).

5.2.5. *Prospective opportunities for the research unit*

Those interviewed were asked to fill out a table compiling several scenarios. The first part of Table 5.1. is divided into four parts:

– fundamental science the person is working on: what are the disciplines the person has the best mastery of, for which he/she is recognized? In most of his/her works, what knowledge does he/she implement most thoroughly?

– The link between teaching and research: if the person is a teacher-researcher, does he/she develop his/her course/tutorials as a function of the discoveries made in his/her field of expertise? Does he/she only teach the basics, the fundamentals (fixed course), or does he/she make students work on recently discovered issues?

– Openness to the international: does the person work with external laboratories? On European projects? Others?

– Multi-disciplinary openness: is the researcher a specialist in his/her domain without any interest in other domains, or on the contrary, does he/she find it difficult to define his/her "basic" domains due to significant multi-disciplinary overlapping? Where is he/she between these two extremes?

In the second part of the table, a list of social issues was established, a non-exhaustive list, the last of which, entitled "other", allowed those interviewed to add what may be missing. The first column concerns the current state of research activities. Which category is the researcher in? On a scale of 1–4, 1 meaning that the researcher was involved in some projects in this domain ("small" projects, simple intervention, work indirectly impacting the domain) and 4 meaning that the person is highly sought after on this issue, that it is his/her primary research work. In the second column, the interviewee was invited to imagine what his/her position could be in the next 5–10 years. Will he/she stick to the same issues, become involved in a new domain or rather deal with one subject while abandoning another?

The third column is about imagining the impact that a constraining policy could have and that of an ecology imposed upon their research work. On the basis of their responses, it was possible to estimate the researcher's dependence on the directives defined by society. What do they imagine with "ecology"? For some, we are already in this scenario: "There would be no impact on the research issues that are already at work within this optic. Moreover, projects that are described as 'green' have a great chance of being chosen for public financing; this will be one of the words to mention when arranging a project." For others, this could have a positive impact by boosting financing from manufacturers who have to bend more and more to regulations and so on. Finally, some considered an extreme ecological scenario, a "return to nature", leading to the complete annihilation of research activities.

In the fourth column, it was a question of an "isolated" society, a society whose borders are closed, limiting imports/exports, exchanges with the outside world. For some teams, this scenario is seen as a "catastrophic scenario", for they exchange extensively with laboratories abroad, or their research work is essentially dedicated to international applications (branches of French companies, etc.). For others, this would have no impact at all, either because their research only takes place in France or because they feel that science has no border and that it would be impossible to stop the flow of knowledge and information.

In the fifth and final column, entitled "will", researchers were invited to discuss what they would like to do (in their career, in a general way), the direction that they would like to give their research, or even their dreams (in relation to their career).

Regardless of the scenario proposed, there is no marked evolution (beyond a factor of 2 for appreciative measures) in the effect of the external on their scientific activity. "A large part of those who deem themselves sociologists or economists are social engineers whose job is to provide income to the managers of private companies and administrations. They offer a rationalization of practical or half-acquired knowledge that the members of the leading class have of the social world. Rulers today need science capable of rationalizing, with a double sense, domination, capable both of reinforcing the mechanisms that ensure this and legitimizing it" [BOU 84]. In light of the results of the interviews, it is obvious that what can be taken for provocation escapes the mission. What appears is that the resolution of social problems and of the mutualization of interests cannot be reduced to the single question of researchers' skills or technical and professional expertise acquired by some. It is rather a matter of a societal issue whose treatment calls for the training of collective expertise elaborated in the framework of a democratic debate (which thus stems from the goal of this work).

	Current state	"Current system" scenario	Ecological society	Isolated society	Will
Information society (Internet of Things (IoT), home automation)	8	11	5	5	8
Reservations (resources)	45	44	48	39	25
Sustainable development, environment, recycling	60	62	43	35	46
Health	30	32	8	12	17
Silver economy (aging population)	3	2	6	4	7
Food processing	14	12	10	7	7
Technological innovations	43	31	20	16	23
Tertiary	3	0	0	0	0
Population ↗ Towns ↗	8	8	1	0	3
Formulation	6	7	4	4	4
Defense	6	0	0	0	0
Entertainment products	1	1	0	0	0
Ethics	5	6	5	6	3
Digital security	4	4	4	4	0
Cosmetics	1	0	0	0	0
Process security	2	2	0	2	2

Table 5.1. *Vision of the future in their research associated with the views of members of the laboratory as a function of different scenarios*

In other words, the question of the aptitude to promote scientific anticipation of utility stems from the internal resolution of local governance problems (influenced by the global) based on collective competences. From this perspective, social engineering must likely integrate collective, experimental logic into its project and methods, and thus contribute to the management of the appropriate, even innovative responses. How then to bring about the evolution of the LT?

5.2.6. *Collective projects? Risky projects?*

Deep down, everything is a matter of beliefs: in the future because no major cloud is darkening the sky, and in the leaders who assume scientific and administrative responsibility. In that case, why attempt to practice activities that stem from divergent thought (while this was not a criterion for entering the laboratory), while there is no obligation to change? But are the time bases of rejuvenation adapted to maintaining the scientific and technological leadership of a research unit? Therein lies the dilemma! Who should be given credit for these changes that have been or will be undergone? The risk apparently has not been evaluated, which does not allow urgency to arise in the necessary changes and their anticipation.

"Creative or innovative periods are precisely those where, under the influence of diverse circumstances, men are led to come together in a more intimate way, where meetings, assemblies are more frequent, relations better maintained, exchanges of ideas more active" [DUR 67]. Apparently, the interviews do not lead to the appearance of encouraged encounters allowing constructive debate. This does not mean, however, that it would take place, but it is then difficult, in the ambient and traditional inertia in units having a recognized quality to their works, to recognize the potentially emerging forces allowing evolutions that should return to the collective ideal. Orléan [ORL 11] wrote on this matter, "This transformation is not the product of intellectual adhesion resulting from a rational analysis of the situation, but rather that of mobilizing the individual desire for a power greater than the individual." The organizational problem consists in seeking, through different forms of perceived outside constraints, present or future, as they are understood and shared, the different means of rejuvenation that correspond to them.

It is thus necessary for these different "manifestations" of external tensions to be shared in order to create a feeling of common action and a sharing of a trusting feeling on the part of the members of the scientific community of the research unit. The intelligible representations on the part of all the staff imply a minimum of desire to share between its members so that they can interact; one of the foreseeable means could be defining a common or partially common original scientific project. This "material" project would not be there to reveal the "mental" state the members are associated to; it would be exteriorization in the movement allowing a link to be created, allowing participation in the evolution of the image, reinforced because all the partners contributed to its creation [DUR 03]. This pooling would then have, in rational anticipation in ambient uncertainty, a renewed attractiveness and reinforced integration power, allowing shared risk-taking.

Those who have ideas to bring about change, and there must be many of them, are divided into two categories [PHI 12]:

– those who do not dare (for different reasons); they have a good understanding of the stakes and the importance of new ideas, but they do not want to take risks and may be afraid to "displease"; having never been "through the mirror", they have never seen failure and may feel exempt of all reproach;

– those who dare to advance by "disturbing" all sorts of conformism, organizations and sometimes "procedures". If they are not supported in their action (in a framework to be determined), they arouse fear, misunderstanding, criticism and rejection (particularly if the conformism belongs to the research's financers).

It is thus difficult, without support, for researchers selected on the criteria of excellence and scientific productivity to become disruptive elements, capable of taking risks and making the community take risks. It will therefore only be possible to escape this sort of impasse by mutualizing the risk.

It is true that the debate between followers of bottom–up policies, useful continuity and those of integrative reflection (top–down) of risk-taking, on open subjects, can appear in a non-stabilized heuristic approach like a discussion without robust foundations by the impossibility of deciding and advancing. The latter must find meaning with a heuristic approach. Debating, however, is attempting to break away from certain paralyzing stereotypes, application desirabilities, competitions, emotions and the search for recognition, even "free-for-alls" as long as this is desired.

The problematic horizon in which this operation is involved extends far beyond this initial contact with the social field of a recognized laboratory. It is easy to note the importance that skilled discourse has taken to give shape to scientific “excellence” policies based on social and developmental relations at the individual, local, national and international levels. This work summarizes forms of human incapacities in an attempt to reflexively accompany any attempt to assist, rejuvenate, competitively train (but also to limit these) and try to cultivate vigilance so as to be able to unmask injustices like the opportunistic union between a possible managerial ideology disconnected from the “field” and the valorization of each member’s skills.

What could be felt during this work is the need to explore the abilities of mobilization of the “troops” in relation to the “inability to act in an organized way”, by situating it in the wake of reflections on the continuity between action “inside” the organization and action “outside” it – this often being already very organized and powerful (ANR, EU, CNRS, etc.). This is an in-depth exploration of the degrees of formalization implied in the organized action that would grant access to both the capacities of this action and its incapacities and thus to a supplementary dimension of the scientific anthropology. This work that remains to be performed should authorize a conceptual clarification of the notion of incapacity to move the lines: first, it would be a matter of phenomenologically performing a series of distinctions between correlative incapacity, relative incapacity and absolute incapacity, only the last signaling the notion of vulnerability for the future of the unity as an organized and competitive structure; second, it would be a matter of unfolding the properly social dimension of “the incapacity” through reflection on the inter-researcher and institutional mediations [FON 14].

In the end, what was attained was a start of the debate on creativity (useful for 3D printing) and active monitoring allowing play on reasons by analogy (see Table 5.2 which presents some comments concerning the role of the actors in these processes).

In this necessary quest for knowledge, there is absolutely no room to oppose bottom–up (convergent) and top–down (divergent) researchers; on the contrary, they should be brought together. The enterprise needs every talent and after having advocated openness and rejuvenation (interdisciplinary, but not only), it does not seem wise to start a dispute between approaches that must be intertwined to make headway. Otherwise, it would be mutilation. Maybe it is necessary to remember Emmanuel Kant’s famous proposition on the obligation “not to treat humans simply as means, but always as ends”.

Accepting to dive into an unknown subject and to continue, even if insecurity is sometimes unsettling	Accepting not mastering all the content in the eyes of others (but with limits in deepening)
When exploring the subject, putting a temporary hold on questions like "what about my program?" "What about my discipline?" "What about me in this operation?"	Trying to explore a subject that is built with a systematic perspective: remembering the goal
Being wary of falling back into already-blazed trails (disciplinary content that could be "hung up"), which can become ruts when they are "ditched" in a research project	In negotiating the time that will be dedicated to the project, not getting caught up with the division of hours between disciplines; being able to imagine the project's life
Being open to the unexpected and ready to get involved in learning as a function of the needs revealed by monitoring or the demands of potential partners	Measuring out interventions as an overseer at new dimensions to avoid them becoming constraints, not responding to a question linked to the subject
For a future team to be enriched by the presence of several disciplines, it is necessary for each member to be sufficiently aware of specifically what he/she is contributing	Learning to share knowledge and accepting that researchers apply sequences that are not proper to their disciplines in order to better manage the project; being open to others
Applying the 3 M rule: Practicing Meta-disciplinary Methodological Mixing	At all times during project design, asking oneself if the specific view, the originality and the subsidiary thought to be supportive are indeed present.

Table 5.2. *For an active monitoring*

To conclude this section, Servent [SER 11] may be onto something when he wrote, "It is understood here that France has always preferred a precise preparation of operations explained by its past experiences – and the strict respect for structured hierarchical instructions that stem from this – to consider changes and new realities, thus to the detriment of innovation". Similarly, should we not remember Edgar Pisani [PIS 56], who in 1956 wrote, "The mission administration is light, it has a taste for doing, it is realistic, moving, it moves towards the event; it is mixed with life, it does not wait for the initiative, it asks for it; it must be able to create legal entities necessary for the creation of its enterprises; it is dynamic and involves risks, while its elder sister must settle for being prudent and imposes the elimination of all risks. It is anonymous; it is discreet by very definition. The functionaries that constitute it are interchangeable. The other needs to bring about adhesion, submission is not enough. It is based on the movement of a man or team's thought.

Management administration is the judge; the mission administration is the actor. The first is Cartesian, the other concrete." Will we know how to be concrete and rebellious? (see also [PEZ 17]).

5.3. Creativity's place in research

In once-developed countries, the consequences of using new means of communication on innovation and creativity are starting to be clearly felt: broken to pieces by the power of instant messaging, the time for individual creative reflection, to alleviate the effects of the pressure on scientific productivity nearly becomes the exception. In this sense, the explosion of communication could even further reduce all forms of free thought, particularly if it is not sought after. For the author, who is engaged in monitoring the subject of additive manufacturing, individual motivation, if we know how to bring it out, must be able to reveal original paths of action and thereby to multiply creative performance. As such, according to Roquette [ROQ 07], "once everyone is invited, even strongly incited, to seek innovation in a society or in an organization, an improvement in the overall efficiency and thus, over time, in productivity can be counted on through sheer numbers. New procedures, uses, circuits, objects, attached through permanent expertise and a constant concern for perfection, keep the concern for progress and the importance of the ideology of progress alive. The future is delivered as a territory to be conquered, rich in promises of abundance and happiness" (see also [GUI 16, GAI 16]). Are we strongly incited to find new paths in 3D printing?

Interdisciplinarity, in this path, nevertheless earned some strips in university walls when it was more or less constrained to solving problems. The State has notably taken it upon itself to remind researchers that their work must be useful in these cases and have practical impacts. At the moment of this injunction, which has become pressing in the past 30 years, the analysis of the swift developments of science and technology, at the point of becoming society's principle productive force [HAB 73], is elaborated through the prism of science for the economy. Nanotechnologies, for example, are also explained by the concert of knowledge produced notably by optics, biology and mechanics in order to draw benefits from each of these disciplines. These examples could intentionally be multiplied to show how much interdisciplinarity spills over (see [LAF 11]).

On the subject of interdisciplinarity, Mahy [MAH 14] wrote, "Above all else, it is a matter of sharing a desire, an intention, wanting to create solidarity that invites the adoption of a generous and trusting posture and once again returning to it when it is momentarily out of sight, an instant replaced by the conquering warrior. The dynamics of such a group are imagined, the maturity that must be proven, its relation to knowledge, in the image of oneself, and the empathy required to establish and preserve a particular quality of confidence." In short, as section 5.2 showed (however, only for a single experience), this is not the end.

The interdisciplinarity exemplified by "science studies" has come to challenge science conceived in epistemological terms as knowledge per item and per concept. In short, according to Granger [GRA 86], "science seeks to transport 'reality' – all reality – in the form of an open object, to the manipulation likely to explain it through concepts capable of producing a representation of it, thanks to which it then becomes possible to have 'precise and penetrative contact' because it is formed under the mode of abstraction."

"Under this head, science, contrary to the interdisciplinarity with which we wish to associate it, is deliberately used to reduce the complexity of what it studies in order to have 'precise and penetrative contact' with its object. Here is the aim of science and it should be noted that it is opposed to that which gives interdisciplinarity its raison d'être: producing a broad view by getting involved in dissolving different disciplines. In other words, the reduction that science performs gives it its strength and precision. Under this head, reducing in no way indicates outrageously fragmenting the object, but partially envisioning it so as to be able to see right through it and be perfectly aware of producing a partial, but precise, view of it. In other terms, science does not seek to comprehensively explain, as does interdisciplinarity, but to reduce reality in the form of objects that it distinguishes, so as to be able to explain them precisely and thus to contribute specialized knowledge about the reality envisioned in this way" [HAM 13].

5.3.1. *Support to creativity?*

Although the notion of change is already old, it imposes itself today as an ideology, strongly linked to the context of immediacy, creation or a network in which the economy of know-how or better "of knowledge" melts away. The idea of permanent change seems to have become a new paradigm, also synonymous with immediacy and instability. However, at the same time, Alexandre [ALE 11] wrote: "In industrial matters, the illusion that the diligence managers will invent railways to concentrate technological support on the preexisting industrial groups that are strong in lobbying, fails to understand future trends." How can we be creative in this décor framed by ANR or H2020 programs, or even National Research Strategy programs?

Moreover, with finances leaving little space for individual initiative, how can we emerge unscathed? What the author has noted is the development of modeling (confirmed by the survey presented in section 5.2). It is associated with Ganascia's reflection [GAN 00]; Ganascia wrote, "Beyond this dematerialiation of supports, scientific practices are undergoing profound transformation. Researchers are deserting the lab bench to sit in front of a computer screen. And with these evolutions, the epistemological status of the experiment is changing radically." Thus, by seeking to resist a constraint, that of producing as much as possible, researchers are being forced to lose contact and confrontation with the matter and, consequently, with the possibility of profoundly validating an idea, which limits the development of industrial activities stemming from ideas.

Finally, to add a layer and to justify this approach, at least partially, a society can only develop in relation to its history and traditions (long for Europe, more recent for the United States, more at technological rupture for Asia), because its know-how is an element of its economic heritage. Having products in line with a company, its positioning and its culture is a requirement for a coherent strategy, because the companies that succeed build their history without real chance and with continuity. When an enterprising strategy is developed, it is necessary to think of the products that it sells and the markets on which they are sold: it is the coherence of everything that constitutes success. And creation is supposed to be a cement between the past and the present, history and the future. Moreover, this is the very definition of modernity: knowing how to identify with the thought of one's time. It is then the classical terrain of the incremental, more or less "disruptive" innovation presented in Volumes 1 and 2. According to Marchesnay [MAR 13], "Herber Marcuse reminded us that the great force of organizations is their inertia. This is all the more significant when it is founded on a broad and long-sighted vision and when it meets with acquired interests."

But the world is changing and will change more when, in the East, the methods of thinking evolve, know-how exists, both ancestral and cutting-edge. In Asia, as in the United States, execution and copy are current business models, but they will quickly become outdated, as the real-time integration of novelties into industrial processes is very fast there (Following the example of manufacturers of rapid prototyping and additive manufacturing machines that are outside of a French company (medium-size enterprise), American, or Eastern companies). The obvious evolution is the implantation of new technologies like 3D printing in local products with high added value, with internationally recognized creativity that is likely to seduce the Western world. How, then, to remain competitive? The business strategist, the university researcher, in his/her mission and with his/her means, after having been a moderate anticipator, a manufacturer, a salesperson, a marketer and a financer in turn, is going to have to integrate all of these notions to become a "creator of new disruptive trends." However, this implies numerous cultural and behavioral evolutions in our country.

5.3.2. *But all the same, strong brakes on creativity...*

First, and this is nothing new, trainers, as their name indicates, are responsible for truly training the young, leading, according to Vincenti [VIN 90], to a certain number of difficulties, particularly concerning the training of engineers, with such difficulties as:

– asking the right questions on a new subject and keeping, as much as possible, the methods and technologies that were at the origin of their training (classical conservatism);

– vision of the technological challenges and of qualitatively modeling emerging situations (while they have good mastery of the methods of quantitative treatment, knowledge demanded by the hierarchy);

– generation of ideas to plan original solutions and so on.

Researchers, for different reasons, are largely ignorant of technologies (the disciplinary approach is generally valued), have no global vision, and find it difficult to think from the general to the study of sub-systems (then to return to the general). More and more, new system management, cultural evolutions are leading to the re-focusing of activities on principal knowledge (with orientation toward the mono-discipline, avoiding investing in a bibliography beyond one's own discipline), which leads more and more to "single solution" brainstorming. It is true that time is taken into account more and more and that is necessary to respect the framing imposed in calls for bids. In short, as shown in section 5.2, creativity originally comes from an operation that takes place "reluctantly".

But when the idea takes shape in terms of proof of concept, this does not mean that everything is over. According to Silverzahn [SIL 16], this is when the hierarchy takes note of its existence and, even worse, becomes aware of its interest, which it then decides to support, even to push it so that it will succeed on the large scale. However, for the price of its support, the general management will demand, understandably, that the project be "well managed". "By good management, the general management actually means, above all else, that order will rule. It must be said that, born clandestinely, the project has often seen its promoters turn to DIY (Do-It-Yourself) and application: starting small, using the available resources, working through co-creation, etc. All of this is not particularly sexy, not particularly popular, not beautiful sea blue, and thus it is too small; how are we going to have a real impact on revenue with this DIY? No, we need to see bigger! But most of all, we need to go from DIY to 'true' management. We need to better control its costs.

"Thus the exit of DIY and the implementation of true management. In the logic of rationalization, management will consist in appropriating everything that creates the specificity of the innovative project. It is necessary to connect to the corporation's computer system and stop using these open platforms [...]. The foundation must be the organization's functions and the project's autonomy must be reduced. The innovative offer must be lined up with the existing offer; it must be more coherent. One must know where he/she is going. He/she must do like the others, simplify. The project, or rather, its degree of breakthrough, dies from the famous Chinese death by a thousand cuts ('Lingchi'), which consists in cutting and successively removing parts and limbs of the condemned person's body before lastly cutting off his/her head. In the end, dozens of small modifications, all with good, perfectly defendable reasons, remodel the project to bring it back into the organization's mold [...]. But these modifications transform the breakthrough project into a project of continuous innovation, that is, into an extension of the current range (the same thing, only better), or more specifically, an extension of the current business model [...]: instead of adapting the organization to innovation, the innovation is adapted to the existing organization by trying to force it into the existing framework, which necessarily aims to denature it" [SIL 16].

The difficulty is that the defense of the project requires considerable energy and financial support that are more and more difficult to find. The creative who innovate risk ending up exhausted, removed from the game by the system in one way or another. Could researchers who expressed themselves in section 5.2. be right?

5.3.3. *What to do?*

The question is knowing how to transform a perceived or anticipated "social interrogation", requested or not by society, into an object of scientific research. On a boundary object like additive manufacturing, reflection for action on the part of the authorities who govern us seems necessary if we do not wish to fail in our scientific and technological operations with application accomplishment, as was the case at the end of the 1980s for a considerable economic market, that of stereolithography, a pioneering activity in additive manufacturing.

With an angelic vision, it could be thought that a research unit is going beyond an average summary of society's functioning. Crazy discoveries that attract media attention are expected, the creative activity in the world is incessant with a rational confrontation with reality, the proactive attitude is permanent and so on. Yet, what is largely noted is that the complexity of the world is, within previously supported paradigms, reduced to advance incrementally with validations by related peers.

Schematically, high-level research laboratories listen to "social demand" (and the significant financing that comes with it) and transform what seems to be a need into generally (but not always) mono-disciplinary scientific orientations that, creating legitimacy, can add research to the continuity. The creators, disturbing the organization, normally cannot join this schema. It is their success that will (maybe) lead to new scientific orientations that are stabilized for a time. Kuhn is thus not yet dead!

It was 75 years ago that Auguste Lumière published his book "Les Fossoyeurs du progrès: les mandarins contre les pionniers de la science." According to Duvivier [DUV 16], this book gives a somber overview of the situation generally experienced by an innovator when he presents a new idea or theory, different from the paradigm in effect. "How many among them were not actually taken for fools, visionaries. The indifference, mockery, and virulent attacks shoving their ideas into oblivion for many years and pushing them to poverty, indifference, isolation, and sometimes even suicide [...]. This phenomena affects every field, from chemistry to medicine, from the unknown researcher to the Nobel Prize winning scientist." Conservatism and perpetuated habits must never be forgotten.

What is expected is a process of inter-comprehension to get involved in a maturation of an issue; it is also leading to the collective expression of a passion for the novelty, fed by the tension between promises and a (some) possible scientist(s): How then to organize the unforeseeable, to propose a trusting, mobile and polymorphous environment? How to bring about risk-taking to support a dynamic of innovation? How to learn by transposing ideas and through reasoning by analogy? How to associate intuition and rational analyses? How to go from the "contemplation" of the unknown to the emergence of a common thread through the acceptance of uncertainty and the possibility of a failure?

It is not necessary for a group engaged in the exploration of a boundary-object, like the fields presented in Volumes 2 and, more specifically, 3, to have a leader who, in order to develop his/her own idea, needs a collective upon whom he/she will impose a vision so that they will contribute their disciplinary skills to the project. The project must build itself up through in-depth exchange between partners. Where no disciplinary scientific talent reigns, an interdisciplinary group is vulnerable through construction and thus, in order to win, must go beyond the boundaries of specialized knowledge. This implies accepting questioning one another within a group, being capable of encouraging one another in a difficult process, associating perspectives: in short, sharing moments of doubt and pleasure (see [EDW 11]).

5.4. Innovation, a consequence of creativity

No product, no factory, no job is created without ideas. This is obvious, although reminders must be given. An economic policy consists in the analysis of the ideas behind products and developing new ideas on robust foundations. On this subject, most economists, politicians and deciders in research and/or its financing are more or less silent, even if there are a number of reports relating to the development of innovations (SNRI (French National Strategy for Research and Innovation) [SNR 09]; SNR (French National Strategy For Research) [SNR 15]; OPECST (French Parliamentary Office of the Scientific and Technological Choices) [OPE 17]; OECD [OEC 07]; Ministère de l'Industrie [MIN 09, MIN 10]; Cour des Comptes (French Court of Audit) [COU 07]; EU – FWP 7 [EU 12]; Eurogroup Consulting [EUR 12]; etc.). "The economic war is a war of ideas, a war that needs idea factories, that has no need for specialists, experts, who are generally sterile, but rather, of new, massive, often fallow fields of crops. Believing, let us recall, is measuring up to the unknown" [MED 12].

If, as Drucker [DRU 06] wrote, "marketing and innovation produce results, the rest is only a cost", many consider innovation to demand only few material investments, at least at the beginning (this must be reflected on, but it must depend on the fields in question). Creativity and imagination, on the contrary, must be at work. It is not when everything is irreversible that everything must be stopped, while the investments become enormous: academic research must contribute, up to its standards and within its mission, to better mastery of techno-scientific risks, and this is the entire stake of this reflection serving as an intermediary between a bibliography and action bases. To do this, it will be useful to determine if the actions of academic research must, as happens in industry, have a time schedule and a budget (e.g. by going to the demonstrator), with possibilities to change the cap, to stop because it is known that an impasse is being approached and so on. This display of time limits should be explained to researchers who invest themselves more in the continuity of the actions linked to the human routines and habits of the research staff, and who are likely to consider a stop to be a failure. Yet, it is known that industrially speaking, approximately 20% of projects are viable and that 95% of new products fail [MUS 07]. Scardigli [SCA 92] defined three times for the spread of a technological innovation:

– beforehand, demonstration of the "prophecy";

– experimentation, feasibility time;

– acculturation of the technology.

It is essentially in the first two domains that an academic researcher should exercise his/her "talents". Moreover, the problem of following up on [scientific] projects is analyzed by Gartner Group [GAR 08], using the hype curve representing the visibility of a technology as a function of time, characterized by:

– the introduction and the first attraction by deciders of a technological basis (example of nanotechnologies). In this phase, the field of possibilities is immense, such as that of the researcher who must be able to use his/her entire imagination. Initial programming must be modest;

– the peak of exaggerated expectations: the emergence of the technology is difficult for different reasons (GMO: media controversy, cost of investment: semiconductors III/V, II/VI, etc.). Only some axes remain promising. Then, there is a need to orient the work toward reinforced operational efficiency and to challenge others;

– industrial applications appear, in this case, outside improvements with a scientific origin; the activities leave the scope of action of academic research;

– society takes over the applications of techno-science by modifying them [GAR 08].

In all these different situations, there is greater and greater interest in the numerical value of things (merchandise, CAC40 (French Stock Exchange Index), evaluation of research and the French National Evaluation of Academic Research System (HCERES)) while forgetting the intelligence of ideas and more or less removing the target for their appearance or reflection. In this exaggerated context, where the supposedly pertinent indicators are nearly legion (as well as quantitative), consideration is often paid only to the later stages, that is, the result, rarely or never beforehand, nor to the failures and experience that come with this. What is the real source of growth early on, then? It is certainly on this basis that it seems useful to reflect in order to find/propose methods of rejuvenation and to depart from the obsessive quest for "economies of scale" and the evasion of risk-taking [BIG 13]. Yet, the results of a global survey on the situation and the perception of innovation have been published in the framework of the "Global Innovation Barometer 2012". This study clearly shows French specificities that are not very favorable to national innovation in a strongly globalized context. It is, although to a lesser degree, at the European level [MAC 12]. According to General Electric (GE) [GEN 12, GEN 15], at the source of the survey, the French environment does not seem to be booming.

Yet, for the author, innovation is one of the keys to the future, which leads to reflection on the necessary transformation of the potential of research to attain economic development goals based on science. In the GE survey, 47% of respondents (world average) define innovation as the means to bring about new processes, products, organizational changes or marketing; 42% see it as a culture, an environment that promotes creativity and permanent change; 60% of French respondents believe that it serves to become a leader and to stay ahead and 51% to bring about solutions that can benefit the greatest number of people.

Yet, it is necessary to recall the hopeless trade balance and unemployment numbers without seeing the lack of exploited ideas hidden behind them (and the lack of their effective transformation). This reflection does not support a global policy of creation *a priori*, but targets a critical analysis to propose ways to put things into practice. Resisting the crisis starts with a well thought out policy of creation, explicitly concerning its criteria and modes of action.

However, several phenomena (at least) are at work in innovation:

– the large size of the centers of industrial research and research units leading, as in academic research, to a subdivision of activities. Gaudin [GAU 78] wrote then that, "the object is cut into pieces, and this division, in and of itself, guides production towards reproduction"; in fact, with "fearful conformism" [GAU 78], a search is also being conducted where something has already been found;

– research is only rarely at the source of innovation; it is the "fact of heterogeneous and improbable initiatives" [GAU 78];

– innovation depends on specific events, results of history, leading to an absence of precise determinism and predictability in the long term (butterfly wing stereotype);

– the technological systems that we use are not isolated from one another; they refer to one another [PUE 08];

– the development of technological artifacts is regulated by their use (appropriation); Puech [PUE 08] described adapting technology to the use and adapting citizens to the technology (idea of co-evolution);

– the elevated cost of investments for new installations is likely to be an obstacle to innovation; "capitalism itself organizes its regulation of the technology by choosing what must be developed and favoring profitable investments. Capitalism moderates the insatiability of technology by its own [financial] insatiability" [CHI 07];

– to understand the production that leads to innovation, it is necessary to resituate independent decisions in the context of the time, the associated irreversibilities, even the fears of undertaking [KID 82]. The conservative trends at all levels tend toward mimicry and the status quo [DUR 12];

– those responsible for production do not demand research whose effects are predicted for the medium term, while they are asked for financial results in the very short term. This phenomenon is amplified by short presence time (a few years) of these people on a single site. They leave the concern to get involved in a risky universe to their successor (hot potato transfer paradigm!).

However, despite these reservations, the technological movement, the means of an attractive financial result, must be the creator of new riches accessible to the citizen. At the same time, the citizen asks for more regulation of the technical change by regulation, the legislation aiming at overall protection of the consumer. Moreover, Kaufmann [KAU 08] reminds us that "the less the norms are obligatory, the more the primary activity of people is to produce new ones until there is an obsession with normality". However, as Landes [LAN 75] expressed it, "the industrialization of the world is pursued for better and for worse, and if there are people in the advanced countries who retch at this cult of material realizations, it is because they have the means to offer themselves a critical attitude, while the large majority of the planet's inhabitants dream of accessing this material ease".

However, in their report, Legait *et al.* [LEG 16] wrote: "The French public R&D system, which has greatly evolved in recent years (creation of the ANR, university autonomy, creation of the Future Investment Program (PIA)'s devices and structures, etc.), has unfortunately become even more complex, and less readable, for many companies consulted and even for some large groups (multiplicity of counters through the PIA, in particular). A lack of professionalism and reactivity in some public actors is often acute, despite the development of valorization devices: lack of knowledge of the needs and expectations of the companies that affect the implementation and quality of collaborative projects, research activities performed in overly narrow scientific fields, sometimes elevated and opaque prices and inadequate communication and project management. The negotiations with public research establishments are, for many reasons, too long: too many speakers, particularly in the case of multi-patron laboratories (for example CNRS and

University; CNRS and Industry), sometimes due to confidentiality problems associated with publications, and very often due to intellectual property clauses. Patents have become strong stakes for public establishments, which often have neither the means nor the skills to promote them according to the number of economic actors consulted. Finally, the human mobility between the private and the public in R&D is very low in France, with the exception of the CIFRE's doctoral students (Industrial Training Conventions for Research), who are highly appreciated by the companies that benefit from them".

However, in the economic context, once we leave the feasibility of the idea of the "bench corner", that is, once we approach technical demonstration, the more the project must consider the more and more elevated number of partners, of the cost that is increasing exponentially, time constraints, the management of complexity and essentially the problems of changing scale. How to recruit people (including those from public research) to attain the targeted goals on the basis of an industrial innovation project? To master the facts and gestures of these (all while positioning them as real allies) to avoid runoffs and divergences. In this context, analyzed by Latour [LAT 05], the success of an innovation activity cannot be spontaneous and free. Moreover, he wrote, "going from science to technology does not amount to going from a paper world to a concrete, dirty, unorganized world. This consists in going from a paper world to another paper world, from one calculation center to another that gathers and processes an even greater number of calculations from more diverse sources (see [BAS 79, FER 92]). The more modern and complex they are, the more the machines need paper to be updated. There is a simple reason here: in the very process of their construction, they disappear from sight because each part chases the other…"

5.4.1. *Academic system*

From a scientific standpoint, the world of French research also comes from an individual approach to once again find itself "magnified" by disciplinary development in Auguste Comte's sense. This development has led to a great deal of technological progress also based on a supply logic. When this progress resulting from a scientific approach coupled with the notion of profit, if possible involved in the short term, becomes an essential goal, this means that there will be interest in problems considered a priority and little, if any, interest in those considered to be of modest interest relative to the principal stake of competition. Often, it is not even known that abnormal situations can exist due to this specialization that is taken to be a label of quality, even of elitist reference (because they are visible?). Thus, through cultural inertia, the system works until social dissatisfaction appears from challenges by others (the "breakers" of the paradigm) creating crises or at least perceptible challenges within the research body.

Thus, the rational method has modeled the scientific and technical world, led to measureless progress translated by an unthinkable hope for life a century ago, through material means unforeseeable 20 years ago. However, at the same time, for different reasons, this success finds itself compromised. Is it, as Forti [FOR 96] wrote, "the fruit of the arrogant Newtonian intelligence" that could have formatted us? Or, on the contrary, as Folscheid [FOL 07] expressed, due to René Descartes: "What is called 'Science', in the sense of pure and selfless knowledge, has officially received its leave in 1637, the publication date of Descartes' 'Discourse on the Method' by the 'Master and possessor of Nature'"?

However, according to Morin [MOR 13], science has rejected complexity, one of the characteristic elements of innovation, for the following reasons:

"– the principle of universal determinism, illustrated by Laplace's demon thanks to its intelligence and extremely developed senses, which not only to knows every past event, but can predict every future event;

– the principle of reduction that consists in knowing a composite whole from the knowledge of the primary elements that constitute it;

– the principle of disjunction that consists in isolating and separating the cognitive difficulties from one another, which has led to the separation between disciplines that have become hermetic to one another".

According to Olivier Godard [GOD 14], "by making complexity one of its watchwords, [science] has brought the ocean of uncertainties and interrogations that constitute it and the fragility of some of its constructions to the forefront, at the same time that the proliferation of invasive new techniques appearing to be so many adventures at the uncertain end for people's concrete life".

Should we reflect better so as to act in the sense of interdisciplinarity, a potential factor of a reunion of academic science and companies? The management of innovation not only demands precise notions of the creative subject; it requires a more general culture comprising law, economy, project management, esthetics and communication, a rather distant culture from that of academic researchers. "Only a spirit of synthesis can overcome the problems, respond to each event, and interpret the relative sense of every phenomenon in its brutality. Common sense is not science, but culture" [ELG 70]. However, on the same subject, Pelluchon [PEL 14] reminds us, and rightly so, that it "is not impossible to undertake a serious interdisciplinary task without a rigorous methodology allowing each specialist to defend his arguments coming from his discipline's field of pertinence."

Finally, an alarm cry (among others), that of Ségalat [SÉG 09], who wrote, "Exercising research that is becoming more and more scholastic, the system is making more space for good students than for artists of science," all while responsibly accepting competition, a price to pay for its freedom of research. Another alarm cry, that of Sperber [SPE 10], who believed that distancing oneself from research with interdisciplinarity is a loss: "This means that the creativity and inventive genius of young researchers are diverted from interdisciplinary research, which slows them down and is less rewarding both intellectually and practically, and thus loops." The interdisciplinarity illustrated in Figure 5.7 remains difficult in this open decor.

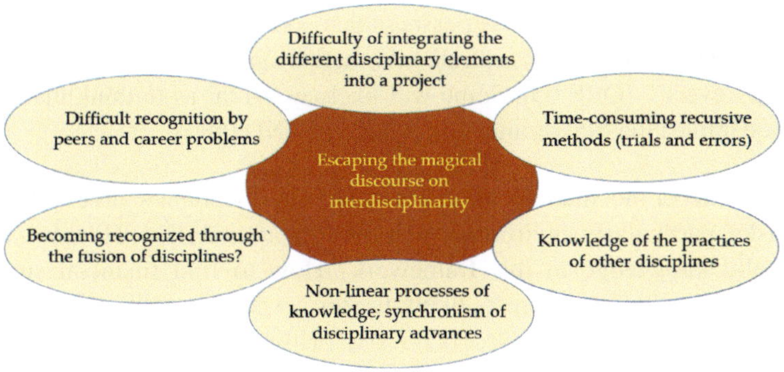

Figure 5.7. *The (small) problems of interdisciplinarity…*

Seeking to understand how researchers get involved in an interdisciplinary process, we investigate their work. Leplat [LEP 97] presented the work as an activity in subjective dimensions. The remarks that each agent "cannot be conceived as a simple system to execute the prescribed task. This task is involved in its history. It not only performs the prescribed task, but also aims, through this performance, at personal goals." The project head's task is among his/her history and the goals that he/she sets. This kind of activity requires a double registry of skills: on the one hand, skills concerning the object to be mastered; on the other, communicational and relational skills (obtaining and keeping the trust of another) [PIO 09], to start the group project. This animation refers to an ability to act by joining a triad of ergonomic psychology [CLO 06]: the subject, the object and others.

5.4.2. *Between productions resulting from science and responsible conscience*

According to Wolf [WOL 09], "[the] disjunction between science and philosophy takes place until the end of the 19th century and leads to the detachment of the culture of the humanities from the scientific culture. This separation will certainly allow considerable scientific advances, but it will also progressively create a problem in a science that no longer knows itself, that no longer has [enough] self-reflective ability, and in a philosophy that, no longer being fed empirically, is incapable of fertilizing science, which is itself incapable of conceiving itself". Yet, for years now, crises have come to transform the context, whether it is a matter of effective sanitary risks such as mad cow disease, asbestos, contaminated blood, or worries (founded or not) about cell phones, electromagnetic waves, GMO, nanotechnologies, NBIC convergence and tomorrow "bio-printing" convergence. In this framework, individuals do not always have the information and, when it exists, at least partially, it is analyzed as a function of the feelings of threat and/or powerlessness that come from it. "This feeling creates that of fear and transforms the 'neutral' information on such and such event into an 'extraordinary event'" [ORF 05]. Some will likely go so far as to think that there is a desire to hide things; the trial of intention is not far [AND 08].

There is little or no increase in the number of researchers invested in works linked to risks, particularly environmental and occupational risks. How to satisfy the need for valid knowledge in this framework? How to find financial support for studies targeting collective interests likely to stop, even "threaten" those who are at the origin of innovation processes [PES 03]? How, in a scientific system very broadly governed by Auguste Comte, engaged in mono-disciplinary excellence, to encourage the demand for disciplinary hybridation [KAL 07]? How to make researchers consider innovations more as ends and less as means? How to be able to prepare the emergence of products, devices corresponding to the future needs of humanity? How to have non-conflicting expertise between scientists and experts in the analysis of possible risks for society [BAR 93]? Here, we have a number of questions awaiting responses in a few sentences.

Or must we accept this first affirmation without reacting: "It is thus necessary to accept that science is no longer at the origin of incontestable knowledge" [MER 06]? Or this second: "Common sense, which is made up of many illusions about the world, must accept being explained by Science without reservation" [HAB 02]? In this unstable context, the debate on research's place in society is not stabilized, particularly for a number of researchers, going from the principle of disinterest [MER 73] to the impossible myth of pure and selfless science, comforted by the implication, even the inclusion, of science in the modern political system

[MOO 06]. This polysemous situation has had numerous advantages, particularly in hiding possible debates, approximations for the exchange between scientists and citizens. However, the tensions are reinforced, leading to evolution.

5.4.3. *Engagement toward a future focused on innovation?*

A situation aiming at engagement to raise new challenges for a better life on our planet is translated today by large goals, all while conserving the traditional economic foundations. It naturally comes from tensions made visible today out of the fear of the future (exhaustion of reservations, global warming, migration, etc.). It must be said that since the Middle Ages, where the collective was the peasant hamlet, we have reached the planetary stage with all its limitations, on the one hand, and the stage of individuals reduced to their insularity, on the other hand [TAG 00]. Assurance (or trust) can thus no longer come from "above". On such heavy foundations, where the cult of the movement for the movement's sake is expressed all over, how can scientists act to contribute responses for our survival (or rather, for the author, that of his descendants)?

Yet, defining a soci(et)al need and the technical and organizational means to satisfy it naturally do not systematically come back into the leadership of techno-scientific research that must maintain its scientific, creative and in-depth character, all while having specific approaches due to its position at the interface with the applicability of the scientific concepts stemming from research. The legitimacy of the engineering sciences is actually built by the dynamic and recursive mutualization of scientific knowledge with an end in sight, an end whose origin comes from the scientific component or that of the appliers or from a demand from the deciders. It constitutes a science of action: "It is a working dynamic [...] through its transformational, manipulative, constructivist power..." [HOT 92].

By granting specific financing, the State and the European Union are motors of the development of science in privileged directions. According to Guellec [GUE 92], "the very nature of technology implies a certain number of market failures: the public authorities must then compensate for this through incentivizing measures or through the allocation of public resources. It thus comes back to the State to support long-term generic R&D whose profitability is insufficient for private agents to be concerned with it". It is indeed on these foundations that the technosciences are developing, with calls for bids from the ANR and/or the European Union. However, and this is well known in industrial research, there is the possibility to anticipate the propositions from the Union or the State. "This policy [...] is a form of structural fraud, certainly light, but it increases as the bureaucracy that pays for official research dominates" [VAN 09]. So long as this statement is and remains valid, there is a possibly perverse effect of this situation; the possibility of better research planning than reality would leave us to believe.

5.4.4. *Caught between two chairs? Between more than two chairs?*

These evolutions lead to a situation halfway between fundamental research and applied R&D "to guarantee", it is said, "a faster economic yield. In the end, research becomes both diversified and interdisciplinary, the decisional processes [...] are more flexible..." [GIE 00]. Recent developments lead to the appearance of numerous and complex couplings between systems, going far beyond disciplines. For researchers in this domain, there is a need to "search alongside" his/her discipline. "Our era is placed under the line of multidisciplinarity. Even if the risk of a loss of discipline linked to the dilution of the specialization is unsettling, the advantages of contact with knowledge cannot be ignored" [SOK 97]. However, as its counterpoint, Merton [MER 73] wrote, "[The system] leads to an accumulation of advantages all the larger as the research domain is stabilized; the founders benefit from cumulative recognition... This is what some call the Saint Matthew Effect".

Scientific production is unevenly distributed over the population of the researchers and teacher-researchers, that is, a small part publishes a very large number of publications while the rest publishes few. Generally, if, for a given period, a bar graph with the number of people N having submitted p publications is made, a rather representative correlation of the facts corresponds to an empirical law in the form:

$$N(p) = K. \exp(-\alpha.p)$$

where K is a constant dependent on the discipline and α is an invariant.

These radical differences in the productivity of the scientific research environments can have several origins:

– differences in talent differentiating the individuals;

– the possibilities given to some to change their activities through administration of research, teaching, paid overtime and so on;

– different treatment by the science operating system: the most reputed researchers (for a whole range of reasons) at a given moment benefit from more advantageous research conditions.

These differences create a cumulative process that affects the competition between scientists in the sense that the researchers who have been the most "productive" can benefit from conditions encouraging greater productivity. The founder of the sociology of science, R.K. Merton [MER 73], provided the first description of the cumulative advantage processes in science that he gathers under the umbrella term "the Saint Matthew Effect". He refers to Saint Matthew due to a passage from his Gospel where it is written, "Whoever has will be given more, and they will have an abundance. Whoever does not have, even what they have will be taken from them."

The hierarchies of universities and large research organizations make recruitment decisions on the basis of a comparison of the candidates' scientific production profiles, using selection methods of peers associated with a conformity of thought from yesterday where the slightest risk is avoided. Yet, beyond partial information, the committees acting for the future employers do not always have in-depth access corresponding to the true quality of the applicant. Several coupled reasons are related to this situation:

– lack of time for detailed judgment and recourse to quantitative analyses;

– lack of critical thinking on the evaluators' parts, even unwholesome favoritism;

– difficulty judging;

– fear of risk-taking.

The employers have a given job (for a long time) with fixed attributes that they are thus happy to provide as best they can. With time, these attributes (the associated levels of remuneration and productivity) are heterogeneous (even if they tend to come together with age). More precisely, new employees compete for positions as "junior" researchers and then "senior" researchers, who are characterized by different levels of remuneration and associated productivity (beyond bonuses and overtime). The agent who was the most productive during the first stage is selected by the organization that is offering the job associated with the greatest expected productive utility. Yet this increases the chances of getting a job that offers a better salary. During the second competition, a dynamic bias gives an advantage to the researchers who were the most productive during the first: we are then speaking of a cumulative advantage or Saint Matthew Effect.

Box 5.1. *Saint Matthew Effect*

"But the object having decomposed and its constituent elements divided and hierarchized by, and due to, these intellectual constructions that are disciplines, the question then begs to be asked of the later convocation of these object portions that, at first, were naturally not judged as fit to constitute the matter of the scientific investigation" [ALV 03]. Thus, to avoid confining the researcher in his/her discipline, a discipline in which he/she exercises his/her freedom (his/her autonomy) and his/her activity allowing him/her to go beyond the limits of knowledge, it is necessary to find paths of "transgression" that must authorize and support new cultural couplings. These must allow the promotion of adapted synergies, authorizing the creation of new artifacts that are useful for society. This is indeed what is expressed by the good idea of coupling between science and technology. By supporting creativity at the interfaces, by developing creative hybridization, it is thus a matter of going beyond other boundaries of knowledge, but possibly using new or revisited methods, by resituating "research and researchers in the context of the city life" [POM 04]. This view is reinforced by the position of Araujo-Jorge [ARA 01], who insists on the fact that, at present, it is necessary "to integrate the knowledge acquired at a higher level to understand the comprehensive functioning of nature in order to truly grasp its complexity". However, if these visions are supported by a number of scientists, reality is certainly not so simple.

There are numerous challenges for the techno-sciences to overcome. In this environment, the approach by demand, which must be worked on responsibly to bring about changes in the cultural foundation, optimization, is no longer done on the cost, but on the quality of the service, which changes many things, if only through the reinforcement of innovation processes and those respecting timeframes. The applications of innovations are not just a matter of uses and technologies. As shown by Akrich *et al.* [AKR 06], the innovation circulates in society through successive "translations" with reciprocal transformations of each. This self-prophesying rhetoric makes the technological adoption necessary and desirable. Among the actor–representatives that effect the social translation of technological innovations, the experts of the trends, are not only the makers of previsions but also the journalists who play a role as informative monitors, even recommendations or lobbying companies. The discourse accompanying the transformation is thus, just as the transformation itself, part of the socio-technical process and short-circuits the long acquisition time through discourse [ALL 15a, ALL 15b].

5.4.5. *Innovation as scientific production: is it born of freedom? What freedom?*

Innovative research is better fed in laboratories that have great financial freedom [AZO 10]. These authors show that scientists are more likely to produce innovative research if they benefit from long-term goals and subsidization than if they are involved in short-term projects. "If you want researchers to diversify and explore new directions, it is important to foresee their horizon in the long term, to give them the time to experiment and possibly fail". The results of this study show that subsidies on projects produce half as many articles with half the quality of those with more open financing. However, at the same time, according to Pestre [PES 14], "Attention must be paid to what comes about, what emerges, but it is not reasonable to forget resilience [...], the weight of the rigidities, of the friction. It is important to keep the hybridization, cross-fertilization, and irreduction process, but also to think about the perpetuated vulnerabilities and the impossibilities of doing, of analyzing the dominations for themselves, their implementations, and their experiences".

Nevertheless, some open questions will have to be discussed:

– How can this freedom serve innovation and then how can it be promoted?

– How to associate the different actors of innovation optimally?

– How to operate the transferability of the results, to bring them closer to real life [ALB 09a, ALB 09b]?

– How to find tomorrow's talents and support them financially and in their careers?

– How to encourage the coming together between academic research and companies to support breakthroughs?

– Who should have leadership to launch an innovative operation?

– What can then be done when the economic fabric in a given domain does not or no longer exists?

Will we know how to find ideas to move forward? Will we be innovative and creative? Yet, what has been noted, particularly upon becoming a senior researcher, is that the national governments (ANR) or European governments (7th PCRDT and today H2020) have gradually imposed an upholding of financial support to research units with cooperative calls for bids on targeted subjects with the economic world and the valorization of research results [BER 08]. According to Heil [HEI 10], the change in research practices is translated by a logic of results (absence of risk-taking) at the expense of a logic of means (see also [BER 08]). New Public Management [DRE 05] thus requires science, normally unforeseeable long-term material during the deepening phases, and the effective obtention of results foreseen in the contract tying researchers to agencies.

Under such conditions, the place for divergence is modest; there is then only a small space for creativity and innovations with high risks. According to Taleb [TAL 07], the system would bring out two behaviors: that of the so-called normal law, that is, that the background paradigm is not challenged, the variables studied are minimal and the approach is classical; whilst the other concerns new phenomena, unknown by "the former art", which corresponds to an uncertain environment, difficult to foresee, where possible breakthroughs can appear (see also [ZAJ 00, SIM 11]).

In a universe of limited financing, the experts consulted take refuge in the fact that it is always better to support credible proposals and ones whose propositions are known to be honest and with a very high scientific level. They will be effectively and successfully explored, which is not necessarily the case when supporting risky operations. Conservatism is thus the obvious result of this. The study conducted by Lesage and Hurteau [LES 16] shows that the effects of reducing financial support for research are translated for subsidiary organizations "to support works with a high potential for concrete results, to the detriment of fundamental research". The upheaval of scientific activities, which the author has not seen since his hiring at the CNRS, but much later (approximately in the 1980s, when additive manufacturing works were being put in place, but still with a rather loose net mesh) this direction against the spirit or the mission of academic research risks slowing down the ideas that still exist on the national level, transforming the aforementioned researchers into producers of scientific papers (whose quality is certainly not questioned) with no global vision nor profit (at least intellectually) that they could make.

It is worth mentioning that the equating of scientific works with numeric variables in a "benchmarking" philosophy, but avoiding the in-depth analysis of this research, leads to a movement to draw closer to the imposed model in evaluation, even if progress to escape the quantitative and simple is being practiced (see e.g. [BRU 08a, BRU 08b]). This strategy, possibly undesired, but detrimental to the support of risk-taking, of behavioral injunctions of researchers led to the acceptance of the model. Juvin and Lipovetsky [JUV 10] remind us on this subject, "the number replaces knowledge, the number kills taste and the aristocracy of judgement [...]. The number tells the truth, the quantity ensures the beautiful, the good, and the true...".

In line with this, Boudon [BOU 03] believed that, "an individual's conformity to a constraint can also be analyzed as the fruit of a will to make his/her behavior intelligible in the eyes of the group he/she belongs to, his/her contact people or, more generally, the public in regards to which he/she acts. By conforming to expectations and prescriptions, he/she makes it possible to anticipate his/her behavior, thereby facilitating the interaction with other individuals and integration into the group…" It is then necessary to expect the flash to come from elsewhere, even if the Ministry responsible for Research, respecting the "Frascati Manual" [OEC 02], explains that, "the program's research organizations form a pedestal for the production of knowledge that must be a reservoir of [incremental?] innovation, supplying our economy". However, the feeling of control in production and the promises to be kept can have the consequence of inhibiting slightly daring initiatives. "[Innovation] needs flexibility and creativity while [feeling] demoralizes operations and exhausts technicians in tasks that lead them away from their goals" [BRU 13]. Moreover, following the example of what has already been analyzed in risk management, an approach is observed; it corresponds to simplification with the risk of forgetting certain results that would force deviation from the agreed-upon, and this to avoid accusatory evaluative expertise (see [BOU 01b]). By not moving too far from "good practices", researchers thereby build their employability (see sections 5.2 and 5.3).

"The concentration of political and economic power around so few people not only brings about material dependence outside the man of science; at the same time, it threatens its profound existence. Indeed, through the development of refined technologies to manage intellectual and moral pressure, it forbids the appearance of new generations of independent valuable human beings" [EIN 99]. "The word 'independent' is cowardly… As the decision to get involved in an adventure on a new subject, without a known market, is not the result of a deep rational evaluation, while a stable job can be kept by reporting to tomorrow the passage to the other side of the 'mirror'…" [MAR 12]. Creating your own company is thus a matter of conviction, of proving that things can be made to advance and, ego in hand, proving that an individual has the right talent to do it. It is still necessary for the system not to stop this obligatory enthusiasm until it is extinguished.

5.5. What solutions to evoke for additive manufacturing?

5.5.1. *General framing*

"The Bloomberg classification of the most innovative countries in the world is based on six tangible parameters: innovation expenses, the number of scientific discoveries, research conducted, the number of researchers, the number of high-tech companies, and the number of patents filed. And the winner is: South Korea, which is definitely first, followed by Sweden, Germany, Switzerland, and Finland. Israel has even surpassed France, which has dropped one place compared to last year" [GIR 17].

Before reaching logics of action (or inaction) as perceived by the author for things concerning additive manufacturing, some data seem useful, particularly by drawing some comparisons. Analyses by the World Economic Forum [WOR 15a, WOR 15b, WOR 15c] show significant differences between the United States and the EU28 in terms of innovations with weakness in France relative to its neighbor, Germany (as well as particularly Sweden, Denmark and Finland). The CERI (Center for Research and Innovation in Education) (2015) highlights significant differences in terms of support for innovation in different developed countries; Figure 5.8 also illustrates large differences within the Union in terms of support for risky operations and thus support for "young sprouts", a field in which France is in a better position than Germany for the number of start-ups created, but with less financing (see also [OEC 15a, OEC 15b, OEC 16]).

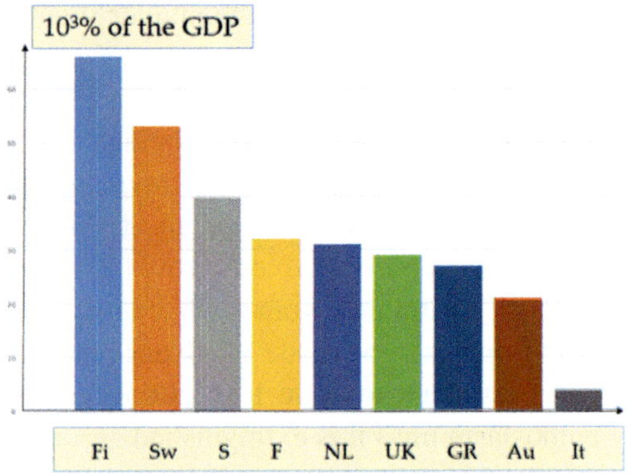

Figure 5.8. *Differed support in the EU for capital-risk operations (2013)*

To continue to "drive the nail", the Deloitte report [GIF 15], which positions the competitiveness of the nations of the world in the field of production, focused on a foreseeable loss in France's competitiveness in the near future. Certainly, other declining European countries give the impression that France is not the only country suffering from this decline, but the question of where we are going is clearly asked (see Table 5.3). This situation is attested to by the World Economic Forum's thoughts [WOR 16] on the risks brought about in development due to governance that is unfavorable to its support.

Union countries and neighbor countries	Current rank	Future rank
Germany	3	3
UK	6	8
Switzerland	12	19
Sweden	13	18
Turkey	16	17
The Netherlands	20	21
France	22	26
Czech Republic	23	20
Finland	24	24
Spain	25	27
Belgium	26	29
Italy	28	30
Ireland	31	31
Romania	33	28
Portugal	35	35
Greece	40	40

Table 5.3. *Average loss in France's competitiveness in the field of production relative to the world*

In an interview published by the German Industrial Federation (BDI), France comes in 18th place out of 35 countries in terms of innovation. More unsettling, it was in 14th place in 2010 and 9th place in 2000 – so just ahead of Germany, which has climbed a few places since then. BDI explains this counter-performance with the absence of ambitious reforms. It recognizes decentralization efforts, both in innovation policy and in the financial instruments put in place, notably in favor of middle-market companies, but it leaves something to be desired. "In total, there has been no success in moving the fixed structures of science and the economy, write the authors of the innovation indicator. France thus hovers and, if we place ourselves in a faraway perspective, it finds itself surpassed in an international comparison" [MAD 15].

In a recent report from the Swiss Council of Science and Innovation [CON 15a], a comparison was made between the funds attributed to innovation and their efficiency (the authors of the reports remaining quite humble in their work considering the imprecision of the indicators and the reliability of the data sources). Nevertheless, this efficiency is an element to be taken into consideration, as shown in Table 5.4. France is not listed in the report, so the efficiency linked to the 5–6 billion €/year of Tax-Research Credit (CIR) in France is not mentioned.

Global innovation index		
Rank	Country	Score
1	Switzerland	648
3	Sweden	623
4	Finland	607
8	Denmark	575
13	Germany	560
20	Austria	534
?	France	?
Relative efficiency of innovation		
6	Switzerland	0.9
19	Germany	0.9
22	Sweden	0.8
41	Finland	0.8
61	Denmark	0.8
69	Austria	0.7

Table 5.4. *Efficiency of support for innovation*

In the technological sector, with start-up companies rapidly invading traditional sectors, large organizations are confronted with the increase of pressure to innovate, according to the last report from Capgemini Consulting [CAP 15]. The challenge lies in the fact that traditional approaches toward innovation could be broken: only 5% of staff engaged in doing R&D feel very motivated to innovate. In some sectors, more than 85% of new products fail and 90% of managers believe that their company is too slow in launching new products and services. However, in this context, innovation, the magic word, is extolled by every community emerging from notable differences between "traditional" companies and "young sprouts" (including those linked to additive manufacturing). Yet, as this is a matter of small but fully evolving groups for young companies in the 3D domain (on the order of 1,000 people for the largest), programs like H2020 for the EU or ANR in France (which are very similar) particularly account for the demands for "heavy weights" in industry and services. In this framework, the place for support of disruptive actions is modest; the incremental remains vigilant.

Weaknesses in traditional approaches to innovate have led some organizations to explore different "avenues" and to seek a new source of inspiration. These organizations have launched or gotten closer to innovation centers in the principal technological poles, with the explicit mandate to accelerate digital innovations in particular. Comprising teams of people and often physical sites, these centers of innovation are established at the world's primary technological poles. The goal is to benefit from the ecosystem of innovation of these centers, such as:

– start-up companies;

– investors in risky capital;

– accelerators;

– providers;

– university establishments.

The world's principal technological poles are the preferred destinations for the establishment of innovation centers. The manufacturing sector is considered the uncontested leader with a proportion of 58%. "Despite the fact that financial services are undergoing an increase in pressure due to digital disturbances, the rate of penetrating innovation centers in those is only 28%. It is extremely difficult to ensure that innovation centers are a success. Among the success factors are clarity on the role of the innovation center, governance to implement innovation, and strong networking with the rest of the industry" [WOR 15a, WOR 15b, WOR 15c].

For manufacturing industries, a report by MENESR (French Ministry in charge of academic research) [MEN 15] illustrates a loss in France's competitiveness relative to certain European countries like Germany (but not only Germany). Figure 5.9 presents these gaps. This context is even amplified in terms of research when the situation report from the CNRS (CNRS-CoNRS [CNR 15]) is examined, where it is difficult to find key words like additive manufacturing or 3D printing, or even bio-printing, a strongly emerging field.

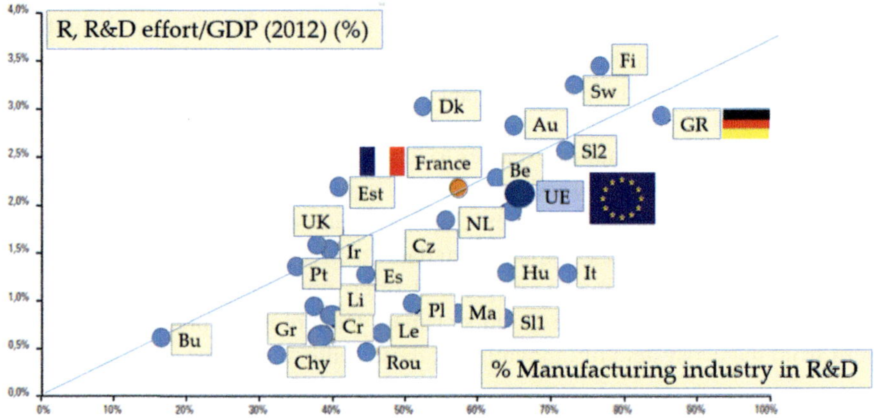

Figure 5.9. *Position of France in research and R&D in manufacturing industries*

For evoREG [EVO 15], by considering innovation systems to be complex systems, the following hypotheses can be formulated to prepare scientific and organizational policies:

– it would be necessary to orient innovation policies toward stakes linked to actor coordination rather than toward optimization and efficiency;

– it would be important to understand that only innovation policies considered to be holistic can be successful in the long term. Consequently, policies should provide the "right" contextual conditions rather than trying to influence individual behaviors;

– due to the nonlinearity of the phenomena considered, if the policies principally lead to unforeseeable events, the incentive measures mobilized by these policies should target marginal systematic effects rather than enormous changes;

– in short, more divergents would be needed in deciders in the upper administration and in the policies that follow them (but would they easily know how to convince their administrative councils?).

In the same direction, in a recent report by the OECD [OEC 15a], the position of Europe (and France) concerning innovation is modest, as attested to by the results presented in Figure 5.10 for advanced materials. The United States, Japan and Korea appear to be the uncontested leaders of a new generation of "breakthrough" technologies in the fields of advanced materials, health, information and communication technologies, likely to upset the existing processes. Attempts at possible changes have been recently expressed [OPE 17].

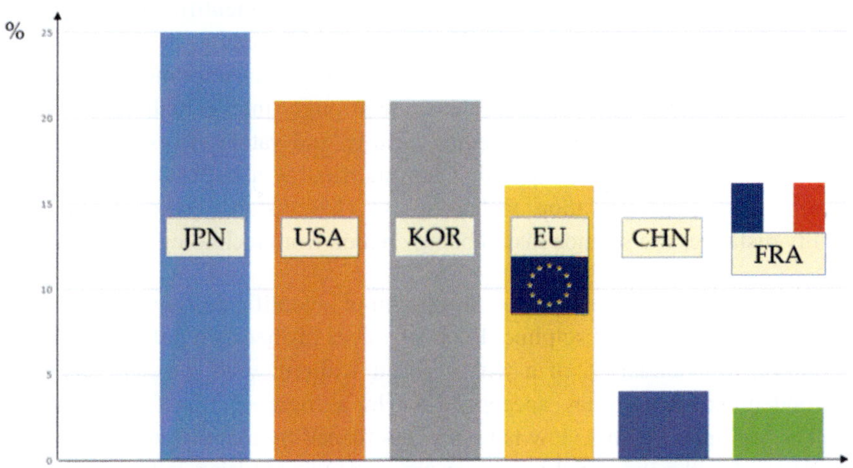

Figure 5.10. *Position of Europe in terms of industrial property in the fields of advanced materials (primary actors; patents filed in the United States and Europe)*

To change things, Louis Schweitzer, quoted by Bottero [BOT 17], wrote, "To end the work in silos, to prefer working together, is one of the conditions *sine non qua* for the economy – and companies – to behave better". So there is nothing to do but follow him on this path…

5.5.2. *And if the history of additive manufacturing in France were examined in light of these comments?*

"We are in a world of 'Sunday drivers'", as George Friedmann writes of men who never leaned over their motor, and for whom things only have, by mystery, to function [BAU 00]. Yet, in the framework outlined in the present work, the author hopes that it will be possible to go against this proposition, that corresponds partially to his vision of research often/sometimes bundled up in the overly conceptual because it is "relevant", conforming to aspirations and adapted to "healthy evolution" conducted according to traditional models from another era or rather another world. In this situation, which is necessary to share and essentially re-focus actions of an individual toward the "basic" mission and so to be more efficient, it will be possible to be recognized as actors of living scientific interdisciplines, autonomous and useful for society. Yet what is happening in the society of researchers is that they are led to undertake complex and abstract mental operations thanks to computer science. In this context, the strongly intellectualized classes are rather distant (on average) from objective reality and rather invest, according to Lasch [LAS 07], in a virtual world that he qualified as a hyper-reality exploiting information and communication technologies. This distance taken from reality potentially distances the science that is being created from technological innovation.

As Henry [HEN 98] recalled, "the evaluation of scientific products always has to do with the 'territory' of a discipline. Each advances his/her own definition, through the definition of boundaries, of a within and a without, with transfers sometimes being condemned, sometimes accepted, but in a logic of transfiguration to be imposed as legitimate". This view is largely confirmed by that of Larrère [LAR 94], who believes that the industrial paradigm proceeds to a division within knowledge, between the natural sciences and human and social sciences, constituting an "anthropocentric epistemic paradigm". In short, there is still a long way to go to promote this responsible indiscipline strongly desired by the author.

To continue proving this point, Nicolas Colin in Marion [MAR 15] noted a sad truth of the French ecosystem: "the companies that are getting involved there want to create the next Google and end up at the head of a small web agency." The French enigma is tragic: "why, when we have a great deal of capital, a great deal of know-how, and a not insignificant amount of rebellion, can't we manage to develop an entrepreneurial economy worthy of this name?" The response has to do with the dosage of the entrepreneurial ecosystem, even in the absence of the mixing of the three primary ingredients. Things cannot take off: "the capitalists, engineers, and rebels do not visit one another, they even resist one another" (see also BPI France [BPI 17]).

Furthermore, according to La Tribune [PIA 16], beyond the technical challenges in globalized competition, it is in terms of vision and coordination that Europe, and France in particular, should progress. "In the United States, additive printing, praised by Obama himself, benefits from a one stop shop; the United Kingdom, the European leader along with Germany, is very present at the academic level. While China, Japan (historic pioneer), but also Singapore and Australia are taking their positions, Europe is spending a great deal of money, but fighting to adopt a strategic vision. France is present at every level, with big purchasers (the large groups of aeronautics and automobile), some specialized companies, and an ecosystem of subcontractors (prototypers, historic actors in plastic processing and metallurgy, specialists in everything additive…)".

However, a feeling of disorder prevails. According to Opinionway [OPI 16]: "The recently formed alliances (Polyshape and Lisi Aerospace, Fives and BeAM, Fives and Michelin) are not enough to mask a lack of cooperation. The authors point out other flaws: strong regionalization of help and projects that lead to costly duplication, atomized subcontracting in the face of strong needs for investment, embryonic training and failed research". Even the French, in surveys, think that we are behind in terms of innovation [OPI 16]. They do not trust policies to support it (this situation can be seen in many other sections of French society). The ability to finance appears to be a brake blocking the expression of the French people's creativity. The major proportion (77%) of them list this as determining, whereas 66% mention the difficulty of finding the right people capable of helping them develop their projects. Doubt, however, persists in the mind of the author's compatriots on the originality and pertinence of their ideas: 48% of them need to be re-assured before getting involved. Finally, 35% of them lack time, above all else (see Figure 5.11).

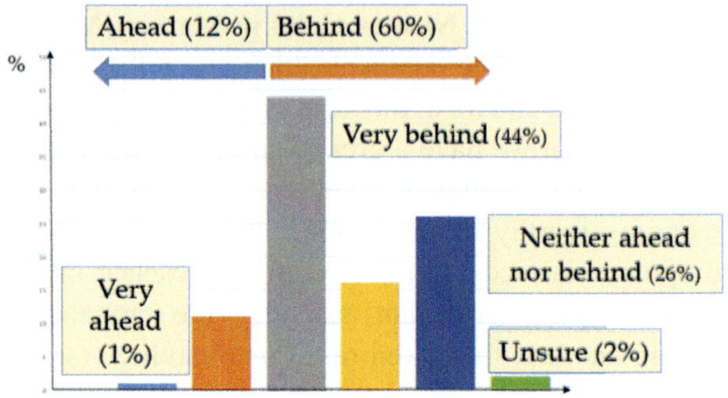

Figure 5.11a. *French perception relative to national innovation (Question: do you believe that France is ahead/behind in terms of innovation in relation to other countries?)*

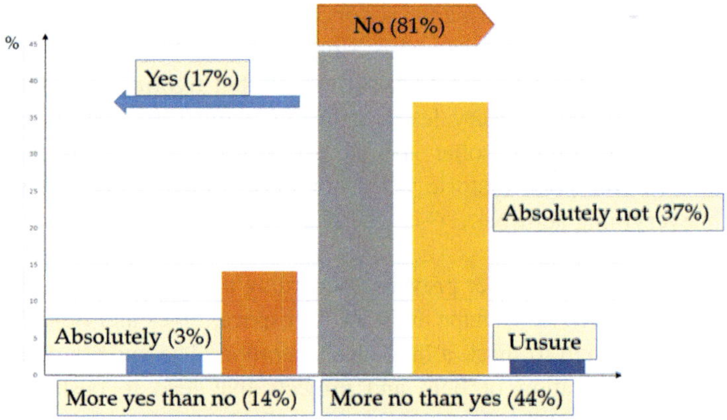

Figure 5.11b. *French perception relative to the State's ability to support innovation*

Table 5.5 uses the "best recipes" from the literature on innovation as entries, put in correspondance with the 3D adventure experienced in Nancy between 1983 and 1995 (the author's transfer to a private structure). As for the other conclusive tables, a + corresponds to an agreement on the match, = to something in between and – to a disagreement. Condensed comments complete this table.

Classical subject in innovation	Correspondence with the 3D experience	Comments
Very new	–	
New method of production	+	Intellectual game based on scientific foundations
Evolution	–	Not enough time to properly anticipate due to a subcritical mass; no prospective analysis performed, but great deal of brainstorming
Creation of a multi-disciplinary team on the project	–	Angelic functioning as a virtuous craftsman who does his best with what he has; absence of industrialization specialists; DIY because specialists from other domains who could have intervened did not want to participate lacking financing
Prototyping/demonstrator	+	Considered indispensible from the start; remarkable help of doctoral students and post-doctorals for those operations that they believed in; motivation is essential
Organization and financing	–	That of the team… and affective links; no adapted financing and unpredictable functioning according to the game of contracts
Association of inductive and deductive thinking	+	All the same, we are in a CNRS-University unit
Opening up to new markets	=	Not at the start of the study; reflection begins only after having realized an initial demonstrator
Placing the user at the center of the process	–	Question not asked
Market analysis	–	The market did not exist
Business model	–	Unknown expression at the time
Existing services	=	Contact with manufacturers; but at first, a certain angelism that our hierarchies did not take into account.

Table 5.5. *Comparison of the 3D experience with the standards of innovation*

Upon analyzing the data collected in Table 5.5, it can be understood why everything was wrong in terms of industrial success: we were not trained and it was not possible to be accompanied. The failure was thus planned from the start of the innovation (but it was nevertheless fun to twist and untwist these processes with the materials that came along with them). However, at the same time, with some disappointments that many start-ups know, Chuck Hall, the founder of 3D Systems in the United States, was able to emerge unscathed. It is difficult to find the difference, maybe because it was not possible to try "playing" together to advance a bit faster in the exploration of additive manufacturing… Maybe we are Neanderthals who he would not have liked to meet?

In an innovative and uncertain environment, the introduction of a management control unknown to the basic researcher is not always necessary when the company is very small, when it is closer to its scientific base than markets themselves, and due to the French system of public aids that makes the intervention of risk capital less urgent [MEY 15]. As long as the aids existed for industrial valorization, this could have not slowed down indispensible researchers. Management according to the author is often limited by the understandable focus of managers on technical and commercial goals and by their primarily scientific culture. The piloting devices implemented are reduced to a psycho-cognitive use to help the decision of the manager who is at the exploration stage or are further developed when the exploitation stage is reached. However, this is generally the case when the role of the management controller is limited to that of safeguarding and interactive functioning, which could have opened up to other forms of knowledge for which we were not prepared.

Astolfi *et al.* [AST 16], to go further in the industrial act, deviated from the views of numerous additive manufacturing actors who "lamented an uncoordinated abundance [of financial support] and the absence of a national strategy or, at the very least, of directory guidelines"… (see also Lubineau [LUB 16]; AIF (French Alliance for the Future of Industry) [AIF 16] and Hourcade [HOU 15]). This is indeed what the report by the Instituts Carnot [INS 17] finds on this subject where the (unfiltered) responses report the existence of unstructured teams with a modest number of people involved in research (user?/creator?) with/on additive manufacturing.

The case of machines and equipment for France corresponds to a negative balance. Yet, the need for high-tech devices in an open scientific and technological space, with abilities to perform at a high level, may be covered by France. It would still be necessary to support this kind of operation. However, at present, France's place in the competition in developing Industry 4.0 remains quite modest, as shown in Figure 5.12 by Berger [BER 14].

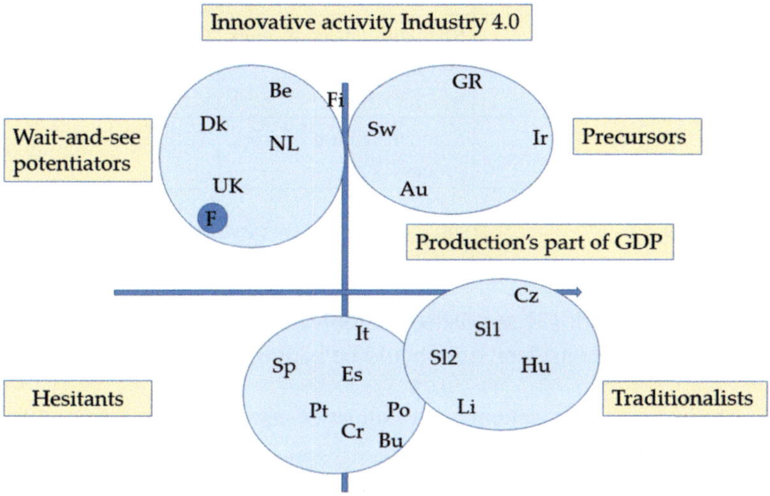

Figure 5.12. *Place of European countries in the "Industry 4.0" dynamic (Sl1: Slovakia; Sl2: Slovenia)*

Hofstede [HOF 01] analyzed the entrepreneurial behaviors in five developed countries, including France. Table 5.6 presents, for the five countries studied, three elements of their specificites, allowing a comparison that is not (due to administrative burdens unfavorable for the development of innovation) in favor of a French dynamic in international competition in new technologies. Moreover, it introduces the quality of the relations between strategy, control, structure and culture, allowing the integration of all of the facts of support for innovation that, for France (even if it is difficult to deal with this subject within national borders) reinforce this weakness (see also [ACK 13]).

Country	Specificity 1	Specificity 2	Specificity 3
USA	Voluntary action by domain	Standardization of production	Not extreme, but effective positioning
UK	Adhocracy	Joint adjustment	Quality of innovation support personnel
China	Simplified structuration	Direct supervision	Top–down incentive strategy
Germany	Professional bureaucracy	Standardization of goals and means	Operationalization of the heart of efficient principal activity
France	Generalized bureaucracy	Standardization of work processes	Technostructure

Table 5.6. *Innovation environments in five developed countries*

Science Europe [SCI 15] suggested, notably, that upon conceiving innovation policies, the following considerations should be kept in mind:

– the challenges that contemporary societies are confronted by will not be resolved by progressive developments. Instead, they demand radically new approaches;

– as research founded on challenges is taking action in relation to the traditional forms of research founded on curiosity, this situation will stimulate the need to elaborate knowledge through disciplines and will thus encourage pioneering approaches;

– when radical innovation emerges, it can change the situation and lead to consequences beyond those brought about by the original challenge;

– radical innovation is an essential element in applying knowledge from research in the real world that stimulates a profound change.

The deciders in charge of innovation programs should:

– take measures through research ecosystems to support the integration of disciplines and stimulate true interaction between all the actors;

– get strongly engaged in the conception, performed cooperatively, of research questions and their framing, between the political deciders and the researchers of a wide scientific range;

– encourage the arrangement of programs centered on innovation that could support all the aspects of innovation and that would also be fertile for the development of nonlinear and immaterial innovation;

– conceive a new appropriate research ecology for the societal challenges that the 21st Century must face.

Beforehand, it may suffice to be convinced of the need to act differently by examining what the United States proposes within their NNMI (National Network for Manufacturing Innovation) [FED 15]. This structuring organization brings together various topical institutes and serves industrial transfer (TRL (technology readiness level) understood between 3 and 7). The functioning organization chart obtained from NNMI (2016) is shown in Figure 5.13.

Figure 5.13. *Development of competitiveness as seen by NNMI*

By linking various bodies of knowledge and skills, this structure reinforces and adds to the efficiency of means to develop competitiveness in the manufacturing domain (see Figure 5.14 from NNMI, 2016). Nevertheless, not everything is black, just gray; there is indeed a certain number of companies, certainly small ones, that are innovative in the additive manufacturing domain in France [NOT 16].

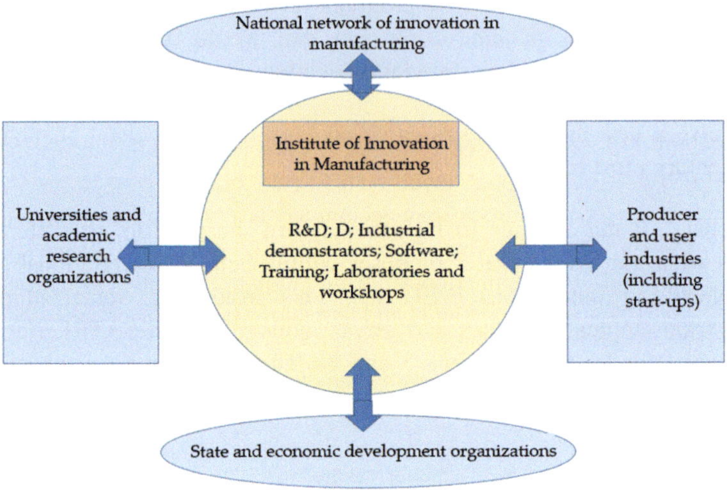

Figure 5.14. *Top–down relations in manufacturing*

For 2016, the planned budget for this operation is US $1.0 billion, including a bio-printing component. At the same time, in the NOE (2015) governmental report, it is "only" written, "Future industry: In order to anticipate the great mutations in the industry, particularly linked to the digital, the National Council of Industry has started a prospective work, branch by branch, to better consider the evolutions of jobs and skills." Not even "fast seconds"! The latest reports from PIPAME, a sector of the French Ministry of Industry [PIP 17], that of ANRT (French National Agency for Research and Technology) [ANR 16] and France Stratégie [FRA 16] or Schmoulder [SCH 16b] or finally from ISF [ISF 16], despite good intentions, are not more proactive. Could Berger, Dorman and Starbuck [BER 16a, BER 16b] be right when they show the complexity of the French innovation system?

On the French side, according to the CESE (French Economical, Social and Environmental Council) [ING 15], these rather unenthusiastic words about 3D manufacturing are still slightly hopeful: "In the vast innovation movement fed by digital tools, questions must be asked about France's difficulties holding its rank. There is an urgency not to let a new opportunity to contribute to the development of our economy and its global competitiveness pass us by". CESE is comforted by the information found in Table 5.7 from Rosenberg *et al.* [ROS 15] presenting a comparison between national efforts for the development of additive manufacturing technologies (see the demand seen by Weller *et al.* [WEL 15]; Earls and Baya [EAR 15], AIF [AIF 15] in connection with the forecast for 3D printing).

Contrary to some foreign countries (UK, Germany or USA), France has trouble bringing out critical-sized actors in the research domain. The assets of many laboratories rely on a small number of people, which makes their support over time uncertain. As in the field of subcontracting, the small size of the research actors is a disadvantage, both in terms of investment ability as attractiveness and maintenance of skills [AST 16].

	USA	GR	China	I/B/PB	GB	Sing.	Korea	Jpn	France
National priority									
Financial support									
Indirect aid									
Academic know-how									
Industrial know-how									
Support structures									
Industrial dynamic and services									
Users dynamic									
Specialized professional associations									
Media									
University training									
Public awareness									
DIY dynamic									
Σ Points	32	26	14	15	23	11	13	16	14

Table 5.7. *Weakness of France's effort in additive manufacturing (dark color: significant effort (3 points); medium: average (2 points); light: weak (1 point); white: insignificant (0 points); GR for German Republic, Sing. for Singapore Jpn for Japan; DIY for "Do-It-Yourself")*

The report by Legait *et al.* [LEG 16] on the improvement of industry–university links notably recommended:

> "– Electing those in charge of R&D in the administration and public establishments according to criteria inspired by the best international examples to guarantee the mobilization of indispensible skills;
>
> – creating scientific director positions within ministries given, on the one hand, functions comparable to those of "chief scientific advisors", established in several other countries and completed, on the other hand, by more significant administrative responsibilities. They will be placed at the top of the administrative hierarchy to participate efficiently in the elaboration of public policies, creating a network of experts within the administration in governance and innovation economy and kindling university work on these issues;
>
> – developing mobility between the public and the private by extending the principle of the CIFRE device to post-doctorals and statutory researchers on the basis of a new program similar to the British "Knowledge Transfer Partnership";
>
> – explicitly displaying, in the statutes of academic researchers and functionary teacher-researchers, a mission of valorization at the same level as their research missions, as this appears for the concerned public establishments (law from July 22, 2013, concerning higher education and research)".

The report is convinced that, in the absence of profound changes in, on the one hand, the governance of innovation in the administration and in public research establishments on the other hand, any attempt to improve "at the margins" is an illusion, as recent decades have shown.

The preceding recommendations are completed by others aiming of:

> "– improving cross-corporation and public research knowledge: implementation of "coupons for innovation" directed at SMEs, designation of "industrial liaison officers" within public research structures, increase in visibility and readability of public research, with a new interface completing the Internet tool, grouping together "small" public calls for projects;
>
> – accelerating the implementation of public/private R&D projects: faster procedures for collaborative research programs, intellectual property clauses, single representative to negotiate contracts;

– bringing the Instituts Carnot revenue contribution to 30% along with a thorough review of their governance and assigning a significant part of the PIA3 to projects associating companies and public laboratories, with exemplary governance;

– improving public R&D management tools to develop relations with companies: notably by the definition of pertinent performance indicators and creation of an "industrial orientation committee" within public research establishments".

5.5.3. *A bit of creativity?*

"Arthur Koestler was not far off when he described the behavior of certain creators in science by comparing them to that of sleepwalkers, stumbling upon major discoveries. However, it is still necessary for the moment to be favorable and the mind prepared, and possibly for chance to come to your aid" [MIL 00]. How to manage the sleepwalkers of 3D printing? How to feed their creativity? [CAS 17].

"The hierarchy should advocate 'disruptive' innovation and get rid of the more traditional 'incremental' kind… and companies can finance on their own because it is often less risky. What the State must do is finance the risk, the disruptive… if we want to be an Innovative Nation tomorrow" [COM 16]. According to Dechamp and Delaunat [DEC 16], "discipline is a means of hierarchizing knowledge corresponding to a cognitive need to categorize and to simplify. However, it is observed that it does not allow links to be made between knowledge and the processing of complex problems that cannot be separated from innovation." To reach this goal, every interdisciplinary operation starts with a goal or vision (e.g. to return to a recent discussion: how to read a book without opening it, as done by MIT [MIT 16]). There is thus a need to have visionaries in a group of operators allowing open and new modes of action to be brought about.

Second, it is necessary for this sort of subject to master the knowledge of the real and the possible, which implies updating knowledge on subjects coming from visionaries. An at least partial means is to interrogate disciplinary specialists whose skills could quickly provide this knowledge. The question posed could be as follows: "Considering your skills in 'such-and-such' domain, how do you see your discipline's contribution to…"? Under these conditions, the reflection group could already have mono-disciplinary bricks for initial reflection on their involvement.

The schema of such an operation is represented in Figure 5.15.

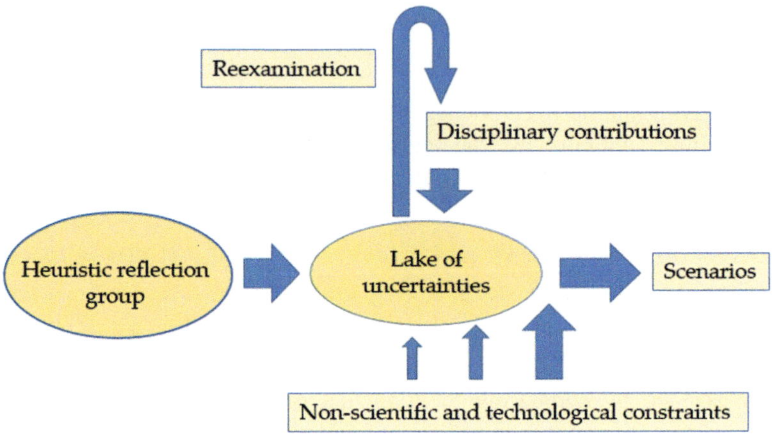

Figure 5.15. *Creativity and study strategy*

It "helps" to create an interdisciplinary group practicing "divergent" thought and having skills on a proposed and selected subject, likely to define, using outside contributions, what is possible and thus known, as well as the areas of the partially or totally unknown. The partisans of the active research model suggest that individuals identify opportunities through motivated, conscious and deliberate research [CHA 14]. As a function of the goals defined by the "pilot", it must be possible to define scientific and technological scenarios adapted to achieving the goal (or to state one's inability to achieve it with the necessary explanation of this impossibility).

Using the best possible knowledge on the diversity of representations (of which an attempt must be made to quantify), there is a possibility of clarifying the role/interests of the different actors in the reflection group in order to valorize diversity and the possibility for more in-depth reflection on particular axes, but also to seek other partners allowing, as much as possible, a balance of the debate relative to the goal. In this sharing of an important part of ignorance, it must be possible to report uncertainties of scientific and technical knowledge, approximations of knowledge, interpretive abuses and skill limitations in order to measure at least the extent of the unresolved questions and the question marks. This approach permits distance to be taken from every ideology, from overly reassuring words and abusive simplifications. By proposing the exploitation of shared territories between disciplines as a zone of freedom, voluntary action and articulation between overly disjointed and even isolated knowledge, it is indeed a matter of defining new practices aiming at operational efficiency.

If there are n (n possibly being equal to the unit) groups for the same goal, it is then possible to consider the uncertainties of proposing n scenarios (S). If there are p goals (not only disjointed ones), the same operation thus leads to n*p scenarios. It is then possible to examine if, in S(i,j) [i ϵ [1,n], j ϵ [1,p]] and bring to light generic questions applicable to the different selected goals (or to a part of them).

Naturally, SWOT analyses are part of the reflection group's work with "measurements" (to be defined semi-quantitatively by multi-criteria approaches) allowing different strategic scenarios to be proposed with a search for optimization of the type People*Means*Duration.

On the basis of these propositions, it is the buyer's job, in his/her decision tree, to make choices (see Table 5.8).

Definition of the goals by the buyer (possibly with an estimation of the time to achieve these goals);
Choice of reflection groups with group leader;
Framing each goal: required knowledge; call for disciplinary specialists;
Return and integration of disciplinary knowledge;
Heuristic approach (midwifery) of scenario definition considering outside constraints;
Search for generic themes among the different goals;
Decisions and search for other ways to spread scientific and technological work into other niches; other specialized research; openings and cooperation, etc.

Table 5.8. *Decision tree*

This frame poses the hypothesis that the buyer has a will to act in this kind of approach that may have quite variable magnitudes (like a proof of concept following MIT's example, mentioned above, expanded to broader actions) as well as financing possibilities adapted to the proposition of credible scenarios to go beyond the currently explored research methods through a more groundbreaking approach. This is obviously one of the methods that could be explored in additive manufacturing matters to bring about some good ideas that are useful for the development of this stimulating technology.

According to France Stratégie [FRA 17]: "Responding to disruptive innovation implies taking risks, thus accepting possible failure, facing their possible negative consequences, and being able to take lessons away from them. This position also concerns citizens and companies as much as the public power. If we do not proactively incorporate innovation, it will end up imposing itself all the same in an even more disruptive way." Let us hope, then, that this call is heard so that there is better development than now of research on the subject of these three volumes in our "Gallic village". According to Charrié and Janin [CHA 17], "to ensure an ability to innovate nationally and to facilitate the adoption of new uses, the State must equip itself with mechanisms to respond to disruption". It will be quite necessary to get involved in the triangle of innovation (see Figure 5.16), as proposed by Potstada and Zybura [POT 14], if we want to play our cards right with additive manufacturing.

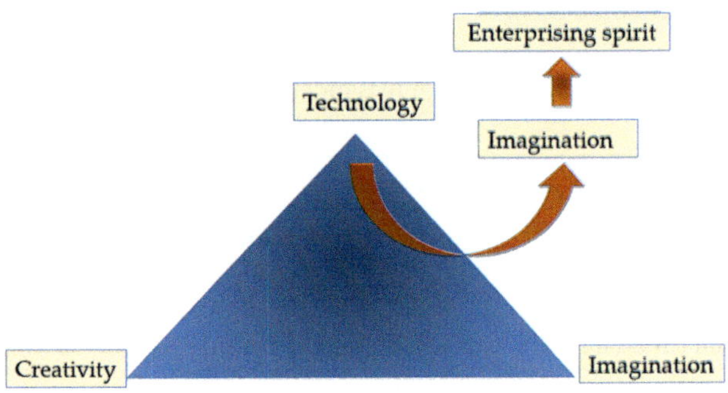

Figure 5.16. *Triptych of developing additive manufacturing*

Everything is (nearly) ready if we follow (nearly all) these recommendations to change the situation beyond the will to move forward, but when and how will this be done? Once again, rejuvenation through the examination of Alain Le Méhauté's words [LE 14] on this adventure may be interesting. The present author has attempted to recall these long-forgotten words in these three volumes: "The issue raised by this clairvoyance [on the concept of additive manufacturing] is that it is opposed to technical blindness and connivance, which subsists in France in particular, among the "financial managers" and those in high technical management, the 'elite' constituting an 'omniscient Platonic aristocracy' – not even arrogant, so sure is it of bearing The Truth. This aristocracy is judged (this is not the right word, it is rather co-opted) from 'short-term' financial choices that obviously are still inevitable, particularly when they are involved in careerist perspectives. The Peter Principle plays entirely on the dynamics constructed for the sole benefit of men of the machine, moreover convinced, thanks to processes of cognitive dissonance well known to psychologists, to act for an obviously imagined common good. As it happens, those responsible for suppressing patent royalties have been promoted and continue to use their 'talents' within 'large enterprises', whose relations with often very innovative subcontracting SMEs are known."

Let us hope that Alain is truly wrong.

5.6. In the form of a conclusion: a summary of the author's point of view

> *"Experimental inflation is no less harmful than economic inflation: there are instruments, they are massively used, and an infinite mass of data is drawn from them, and in the end, we do not know how to take anything away from this. The data fill entire libraries, then sleep in dusty archives, and no one is concerned with them anymore". [THO 83]*

> *"I have forced myself to describe the world not as it is, but as it is when I add myself to it, which, obviously, does not simplify it". [GIO 79]*

> *"It is in this opposition between gradual and disruptive innovations that we can find the stakes: must innovation prolong the active cycles, support the existing structures, and make our lifestyle sustainable? Or rather, must it open radically new paths, substitute unprecedented techniques with those inherited from the past, and revolutionize our societies? Due to this ambivalence, the rhetoric of technological innovation ends up taking the shape of a paradoxical injunction: 'let everything change so that nothing changes!' We must recognize that this is hardly stimulating...". [KLE 15]*

"Interdisciplinarity can and must be evaluated on the dimensions that it builds. Its epistemology must be based on the intelligence of the differences in practices: are we dealing with an 'internal' interdisciplinarity where it is supposed that experimental science can be equipped by mathematics and computer science? Or rather, are we dealing with external and open interdisciplinarity where autonomous disciplines bring about crosses in flows of knowledge in an unexpected way?". [MAT 14]

"The pessimist sees difficulty in every opportunity. The optimist sees the opportunity in every difficulty". [LAP 17]

"Nothing is more dangerous than an idea, when one has only one idea". [WIK 17]

"Those who tell you that the only alternative is to give up are imposters. To give up, it is really best to invest energy that is opposed to the acceleration of time". [SMA 11]

"Without talent, ingenuity, or curiosity, neither chance nor need will be enough to produce innovation". [FOR 02]

"Results are obtained by exploiting opportunities, not by solving problems". [DRU 06]

"Behavior is motivation passed through the filter of opportunity". [SHI 11]

"Crossing the boundaries between disciplines [is] a subversive undertaking, for it has every chance of violating the sanctuaries of commonly accepted ways of thinking". [GRE 90]

"Foxes are interested in everything and have no difficulty going from one issue to another. Hedgehogs are only interested in a small number of issue, which they hold to be fundamental, and they dedicate themselves to the same issues for years, if not decades. Most large discoveries are made by foxes, most small discoveries by hedgehogs. To progress, science needs hedgehogs and foxes". [DYS 11]

> *"The world is very complicated and it is clearly impossible for the human mind to understand it completely. That is why humanity has found an artifice that allows us to reproach the world's complex nature for its chance character and thus to allow abstraction of a domain where simple laws are found. Complications are then qualified as initial conditions, and domains as regularities of the laws of Nature". [WIG 49]*

Figure 5.17 can guide research and R&D works in additive manufacturing. The activities proposed in this figure find their origins in different presentations made in Volumes 1 and 2. It is also based on the propositions of the "Consortium for Additive Manufacturing Materials" [CON 15b].

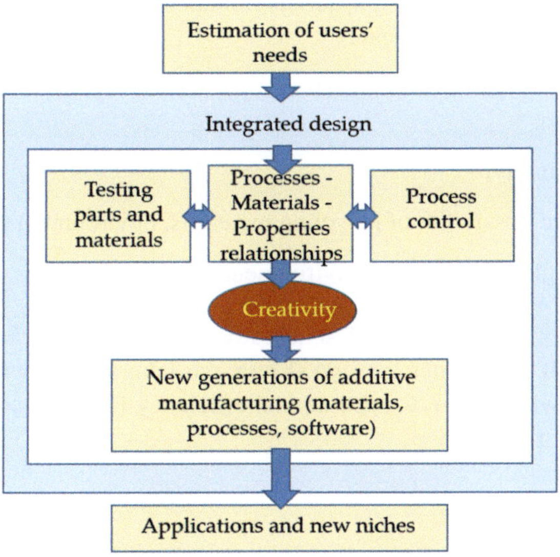

Figure 5.17. *Incremental innovations in additive manufacturing*

In the course of these two volumes, recurrent questions have emerged. These are summarized below:

– improving material couplings (including initially shaping them) – processes – software;

– developing new, more ergonomic processes (software), more robust (quality) processes with better homogeneity of the properties within the material making up the object;

– broadening the range of usable materials;

– optimizing the re-design of 3D objects by considering property–structure relationships;

– treating the "inverse problem" to obtain the object desired in one shot, consistent with the given instructions (considering deformations induced by the process and the complete history linked to manufacturing the object);

– developing processes with variable sizes of voxels to have a good surface finishing and a reduced manufacturing time;

– updating multi-material and/or material gradient processes with different mechanical properties;

– examining the integration of additive manufacturing technologies into all (subtractive) manufacturing processes;

– going more in-depth into the normalization of the subject, following the USA's example [ANS 17];

– controlling quality;

– examining the possibility of recycling materials; environmental conditions;

– following hygiene, safety and environment studies.

This list is not exhaustive and it already corresponds to an important and significant work. If France has very good research laboratories, will we know how to mobilize them with an aim for operational efficiency in the significant domain of additive manufacturing?

5.7. Bibliography

[ACK 13] ACKERMANN G., "Le système territorial d'innovation", Cahiers du LABRII, available at: http://riien.univ.littoral.fr/wp-content/uploads/2013/01/doc-261-pdf, 2013.

[AIF 15] AIF (ALLIANCE POUR L'INDUSTRIE DU FUTURE), Lancement officiel de l'Alliance Industrie du Futur!, available at: https://www.imt.fr/lancement-officiel-de-lalliance-industrie-du-futur/-, 2015.

[AIF 16] AIF ALLIANCE POUR L'INDUSTRIE DU FUTUR, "Construire ensemble une filière Fabrication Additive Française forte", available at: http://aif.diwi.org/, 2016.

[AKR 06] AKRICH M., LATOUR B., CALLON M., *Sociologie de la traduction*, Presses des Mines, Paris, 2006.

[ALB 09a] ALBALADEJO C., GESLIN P., MAGDA D. *et al.* (eds), *La mise à l'épreuve: le transfert des connaissances scientifiques en questions*, Quae, Versailles, 2009.

[ALB 09b] ALBALADEJO C., HUBERT B., ROCHE B., "Chercheurs en situation de partenariat: prescrire la subjectivité – Réflexion à partir du programme de pérennisation de l'agropastoralisme au Pays Basque", in BEGUIN P., CERF M. (eds), *Dynamique des savoirs, dynamique des changements*, Octarès, Toulouse, 2009.

[ALE 11] ALEXANDRE L., *La mort de la mort, comment la techno-médecine va bouleverser l'humanité?*, J.C. Lattès, Paris, 2011.

[ALL 15a] ALLARD L., "Numérique: les "nouveaux" usages sont-ils si nouveaux?", available at: http://www.inaglobal.fr/numerique/article/numerique-les-nouveaux-usages-sont-ils-si-nouveaux-8442, 2015.

[ALL 15b] ALLARD L., "Internet tue-t-il la créativité ou l'améliore-t-il?", available at: http://www.atlantico.fr/decryptage/internet-tue-t-creativite-ou-ameliore-t-laurence-allard-2324526.html, 2015.

[ALV 03] ALVAREZ-PEREYRE F., *L'exigence interdisciplinaire*, MSH, Paris, 2003.

[AND 08] ANDRE J.C., "Newton et l'expansion de l'univers des risques", in MUR J.-M (ed.), *Emergence des risques*, EDP Sciences, Paris, 2008.

[AND 13a] ANDRÉ J.C., "Towards a Socially Responsible Research (SRR) charter in Engineering Sciences at CNRS level", *International Journal of Technoethics*, vol. 4, pp. 39–51, 2013.

[AND 13b] ANDREEWSKY E., "Mais la carte parfois transforme le territoire", in LE MOIGNE J.L., MORIN E. (eds), *Intelligence de la complexité*, Herrmann, Paris, 2013.

[ANR 16] ANRT, "Regards croisés sur la recherche partenariale – Comment se construisent les relations entre les établissements publics de recherche et les entreprises?", available at: http://www.anrt.asso.fr/fr/futuris/pdf/rapport-rcrp.pdf, 2016.

[ANS 17] ANSI, "AMSC Standardization Roadmap For Additive Manufacturing", available at: https://www.ansi.org/standards_activities/standards_boards_panels/amsc/amsc-roadmap, 2017.

[ARA 01] ARAUJO-JORGE M.M., "L'impact épistémologique de la complexité", in CROZON M., JACQUIN Y. (eds), *L'élémentaire et le complexe*, EDP Sciences, Paris, 2001.

[AST 16] ASTOLFI C.P., CONSTANTIN E., MOULET A., "Fabrication additive; mobiliser les forces françaises", Les Docs de La Fabrique/Presses des Mines-Transvalor, Paris, 2016.

[AZO 10] AZOULAY P., GRAFF-ZIVIN J.S., MANSO G., "Incentives and creativity: evidence from the academic life sciences", available at: http://pazoulay.scripts.mit.edu/docs/hhmi.pdf, 2010.

[BAR 93] BARKE R.P., JENKINS-SMITH H.C., "Politics and Scientific expertise: scientists, risk perception and nuclear waste policy", *Risk Analysis*, vol. 13, pp. 425–439, 1993.

[BAR 04] BARRE R., "Le discours sur le déclin de la recherche: un indicateur du besoin de connaissances et de débat", *Le Banquet*, vol. 19, p. 6, 2004.

[BAS 79] BASKER P.J., *A History of Engineering Drawing*, Northgate Publishing, London, 1979.

[BAU 00] BAUDRILLARD J., *Mots de passe*, Fayard, Paris, 2000.

[BEN 15] BENSAUDE-VINCENT B., "Promesses et régimes d'historicité en technoscience", in AUDETAT M, BARAZZETTI G., DORTHE G. (eds), *Sciences et technologies émergentes: pourquoi tant de promesses?*, Herrmann, Paris, 2015.

[BER 08] BERTHELOT J.M., *L'emprise du vrai – Connaissance scientifique et modernité*, PUF, Paris, 2008.

[BER 13] BERGER R., "Industrie 4.0: L'usine connectée", available at: http://www.gimelec.fr/Actualites/Actualites-Gimelec/Le-Gimelec-publie-Industrie-4.0-les-leviers-de-la-transformation, 2013.

[BER 14] BERGER R., "Industry 4.0: the new industrial revolution; how Europe will succeed", available at: https://www.rolandberger.com/media/pdf/Roland_Berger_TAB_Industry_4_0_20140403.pdf, 2014.

[BER 16a] BERGER S., DORMAN R., STARBUCK H., "Reforms in the French industrial ecosystem", available at: http://www.sf2m.asso.fr/CommissionsThematiques/DocComThematiques/RNM_MIT_Final_Summary.pdf, 2016.

[BER 16b] BERGER S., DORMAN R., STARBUCK H., "Reforms in the French industrial ecosystem", available at: http://cache.media.enseignementsup-recherche.gouv.fr/file/Actus/73/9/Reforms_in_the_French_Industrial_Ecosystem_(fr)_539739.pdf, 2016.

[BIG 13] BIGOT G., *La trahison des chefs*, Fayard, Paris, 2013.

[BOT 17] BOTTERO L., "Louis Schweitzer: "L'interdisciplinarité doit entrer dans les mentalités", available at: http://marseille.latribune.fr/economie/2017-03-30/louis-schweitzer-l-interdisciplinarite-doit-entrer-dans-les-mentalites.html, 2017.

[BOU 84] BOURDIEU P., *Questions de sociologie*, Les Editions de Minuit, Paris, 1984.

[BOU 96] BOURDIEU P., "Le champ scientifique", *Actes de la Recherche en Sciences Sociales*, vol. 2, nos. 2–3, pp. 88–104, 1996.

[BOU 01a] BOURG D., SCHLEGEL J.L., *Parer aux risques de demain: le principe de précaution*, Le Seuil, Paris, 2001.

[BOU 01b] BOURRIER M. (ed.), *Organiser la fiabilité*, L'Harmattan, Paris, 2001.

[BOU 03] BOUDON R., *Raisons, bonnes raisons*, PUF, Paris, 2003.

[BOU 09] BOUDON R., *La rationalité*, PUF, Paris, 2009.

[BPI 17] BPI FRANCE, "Créer et innover en Ile-de-France: le rôle des plateformes d'innovation", available at: http://www.bpifrance.fr/A-la-une/Actualites/Creer-et-innover-en-Ile-de-France-le-role-des-plateformes-d-innovation-31338, 2017.

[BRU 08a] BRUNO I., *A vos marques, prêts, partez... Cherchez*, Editions du Croquant, Bellecombe-en-Bauge, 2008.

[BRU 08b] BRUNO I., "Y-a-t-il un pilote dans l'avion? Tableaux de bord, indicateurs, cibles chiffrées: les balises de la décision", *Boeck University Politix*, vol. 82, pp. 95–117, 2008.

[BRU 13] BRUNET-LABBEZ J., *Innover avec la génération Y*, Studyrama, Paris, 2013.

[CAP 15] CAPGEMINI CONSULTING, "The innovation game: why and how business are investing in innovation centers", available at: https://www.capgemini-consulting.com/resource-file-access/resource/pdf/innovation_center_v14.pdf, 2015.

[CAS 17] CASTILLO A., "La créativité ou l'art d'exhumer des analogies cachées", available at: https://www.letemps.ch/economie/2017/02/16/creativite-lart-dexhumer-analogies-cachees, 2017.

[CEL 72] CÉLINE L.F., *Voyage au bout de la nuit*, Folio, Paris, 1972.

[CHA 14] CHANUT-GUIEU C., GUIEU G., "Créativité, PME et entrepreneuriat: des zones d'ombre et de lumière", *Revue internationale P.M.E.: économie et gestion de la petite et moyenne entreprise*, vol. 27, pp. 35–63, 2014.

[CHA 17] CHARRIE J., JANIN L., "Répondre à l'innovation disruptive", available at: http://www.strategie.gouv.fr/sites/strategie.gouv.fr/files/atoms/files/2017-2027-actions-critiques-innovation-disruptive-ok.pdf, 2017.

[CHI 07] CHIAPELLO E., "Dynamique du capitalisme et régulation de la technique", in FELTZ B., GOUJON P., LAVELLE S. *et al.* (eds), *Ethique, technique et démocratie*, Brulant-Academia, Louvain-la-Neuve, Belgium, pp. 87–99, 2007.

[CLO 06] CLOT Y., *La fonction psychologique du travail*, PUF, Paris, 2006.

[CNR 15] CNRS-CoNRS, "Rapport de conjoncture 2014", available at: http://rapports-du-comite-national.cnrs.fr/rapport-conjoncture/rapport-de-conjoncture-2014/iii-surfaces-et-interfaces, 2015.

[COM 05] COMMISSION EUROPEENNE, Charte européenne du chercheur, EUR 21620, available at: http://www.europa.eu.int/eracareers/europeancharter, 2005.

[COM 16] COMTESSE X., "Une innovation "disruptive"", available at: http://www.hebdo.ch/les-blogs/un-tsunami-num%C3%A9rique-r%C3%A9volutionne-la-sant%C3%A9/une-innovation-disruptive, 2016.

[CON 15a] CONSEIL SUISSE DE LA SCIENCE ET DE L'INNOVATION, "Measuring Innovation – A Discussion of Innovation Indicators at the National Level", available at: http://www.swir.ch/images/stories/pdf/en/Arbeitsdokument_3_2015_Measuring_Innovation.pdf , 2015.

[CON 15b] CONSORTIUM FOR ADDITIVE MANUFACTURING MATERIALS, "Strategic Roadmap for the Next Generation of Additive Manufacturing Materials", available at: http://news.psu.edu/tag/strategic-roadmap-next-generation-additive-manufacturing-materials, 2015.

[COU 07] COUR DES COMPTES, Le rôle et la stratégie du CNRS, Report, Paris, pp. 113–153, 2007.

[DEG 12] DE GAULEJAC V., *La recherche malade du management*, Quae, Versailles, 2012.

[DEC 16] DECHAMP G., DELAUNAY C., "Favoriser et gérer l'interdisciplinarité dans l'entreprise: la notion de territoires partagés", *Management & Avenir*, vol. 88, pp. 37–58, 2016.

[DRE 05] DRESCHLER W., "The rise and the denise of the New Public Management", *Post-autistic Economics Review*, vol. 33, pp. 104–112, 2005.

[DRU 06] DRUCKER P., *Innovation and Entrepreneurship*, Collins, New York, 2006.

[DUR 67] DURKHEIM E., *Jugements de valeur et jugements de réalité*, PUF, Paris, 1967.

[DUR 03] DURKHEIM E., *Les formes élémentaires de la vie religieuse; le système totémique en Australie*, PUF, Paris, 2003.

[DUR 12] DURAND D., *Des outils, des machines et des Hommes*, PUF, Paris, 2012.

[DUV 16] DUVIVIER D., "Les novateurs, ces déviants", available at: http://www.agoravox.fr/tribune-libre/article/les-novateurs-ces-deviants-180824, 2016.

[DYS 11] DYSON F., *Portrait du scientifique en rebelle*, Actes Sud, Paris, 2011.

[EAR 15] EARLS A., BAYA V., "The road ahead for 3-D printers", available at: http://www.pwc.com/us/en/technology-forecast/2014/3d-printing/features/future-3d-printing.html, 2015.

[EDW 11] EDWARDS D., *Le manifeste du laboratoire*, Odile Jacob, Paris, 2011.

[EIN 99] EINSTEIN A., *Comment je vois le monde*, Flammarion, Paris, 1999.

[ELG 70] ELGOZY G., *Les damnés de l'opulence*, Calmann-Lévy, Paris, 1970.

[EU 12] EU FWP 7, 7ème Programme Cadre de l'UE, available at: http://cordis.europa.eu/fp7/home_fr.html, 2012.

[EUR 12] EUROGROUP CONSULTING, "L'art du management de l'innovation dans le service public", available at: http://www.fonction-publique.gouv.fr/publications/publications-hors-collection-1, 2012.

[EVO 15] EVOREG, "Are innovation systems complex systems?", available at: http://www.evoreg.eu/docs/files/shno/Note_evoREG_33.pdf, 2015.

[FED 15] FEDERAL RESEARCH & DEVELOPMENT, "Federal Research and Development Funding: FY 2016", available at: http://www.fas.org/sgp/crs/misc/R43944.pdf, 2015.

[FER 92] FERGUSSON E., *Engineering and the Mind's Eye*, MIT Press, Cambridge, 1992.

[FEY 98] FEYNMAN R.P., *Vous y comprenez quelque chose Monsieur Feynman?*, Odile Jacob, Paris, 1998.

[FIT 04] FITREMAN J., "États généraux de la recherche 2004", available at: http://cip-etats-generaux.apinc.org/article.php3?id_article=475, 2004.

[FOL 07] FOLSCHEID D., "Les changements du monde actuel et les relations nouvelles entre Science et Société", *Les cahiers du MURS "Sciences et devenir de l'Homme", Symposium Proceedings 13 November 2007*, Paris, France, pp. 125–134, 2007.

[FON 14] FONDS RICOEUR, "L'incapacité dans la philosophie et les sciences sociales", available at: http://cems.ehess.fr/index.php?3155, 2014.

[FOR 96] FORTI A., "La mort de Newton", in FORTI A. (ed.), *La mort de Newton*, Maisonneuve et Larose, Paris, 1996.

[FOR 02] FORAY D., "Ce que l'économie néglige ou ignore en matière d'analyse de l'innovation", in ALTER N. (ed.), *Les logiques de l'innovation*, La Découverte, Paris, 2002.

[FRA 16] FRANCE STRATEGIE, "Quinze ans de politiques d'innovation en France", available at: http://www.strategie.gouv.fr/publications/quinze-ans-de-politiques-dinnovation-france, 2016.

[FRA 17] FRANCE STRATEGIE, "2017/2027 – Répondre à l'innovation disruptive – Actions critiques", available at: http://www.strategie.gouv.fr/publications/20172027-repondre-linnovation-disruptive-actions-critiques, 2017.

[GAI 16] GAIGNAIRE A., "Ingénieurs du futur recherchés", available at: https://www.lesaffaires.com/dossier/profession-ingenieur/ingenieurs-du-futur-recherches/591065, 2016.

[GAN 00] GANASCIA J.G., "Science x.0", in HAIGNERE C. (ed.), *Plaidoyer pour réconcilier la science avec la culture*, Le Pommier, Paris, 2000.

[GAR 08] GARTNER GROUP, available at: http://www.gartner.com, 2008.

[GAU 78] GAUDIN T., *L'écoute des silences*, Editions 10/18, Paris, 1978.

[GEN 12] GENERAL ELECTRIC, "Global Innovation barometer", available at: http://www.ge.com/innovationbarometer/, 2012.

[GEN 15] GENERAL ELECTRIC, "General Electric imprime en 3D le premier mini réacteur fonctionnel", available at: http://www.additiverse.com/actualites/2015-05-18/general-electric-imprime-en-3d-le-premier-mini-reacteur-fonctionnel, 2015.

[GIE 00] GIESEN K.G., "Production d'artéfacts et ingénierie sociale; les institutions internationales comme agents paradigmatiques", *AFRI*, vol. 1, pp. 70–86, available at: http://www.afri-ct.org/IMG/pdf/giesen2000.pdf, 2000.

[GIF 15] GIFFI C.A., "Global manufacturing competitiveness forum", available at: http://www.compete.org/storage/documents/2016_GMCI_NCF_Slides_v8_Giffi_120415.pdf, 2015.

[GIO 79] GIONO J., *Voyage en Italie*, Poche, Paris, 1979.

[GIR 17] GIRET V., "Le décryptage éco. La France à la traîne de l'innovation", available at: http://www.francetvinfo.fr/replay-radio/le-decryptage-eco/le-decryptage-eco-la-france-a-la-traine-de-l-innovation_2015094.html, 2017.

[GOD 14] GODARD O., "Contexte et enjeu d'une éthique de l'expertise scientifique; les risques collectifs environnementaux et sanitaires", in MARTIN T. (ed.), *Ethique de la recherche et risques humains*, PUFC, Besançon, France, 2014.

[GRA 86] GRANGER G.G., "Pour une épistémologie du travail scientifique", in HAMBURGER J. (ed.), *La philosophie des sciences aujourd'hui*, Gauthier-Villars, Paris, 1986.

[GRE 90] GREENBERG V.D., Cited by STAUNE J. (2015), *Les clés du futur; réinventer ensemble la société, l'économie et la science*, Plon, Paris, 1990.

[GUE 92] GUELLEC D., "Croissance endogène: les principaux mécanismes", *Économie & Prévision*, vol. 106, pp. 41–50, 1992.

[GUE 09] GUELLEC D., *Economie de l'Innovation*, La Découverte, Paris, 2009.

[GUI 16] GUILLARD J.C., "L'innovation appliquée, l'antidote contre le Darwinism Digital", available at: http://www.journaldunet.com/solutions/expert/65558/l-innovation-appliquee--l-antidote-contre-le-darwinism-digital.shtml, 2016.

[GUO 13] GUO N., LEU M.C., "Additive manufacturing: technology, applications and research needs", *Frontiers of Mechanical Engineering*, vol. 8, pp. 215–243, 2013.

[HAB 73] HABERMAS J., *La technique et la science comme "idéologie"*, Gallimard, Paris, 1973.

[HAB 02] HABERMAS J., *L'avenir de l'aventure humaine: vers un eugénisme libéral?*, Gallimard, Paris, 2002.

[HAL 16] HALF R., "Techniques de management pour doper la créativité des équipes", available at: https://www.roberthalf.fr/blog/techniques-de-management, 2016.

[HAM 13] HAMEL J., "L'interdisciplinarité, manière de faire ou de dire la science?", EspacesTemps.net, available at: http://www.espacestemps.net/articles/linterdisciplinarite-maniere-de-faire-ou-de-dire-la-science-2/, 2013.

[HEI 10] HEIL C., De la science à la technoscience, du chercheur au technologue, Thèse de l'Université d'Evry, Val d'Essonne, Evry, 2010.

[HEN 98] HENRY C., "Le 'je' intellectuel et le 'jeu' interdisciplinaire", *Le Genre Humain*, vol. 33, pp. 155–170, Le Seuil, Paris, 1998.

[HIR 05] HIRSCH J.E., "An index to quantify an individual's scientific research output", *Proceedings of the National Academy of Sciences*, vol. 102, pp. 16569–16572, 2005.

[HOF 01] HOFSTEDE G., *Culture's Consequences: Comparing Values, Behaviors, Institutions and Organizations Across Nations*, SAGE Publications, Thousand Oaks, 2001.

[HOR 16] HORCKMANS M., Internship report, University of Lorraine, Nancy, 2016.

[HOT 92] HOTTOIS G., "Le règne de l'opératoire", in PRADES J. (ed.), *La technoscience: les fractures des discours*, L'Harmattan, Paris, 1992.

[HOU 15] HOURCADE J.C., "L'impression 3D vue au prisme des dépôts de brevets", available at: http://www.industrie-techno.com/l-impression-3d-vue-au-prisme-des-depots-de-brevets.36223, 2015.

[ING 15] INGELAERE R., "Innovation technologique et performance industrielle: l'exemple de l'impression 3D", available at: http://www.lecese.fr/sites/default/files/pdf/Avis/2015/2015_07_innovations_techno_impression_3d_.pdf, 2015.

[INS 17] INSTITUTS CARNOT, "Cartographie des acteurs clés de la R&D en Fabrication Additive en France", available at: http://www.cetim.fr/Mecatheque/Resultats-d-actions-collectives/Cartographie-des-acteurs-cles-de-la-R-D-en-fabrication-additive, 2017.

[ISF 16] ISF, "Relever les défis d'une économie prospère et responsable", available at: http://home.iesf.fr/752_p_43968/parution-des-propositions-des-iesf.html, 2016.

[JUV 10] JUVIN A., LIPOVETSKY G., *L'occident mondialisé: controverses sur la culture planétaire*, Grasset, Paris, 2010.

[KAL 07] KALUSZYNSKI M., "A qui appartient la République? Objets disciplinaires, objets disciplinés – De l'invitation à l'hybridation disciplinaire", in FAVRE P., FILLIEULE O., JOBARD F. (eds), *L'Atelier du politiste*, La Découverte, Paris, 2007.

[KAU 08] KAUFMANN J.C., *Quand Je est un autre: Pourquoi et comment ça change en nous*, Armand Collin, Paris, 2008.

[KAU 15] KAUFMAN S.B., "The Emotions That Make Us More Creative", Harvard Business Review available at: https://hbr.org/2015/08/the-emotions-that-make-us-more-creative, 2015.

[KID 82] KIDDER J.T., *Project Eagle*, Flammarion, Paris, 1982.

[KLE 15] KLEIN E., "Comment faire progresser l'idée de progrès?", available at: http://www.up-magazine.info/index.php?option=com_content&view=article&id=5305:comment-faire-progresser-l-idee-de-progres&catid=103:analyses&Itemid=542&utm_source=newsletter_175&utm_medium=email&utm_campaign=up-nl-vnew, 2015.

[KOS 06] KOSCIUSKO-MORIZET N., "Tel(s) quel(s): aux frontières de la Science et de la Société", *Passages*, vol. 148, p. 15, 2006.

[LAF 11] LAFLAMME S., "Recherche interdisciplinaire et réflexion sur l'interdisciplinarité", *Nouvelles Perspectives en Sciences Sociales*, vol. 7, pp. 49–64, 2011.

[LAN 75] LANDES D.S., *L'Europe technicienne: Révolution technique et libre essor industriel en Europe occidentale de 1750 à nos jours*, Gallimard, Paris, 1975.

[LAP 17] LA PHILOSOPHIE, "Citations de Churchill", available at: http://la-philosophie.com/citation-churchill, 2017.

[LAR 94] LARRERE L., "Ordre biologique, ordre technologique; le cas de l'écologie", in TINLAND F. (ed.), *Ordre biologique, ordre technologique*, Champ-Vallon, Seyssel, 1994.

[LAR 17] LARRERE C, LARRERE R., *Bulles Technologiques*, Wildproject, Paris, 2017.

[LAS 07] LASCH C., *La révolte des élites et la trahison de la démocratie*, Champs-Essais, Paris, 2007.

[LAT 05] LATOUR B., *La science en action; introduction à la sociologie des sciences*, La Découverte, Paris, 2005.

[LAT 07] LATOUR B., *L'espoir de Pandore*, La Découverte, Paris, 2007.

[LE 02] LE MAREC J., Ce que le terrain fait aux concepts: vers une théorie des composites, HDR, Paris Diderot University, 2002.

[LEM 14] LE MEHAUTE A., "Interview d'Alain Le Méhauté – L'un des pères de l'impression 3D", available at: http://www.priximprimante3d.com/inventeur/, 2014.

[LEG 16] LEGAIT B., RENUCCI A., SIKORAV J.L., Les relations entre les entreprises et la recherche publique; lever des obstacles à l'innovation en France, Report addressed to the French Minister of Education, available at: http://cache.media.enseignementsup-recherche.gouv.fr/file/2015/94/7/rapport-071_542947.pdf, 2016.

[LEP 97] LEPLAT J., *Regards sur l'activité en situation de travail*, PUF, Paris, 1997.

[LES 16] LESAGE S.E., HURTEAU P., "L'innovation en recherche universitaire: Austérité et arrimage aux intérêts du secteur privé", available at: http://iris-recherche.s3.amazonaws.com/uploads/publication/file/web_NouveauTexteEtude_RechercheUniversitaire_vli_02.pdf, 2016.

[LUB 16] LUBINEAU P., "La Fabrication Additive en France: cartographie des acteurs", available at: http://aif.diwi.org/wp-content/uploads/2016/07/20160719-Etude-Carnot-FIMP-Fab-Add-Cartographie-des-compétences-Diff-AIF.pdf, 2016.

[MAC 12] MACHECOURT A., "L'innovation européenne est à la traîne", available at: http://www.latribune.fr/entreprises-finance/20120511trib000698101/l-innovation-europeenne-a-la-traine.html, 2012.

[MAD 15] MADELIN T., "L'industrie allemande critique le cadre d'innovation français", available at: http://www.lesechos.fr/idees-debats/sciences-prospective/156032-lindustrie-allemande-critique-le-cadre-dinnovation-francais-86495.php?dPM9t7dQ8qWC8I1u.99, 2015.

[MAH 02] MAHE A., La communication scientifique en (r)évolution, Thesis, Université Claude Bernard, Lyon, 2002.

[MAH 14] MAHY I., "L'art de l'inter à travers l'oeil du coeur", in LEMAY V., DARBELLAY F. (eds), *L'interdisciplinarité racontée; chercher hors frontières, vivre l'interdisciplinarité*, Peter Lang, Bern, 2014.

[MAR 12] MARTINAUD B., *Start-up: anti-bible à l'usage des fous et des futures entrepreneurs*, Pearson, Orléans, 2012.

[MAR 13] MARCHESNAY M., "Le système 1 "Parva sed apta": l'innovation dans les PME", in BOUTEILLER S., DJELLAL F., UZUNIDIS D. (eds), *L'innovation: analyser, anticiper, agir*, Peter Lang, Brussels, 2013.

[MAR 15] MARION F., "Pourquoi les startups françaises ne deviennent-elles pas des géants?", available at: http://www.up-magazine.info/index.php?option=com_content&view=article&id=4976:pourquoi-les-startups-francaises-ne-deviennent-elles-pas-des-geants&catid=103:analyses&Itemid=542&utm_source=newsletter_123&utm_medium=email&utm_campaign=up-nl-vnew, 2015.

[MAT 14] MATHIEU N., SCHMID A.F., "Reconsidérer le lien entre modélisation et interdisciplinarité", in MATHIEU N., SCHMID A.F. (eds), *Modélisation et interdisciplinarité; six disciplines en quête d'épistémologie*, Quae, Paris, 2014.

[MED 12] MEDIAPART, "La croissance en cul de sac", available at: http://blogs.mediapart.fr/blog/ujamaa/090512/la-croissance-en-cul-de-sac, 2012.

[MEN 15] MENESR, "Note d'information: Les efforts de recherche en Union Européenne", available at: http://cache.media.enseignementsup-recherche.gouv.fr/file/2015/09/4/NI_efforts_RetD_UE_-_15.03_453094.pdf, 2015.

[MER 42] MERTON R.K., *The Sociology of Science*, Chicago University Press, Chicago, 1942.

[MER 68] MERTON R.K., *Social Theory and Social Structure*, Free Press, New York, 1968.

[MER 73] MERTON R.K., *The Sociology of Science: Theoretical and Empirical Investigations*, University of Chicago Press, Chicago, 1973.

[MER 06] MERZ M., MAASEN S., "TA goes STS: l'évaluation des choix technologiques change d'orientation", in LE RESCHE J.P., BENNINGHOFF M., CRETTAZ VON ROTEN F. *et al.* (eds), *La fabrique des Sciences*, Presses Polytechniques et Universitaires de Lausanne, Lausanne, 2006.

[MEY 15] MEYSONNIER F., "Les dispositifs de pilotage de la performance en environnement innovant et incertain: étude comparative de 8 startups", available at: https://hal.archives-ouvertes.fr/hal-01147206/document, 2015.

[MIL 00] MILLER A.I., *Intuitions de génie*, Flammarion, Paris, France, 2000.

[MIL 15] MILLERAND F., "Les imaginaires de la science 2.0; de l'idéal de la science ouverte au marketing 2.0", *Communication*, vol. 33, available at: http://communication.revues.org/6070, 2015.

[MIN 09] MINISTERE DE L'INDUSTRIE (MORAND P., MANCEAU D.), "Pour une nouvelle vision de l'Innovation", available at: http://archives.entreprises.gouv.fr/2012/www.industrie.gouv.fr/enjeux/innovation/rapport-morand-2009.pdf, 2009.

[MIN 10] MINISTERE DE L'INDUSTRIE, "Technologies clés 2010", available at: http://www.industrie.gouv.fr/techno_cles_2010/html/sommaire.php, 2010.

[MIT 16] MIT, "Convergence: The Future of Health (A novel research strategy that integrates biomedical knowledge with advanced engineering skills and expertise from physical, computational, and mathematical sciences)" available at: http://static1.squarespace.com/static/570561df4d088ef84160338a/t/576bf712ff7c50b7fcb48482/1466693398318/Convergence-The-Future-of-Health-2016-Report.pdf, 2016.

[MOO 06] MOORE K., FRICKEL S. (eds), *The New Political Sociology of Science: Institutions, Networks and Power*, University of Wisconsin Press, Madison, 2006.

[MOR 13] MORIN E., "Complexité restreinte, gomplexité générale", in LE MOIGNE J., MORIN E. (eds), *Intelligence de la complexité*, L'Harmattan, Paris, 2013.

[MUS 07] MUSSO P., PONTHON L., SEUILLET E., *Fabriquer le futur 2; l'imaginaire au service de l'innovation*, Village Mondial, Paris, 2007.

[NOT 16] NOTTEAU Y., "Impression 3D: Les entreprises françaises innovantes en 2016", available at: http://www.additiverse.com/impression-3d-les-entreprises-francaises-innovantes/, 2016.

[OEC 02] OECD, *Manuel de Frascati*, OECD, Paris, 2002.

[OEC 07] OECD, "Stratégie de l'OCDE pour l'innovation", available at: http://www.oecd.org/pages/0,3417,fr_41462537_41454856_1_1_1_1_1,00.html, 2007.

[OEC 15a] OECD, "Principaux indicateurs de la science et de la technologie", available at: http://www.keepeek.com/Digital-Asset-Management/oecd/science-and-technology/principaux-indicateurs-de-la-science-et-de-la-technologie/volume-2015/issue-1_msti-v2015-1-fr, 2015.

[OEC 15b] OECD, "Les pays doivent intensifier leurs efforts de R&D dans les technologies de pointe", available at: http://www.oecd.org/fr/innovation/les-pays-doivent-intensifier-leurs-efforts-de-r-d-dans-les-technologies-de-pointe.htm, 2015.

[OEC 16] OECD, G20: Innovation Report 2016 – Report prepared for the G20 Science, Technology and Innovation Ministers Meeting, available at: http://www.oecd.org/science/inno/G20-innovation-report-2016.pdf, 2016.

[OPE 17] OPECST, "L'évaluation de la stratégie nationale de recherche France-Europe 2020", available at: http://www2.assemblee-nationale.fr/content/download/51750/501173/version/1/file/rapport+SNR_version+provisoire.pdf, 2017.

[OPI 16] OPINIONWAY, "Les français et l'innovation", available at: http://www.youscribe.com/catalogue/tous/les-francais-freines-dans-leur-creativite-2733171, 2016.

[ORF 05] ORFALI B., *La Société face aux événements extraordinaires*, Zagros, Paris, 2005.

[ORL 11] ORLEAN A., *L'empire de la valeur; refonder l'économie*, Le Seuil, Paris, 2011.

[ORW 72] ORWELL G., *1984*, Gallimard, Paris, 1972.

[PEL 14] PELLUCHON C., "Penser la place de l'expertise et de la délibération éthique dans la politique; réflexion sur les conditions d'une plus grande innovation en matière d'action environnementale et en bioéthique", in MARTIN T. (ed.), *Ethique de la recherche et risques humains*, PUFC, Besançon, 2014.

[PES 03] PESTRE D., *Science, argent et politique*, INRA, Paris, 2003.

[PES 14] PESTRE D., "Du gouvernement du progrès technique et de ses effets", in PESTRE D. (ed.), *Le gouvernement des technosciences; gouverner le progrès et ses dégâts depuis 1945*, La Découverte, Paris, 2014.

[PEZ 17] PEZZIARDI P., VERDIER H., "Des startups d'état à l'état plateforme", Fondation pour l'innovation politique, available at: http://www.fondapol.org/wp-content/uploads/2017/01/096-PEZZIARDI_2016-12-22-web.pdf, 2017.

[PHI 12] PHILIPPE J., "L'innovation managériale, comment innover dans l'univers bancaire?", in EUROGROUP CONSULTING, *L'art du management de l'innovation dans le service public*, Eurogroup, Paris, 2012.

[PIA 16] PIALOT D., "La France veut profiter des promesses de l'impression additive", available at: http://www.latribune.fr/entreprises-finance/industrie/la-france-veut-profiter-des-promesses-de-l-impression-additive-573314.html, 2016.

[PIG 04] PIGANIOL P., "Le chercheur est-il apte à la recherche?", *Le Banquet*, vol. 19, p. 4, 2004.

[PIO 09] PIOT T., "Quels indicateurs pour mesurer le développement professionnel dans les métiers adressés à autrui?", *Questions vives*, vol. 5, 2009.

[PIP 17] PIPAME, "Prospective; futur de la fabrication additive, Final Report, Direction Général des Entreprises, Paris, 2017.

[PIS 56] PISANI E., "Administration de gestion, administration de mission", *Revue française de science politique*, vol. 6, pp. 315–330, 1956.

[POM 04] POMPIDOU A., "Pour une renaissance de la culture scientifique et technique", *Assises nationales de la culture scientifique et technique*, Axiales, pp. 26–31, 2004.

[POP 11] POPE A., *La guerre des sciences*, Les Empêcheurs de Penser en Rond, Paris, 1711.

[POT 14] POTSTADA M., ZYBURA J., "The role of context in science fiction prototyping: the digital inustrial revolution", *Technological Forecasting and Social Change*, vol. 84, pp. 101–114, 2014.

[PUE 08] PUECH M., *Homo-sapiens technologicus: philosophie de la technologie contemporaine; philosophie de la sagesse contemporaine*, Le Pommier, Paris, 2008.

[ROQ 07] ROQUETTE M.L., *La créativité*, PUF, Paris, 2007.

[ROS 13] ROSA H., *Accélération, une critique sociale du temps*, La Découverte, Paris, 2013.

[ROS 15] ROSENBERG J., MORAND P., TURCQ D., "L'impression 3D: porte d'entrée dans l'industrie du 21ème siècle", available at: http://www.adverbe.com/wp-content/uploads/2015/09/rapport-impression3d.pdf, 2015.

[SAI 13] SAINT-MARTIN A., *La sociologie de Robert K. Merton*, La Découverte, Paris, 2013.

[SCA 92] SCARDIGLI V., "Les producteurs de sens", available at: http://documents.irevues.inist.fr/bitstream/handle/2042/32724/C&T_1992_24_209.pdf?sequence=1, 1992.

[SCH 16a] SCHMOUKER O., "Comment savoir si votre idée est LA bonne?", available at: http://www.lesaffaires.com/blogues/olivier-schmouker/comment-savoir-si-votre-idee-est-la-bonne/590981, 2016.

[SCH 16b] SCHMOULDER O., "Les 10 commandements de l'hypercréativité selon Ubisoft", available at: https://www.lesaffaires.com/blogues/olivier-schmouker/les-10-commandements-de-l-hypercreativite-selon-ubisoft/591280, 2016.

[SCI 15] SCIENCE EUROPE, "Humanities scientific committee opinion paper: Radical innovation: humanities research crossing knowledge boundaries and fostering deep change", available at: http://www.scienceeurope.org/uploads/PublicDocumentsAndSpeeches/SCsPublicDocs/151222_HUMAN_OP_Radical_Innovation_web.pdf, 2015.

[SÉG 09] SEGALAT L., *La science à bout de souffle?*, Le Seuil, Paris, 2009.

[SER 11] SERVENT P., *Le complexe de l'autruche*, Perrin, Paris, 2011.

[SHI 11] SHIRKY C., *Cognitive Surplus: How Technology Makes Consumers into Collaborators*, Penguin, New York, 2011.

[SIL 16] SILVERZAHN P., "Comment l'innovation est étouffée par ceux qui la défendent", available at: https://www.contrepoints.org/2016/05/03/251100-comment-innovation-est-tuee-par-ceux-qui-la-defendent#FQQiKgW5FRimzZ9f.99, 2016.

[SIM 11] SIMS P., *Little Bets: How Breakthrough Ideas Emerge from Small Discoveries*, Business Books, London, UK, 2011.

[SMA 11] SMATI R., *Eloge de la vitesse*, Eyrolles, Paris, 2011.

[SNR 09] SNRI, "Stratégie Nationale de Recherche et d'Innovation", available at: http://media.enseignementsup-recherche.gouv.fr/file/SNRI/69/8/Rapport_general_de_la_SNRI_-_version_finale_65698.pdf, 2009.

[SNR 15] SNR, "Stratégie Nationale de la Recherche", available at: http://www.enseignementsup-recherche.gouv.fr/pid24538/strategie-nationale-de-recherche-s.n.r.html, 2015.

[SOK 97] SOKAL A., BRICHMONT J., *Impostures Intellectuelles*, Odile Jacob, Paris, 1997.

[SPE 10] SPERBER D., "Pourquoi repenser l'interdisciplinarité?", in ORIGGI G., DARBELLAY F. (eds), *Repenser l'interdisciplinarité*, Slatkine, Geneva, 2010.

[STE 97] STENGERS I., *La guerre des sciences*, La Découverte, Paris, 1997.

[TAG 00] TAGUIEFF P.A., *L'effacement de l'avenir*, Galilée, Paris, 2000.

[TAL 07] TALEB N.N., *The Black Swan – The Impact of the Highly Improbable*, Random House, New York, 2007.

[THO 83] THOM R., *Paraboles et catastrophes*, Flammarion, Paris, 1983.

[VAN 09] VAN ANDEL P., BOURCIER D., De la sérendipité dans la science, la technique, l'art et le droit : Leçons de l'inattendu, L'ACT MEM, Libres Sciences, Paris, 2009.

[VIN 90] VINCENTI W., *What Engineers Know and How They Know It*, John Hopkins University Press, Baltimore, 1990.

[WEL 15] WELLER C., KLEER R., PILLER F.T., "Economic implications of 3D printing: markets tructure models inlight of additive manufacturing revisited", *International Journal of Production Economics*, vol. 164, pp. 43–56, 2015.

[WIG 49] WIGNER E.P., "Invariance in physical theory", *Proceedings of the American Philosophical Society*, vol. 93, pp. 521–526, 1949.

[WIK 17] WIKIQUOTE, "Citations Emile-Auguste Chartier", available at: https://fr.wikiquote.org/wiki/%C3%89mile-Auguste_Chartier, 2017.

[WOL 09] WOLF M., "Consentement, autonomie, complexité", available at: http://www.ethique.inserm.fr, 2009.

[WOR 12] WORLD ECONOMIC FORUM, The Global Competitiveness Report 2012–2013, available at: http://www3.weforum.org/docs/WEF_GlobalCompetitivenessReport_2012-13.pdf, 2012.

[WOR 15a] WORLD ECONOMIC FORUM, "Collaborative innovation: transforming business, driving growth", available at: http://www3.weforum.org/docs/WEF_Collaborative_Innovation_report_2015.pdf, 2015.

[WOR 15b] WORLD ECONOMIC FORUM, "Deep shift technology tipping points and societal impacts", available at: http://www3.weforum.org/docs/WEF_GAC15_Technological_Tipping_Points_report_2015.pdf, 2015.

[WOR 15c] WORLD ECONOMIC FORUM, How 21st Century longevity can create markets and drive economic growth, White paper, available at: http://www3.weforum.org/docs/WEF_GAC_Ageing_White_Paper.pdf, 2015.

[WOR 16] WORLD ECONOMIC FORUM, Annual Meeting, available at: https://www.weforum.org/events/world-economic-forum-annual-meeting-2016, 2016.

[ZAJ 00] ZAJDENWEBER D., *L'économie des extrêmes*, Flammarion, Paris, 2000.

Conclusion

Tower of Babel (photograph taken at the Museum of Ancient Art in Brussels; Painting by Joos de Momper)

"The definition of a problem tends to be fixed in the position defended by bureaucratic agencies and thus to resist all transformation. Cyert and March [CYE 63], organization theoreticians, wrote that organizations seek to bypass uncertainty by following routine procedures; they do not anticipate problems, but respond to the effects in return induced by their own behavior. From this fact, they tend to "go from one crisis to another" based on standardized procedures to make decisions. This method of functioning, seen in both private and public institutions, allows explanation of the fact that when a new problem shows up, it can be badly understood and be the object of inappropriate treatment". [REI 10]

"The event undoes the time required to invent another time, from this another world, another view is inaugurated". [DUF 11]

"In this society of ferrymen, we are capable of bringing about the unexpected, of fertilizing and transforming ideas into desires, then into projects, and, finally, into realizations. We can live with greater efficiency and harmony. This is a more just society, which again questions the criteria for evaluating success, where we are happier and less superficial, and a society where we manage to organize and share abundance. Doing this together becomes a free pleasure. This is a society where pleasure, far from being absent, is a fundamental principle, an essential right". [NOV 13]

"Once everyone is invited, even strongly incited, to seek innovation in a society or in an organization, an improvement in the overall efficiency and thus, over time, in productivity can be counted on through sheer numbers. New procedures, uses, circuits, objects, attached through permanent expertise and a constant concern for perfection, keep the concern for progress and the very ideology of progress alive. The future is delivered as a territory to be conquered, rich in promises of abundance and happiness". [ROU 07]

"Totally freeing a part of the available energy to resolutely explore radically different territories. The element of surprise, of changing views, of creative reflection must be sharpened. New perspectives, new action levers that create meaning and can embark upon the immediately elusive must be updated". [LAG 13]

"If you want to build a ship, teach people to long for the endless immensity of the sea". [SAI 00]

"The large domain of discoveries is always the unlabeled residue. All around the known and disciplined facts of each science floats the dust of exceptional observations, miniscule, rare, irregular events, which is always easier to leave aside than study". [KOE 11]

"We need a new paradigm shift that sees the world as an integrated whole, not a scattered collection of parts". [PAU 07]

"All governance of enterprise supposes having directives in terms of an appetite for risks and it is the administration council's job to give clear directives to the general management, and through it, to all of the collaborators, to explain what appetite for risks is best for the organization". [LOU 17]

"For nothing is lost. From the crisis, from the political will to bank on an innovative ecosystem, a new entrepreneurial will is felt. The youth has no fear of getting involved. And they surprise. We can see the emergence of the digital, biotech, and medtech. Nothing that compensates the destruction of jobs, but one feels a new breath to twist the neck of the announced death of industry". [SCH 17]

Bibliography

[DUF 11] DUFOURMENTELLE A., *Eloge du risque*, Payot, Paris, 2011.

[KOE 11] KOESTLER A., *Le cri d'Archimède: la découverte de l'art et l'art de la découverte*, Les Belles Lettres, Paris, 2011.

[LAG 13] LAGADEC P., *Piloter en univers inconnu*, Préventique, Bordeaux, 2013.

[LOU 17] LOUISOT J.M., "Optimiser la prise de risques", *Risk-Assur*, vol. 474, pp. 28–37, available at: https://www.riskassur-hebdo.com, 2017.

[NOV 13] NOVEL A.S., RIOT S., *Vive la co-révolution; pour une société collaborative*, Manifestô-Gallimard, Paris, 2013.

[PAU 07] PAULI G., *Croissance sans limites: objectif zéro pollution – Croissance économique et régénération de la nature*, Quintessence Editions, Paris, 2007.

[REI 10] REICH M.R., "Le cas de la pollution du Michigan par le PBB", in AKRICH M., BARTH Y., RÉMY C. (eds), *Sciences, politique et environnement*, MINES ParisTech Editions, Paris, 2010.

[ROU 07] ROUQUETTE M.L., *Propagande et citoyenneté*, PUF, Paris, 2007.

[SAI 00] SAINT EXUPÉRY A., *Citadelle*, Folio, Paris, 2000.

[SCH 17] SCHMITT L., "L'industrie en pleine révolution", available at: http://www.estrepublicain.fr/actualite/2017/01/19/l-industrie-en-pleine-revolution, 2017.

Index

M, N, O

P, S, V

Other titles from

iSTE

in

Systems and Industrial Engineering – Robotics

2017

ANDRÉ Jean-Claude
From Additive Manufacturing to 3D/4D Printing 1

ARCHIMÈDE Bernard, VALLESPIR Bruno
Enterprise Interoperability: INTEROP-PGSO Vision

CAMMAN Christelle, FIORE Claude, LIVOLSI Laurent, QUERRO Pascal
Supply Chain Management and Business Performance: The VASC Model

FEYEL Philippe
Robust Control, Optimization with Metaheuristics

MARÉ Jean-Charles
Aerospace Actuators 2: Signal-by-Wire and Power-by-Wire

POPESCU Dumitru, AMIRA Gharbi, STEFANOIU Dan, BORNE Pierre
Process Control Design for Industrial Applications

RÉVEILLAC Jean-Michel
Modeling and Simulation of Logistics Flows 1: Theory and Fundamentals
Modeling and Simulation of Logistics Flows 2: Dashboards, Traffic Planning and Management
Modeling and Simulation of Logistics Flows 3: Discrete and Continuous Flows in 2D/3D

2016

ANDRÉ Michel, SAMARAS Zissis
Energy and Environment
(Research for Innovative Transports Set - Volume 1)

AUBRY Jean-François, BRINZEI Nicolae, MAZOUNI Mohammed-Habib
Systems Dependability Assessment: Benefits of Petri Net Models (Systems Dependability Assessment Set - Volume 1)

BLANQUART Corinne, CLAUSEN Uwe, JACOB Bernard
Towards Innovative Freight and Logistics (Research for Innovative Transports Set - Volume 2)

COHEN Simon, YANNIS George
Traffic Management (Research for Innovative Transports Set - Volume 3)

MARÉ Jean-Charles
Aerospace Actuators 1: Needs, Reliability and Hydraulic Power Solutions

REZG Nidhal, HAJEJ Zied, BOSCHIAN-CAMPANER Valerio
Production and Maintenance Optimization Problems: Logistic Constraints and Leasing Warranty Services

TORRENTI Jean-Michel, LA TORRE Francesca
Materials and Infrastructures 1 (Research for Innovative Transports Set - Volume 5A)
Materials and Infrastructures 2 (Research for Innovative Transports Set - Volume 5B)

WEBER Philippe, SIMON Christophe
Benefits of Bayesian Network Models
(Systems Dependability Assessment Set – Volume 2)

YANNIS George, COHEN Simon
Traffic Safety (Research for Innovative Transports Set - Volume 4)

2015

AUBRY Jean-François, BRINZEI Nicolae
Systems Dependability Assessment: Modeling with Graphs and Finite State Automata

BOULANGER Jean-Louis
CENELEC 50128 and IEC 62279 Standards

BRIFFAUT Jean-Pierre
E-Enabled Operations Management

MISSIKOFF Michele, CANDUCCI Massimo, MAIDEN Neil
Enterprise Innovation

2014

CHETTO Maryline
Real-time Systems Scheduling
Volume 1 – Fundamentals
Volume 2 – Focuses

DAVIM J. Paulo
Machinability of Advanced Materials

ESTAMPE Dominique
Supply Chain Performance and Evaluation Models

FAVRE Bernard
Introduction to Sustainable Transports

GAUTHIER Michaël, ANDREFF Nicolas, DOMBRE Etienne
Intracorporeal Robotics: From Milliscale to Nanoscale

MICOUIN Patrice
Model Based Systems Engineering: Fundamentals and Methods

MILLOT Patrick
Designing Human–Machine Cooperation Systems

NI Zhenjiang, PACORET Céline, BENOSMAN Ryad, RÉGNIER Stéphane
Haptic Feedback Teleoperation of Optical Tweezers

OUSTALOUP Alain
Diversity and Non-integer Differentiation for System Dynamics

REZG Nidhal, DELLAGI Sofien, KHATAD Abdelhakim
Joint Optimization of Maintenance and Production Policies

STEFANOIU Dan, BORNE Pierre, POPESCU Dumitru, FILIP Florin Gh., EL KAMEL Abdelkader
Optimization in Engineering Sciences: Metaheuristics, Stochastic Methods and Decision Support

2013

ALAZARD Daniel
Reverse Engineering in Control Design

ARIOUI Hichem, NEHAOUA Lamri
Driving Simulation

CHADLI Mohammed, COPPIER Hervé
Command-control for Real-time Systems

DAAFOUZ Jamal, TARBOURIECH Sophie, SIGALOTTI Mario
Hybrid Systems with Constraints

FEYEL Philippe
Loop-shaping Robust Control

FLAUS Jean-Marie
Risk Analysis: Socio-technical and Industrial Systems

FRIBOURG Laurent, SOULAT Romain
Control of Switching Systems by Invariance Analysis: Application to Power Electronics

GROSSARD Mathieu, REGNIER Stéphane, CHAILLET Nicolas
Flexible Robotics: Applications to Multiscale Manipulations

GRUNN Emmanuel, PHAM Anh Tuan
Modeling of Complex Systems: Application to Aeronautical Dynamics

HABIB Maki K., DAVIM J. Paulo
Interdisciplinary Mechatronics: Engineering Science and Research Development

HAMMADI Slim, KSOURI Mekki
Multimodal Transport Systems

JARBOUI Bassem, SIARRY Patrick, TEGHEM Jacques
Metaheuristics for Production Scheduling

KIRILLOV Oleg N., PELINOVSKY Dmitry E.
Nonlinear Physical Systems

LE Vu Tuan Hieu, STOICA Cristina, ALAMO Teodoro,
CAMACHO Eduardo F., DUMUR Didier
Zonotopes: From Guaranteed State-estimation to Control

MACHADO Carolina, DAVIM J. Paulo
Management and Engineering Innovation

MORANA Joëlle
Sustainable Supply Chain Management

SANDOU Guillaume
Metaheuristic Optimization for the Design of Automatic Control Laws

STOICAN Florin, OLARU Sorin
Set-theoretic Fault Detection in Multisensor Systems

2012

AÏT-KADI Daoud, CHOUINARD Marc, MARCOTTE Suzanne, RIOPEL Diane
Sustainable Reverse Logistics Network: Engineering and Management

BORNE Pierre, POPESCU Dumitru, FILIP Florin G., STEFANOIU Dan
Optimization in Engineering Sciences: Exact Methods

CHADLI Mohammed, BORNE Pierre
Multiple Models Approach in Automation: Takagi-Sugeno Fuzzy Systems

DAVIM J. Paulo
Lasers in Manufacturing

DECLERCK Philippe
Discrete Event Systems in Dioid Algebra and Conventional Algebra

DOUMIATI Moustapha, CHARARA Ali, VICTORINO Alessandro, LECHNER Daniel
Vehicle Dynamics Estimation using Kalman Filtering: Experimental Validation

GUERRERO José A, LOZANO Rogelio
Flight Formation Control

HAMMADI Slim, KSOURI Mekki
Advanced Mobility and Transport Engineering

MAILLARD Pierre
Competitive Quality Strategies

MATTA Nada, VANDENBOOMGAERDE Yves, ARLAT Jean
Supervision and Safety of Complex Systems

POLER Raul *et al.*
Intelligent Non-hierarchical Manufacturing Networks

TROCCAZ Jocelyne
Medical Robotics

YALAOUI Alice, CHEHADE Hicham, YALAOUI Farouk, AMODEO Lionel
Optimization of Logistics

ZELM Martin *et al.*
Enterprise Interoperability –I-EASA12 Proceedings

2011

CANTOT Pascal, LUZEAUX Dominique
Simulation and Modeling of Systems of Systems

DAVIM J. Paulo
Mechatronics

DAVIM J. Paulo
Wood Machining

GROUS Ammar
Applied Metrology for Manufacturing Engineering

KOLSKI Christophe
Human–Computer Interactions in Transport

LUZEAUX Dominique, RUAULT Jean-René, WIPPLER Jean-Luc
Complex Systems and Systems of Systems Engineering

ZELM Martin, *et al.*
Enterprise Interoperability: IWEI2011 Proceedings

2010

BOTTA-GENOULAZ Valérie, CAMPAGNE Jean-Pierre, LLERENA Daniel, PELLEGRIN Claude
Supply Chain Performance / Collaboration, Alignement and Coordination

BOURLÈS Henri, GODFREY K.C. Kwan
Linear Systems

BOURRIÈRES Jean-Paul
Proceedings of CEISIE'09

CHAILLET Nicolas, REGNIER Stéphane
Microrobotics for Micromanipulation

DAVIM J. Paulo
Sustainable Manufacturing

GIORDANO Max, MATHIEU Luc, VILLENEUVE François
Product Life-Cycle Management / Geometric Variations

LOZANO Rogelio
Unmanned Aerial Vehicles / Embedded Control

LUZEAUX Dominique, RUAULT Jean-René
Systems of Systems

VILLENEUVE François, MATHIEU Luc
Geometric Tolerancing of Products

2009

DIAZ Michel
Petri Nets / Fundamental Models, Verification and Applications

OZEL Tugrul, DAVIM J. Paulo
Intelligent Machining

PITRAT Jacques
Artificial Beings

2008

ARTIGUES Christian, DEMASSEY Sophie, NERON Emmanuel
Resources–Constrained Project Scheduling

BILLAUT Jean-Charles, MOUKRIM Aziz, SANLAVILLE Eric
Flexibility and Robustness in Scheduling

DOCHAIN Denis
Bioprocess Control

LOPEZ Pierre, ROUBELLAT François
Production Scheduling

THIERRY Caroline, THOMAS André, BEL Gérard
Supply Chain Simulation and Management

2007

DE LARMINAT Philippe
Analysis and Control of Linear Systems

DOMBRE Etienne, KHALIL Wisama
Robot Manipulators

LAMNABHI Françoise *et al.*
Taming Heterogeneity and Complexity of Embedded Control

LIMNIOS Nikolaos
Fault Trees

2006

FRENCH COLLEGE OF METROLOGY
Metrology in Industry

NAJIM Kaddour
Control of Continuous Linear Systems

Printed and bound by CPI Group (UK) Ltd, Croydon, CR0 4YY

16/09/2026

14963074-0003